Lecture Notes in Computer Science 14784

Founding Editors

Gerhard Goos
Juris Hartmanis

The series Lecture Notes in Computer Science (LNCS), including its subseries Lecture Notes in Artificial Intelligence (LNAI) and Lecture Notes in Bioinformatics (LNBI), has established itself as a medium for the publication of new developments in computer science and information technology research, teaching, and education.

LNCS enjoys close cooperation with the computer science R & D community, the series counts many renowned academics among its volume editors and paper authors, and collaborates with prestigious societies. Its mission is to serve this international community by providing an invaluable service, mainly focused on the publication of conference and workshop proceedings and postproceedings. LNCS commenced publication in 1973.

Guangdong Bai · Fuyuki Ishikawa ·
Yamine Ait-Ameur · George A. Papadopoulos
Editors

Engineering of Complex Computer Systems

28th International Conference, ICECCS 2024
Limassol, Cyprus, June 19–21, 2024
Proceedings

 Springer

Editors

Guangdong Bai 🆔
School of Electrical Engineering
and Computer Science, Faculty
of Engineering, Architecture and Information
Technology
University of Queensland
Brisbane, QLD, Australia

Yamine Ait-Ameur 🆔
IRIT
University of Toulouse
Toulouse, France

Fuyuki Ishikawa 🆔
Information Systems Architecture Science
Research Division
National Institute of Informatics
Chiyoda-ku, Tokyo, Japan

George A. Papadopoulos 🆔
Department of Computer Science
University of Cyprus
Nicosia, Cyprus

ISSN 0302-9743   ISSN 1611-3349 (electronic)
Lecture Notes in Computer Science
ISBN 978-3-031-66455-7   ISBN 978-3-031-66456-4 (eBook)
https://doi.org/10.1007/978-3-031-66456-4

Preface

Welcome to the 28th International Conference on Engineering of Complex Computer Systems (ICECCS 2024), held in Limassol, Cyprus, from June 19th to June 21st. This edition was part of the well-established event that has been held around the world for the past 27 years. The goal of this conference is to bring together industrial, academic, and government experts from a variety of application domains and software disciplines, to discuss how the disciplines' problems and solution techniques interact within the whole system.

This year the conference expanded beyond the traditional focus on complex computer systems to include new topics regarding complex systems that process vast amounts of data by leveraging emerging artificial intelligence (AI), large language models, and machine learning techniques. We invited two recognized researchers in the field to give keynote addresses: Constantine Dovrolis is a Professor and Director of the Centre for Computational Science and Technology at Cyprus Institute and Christos Panayiotou is a Professor in the Department of Electrical and Computer Engineering at the University of Cyprus. Professor Dovrolis delivered a keynote talk on the emergence and evolution of hierarchical structure in complex systems and Professor Panayiotou delivered a keynote talk on intelligent transportation systems.

We received 68 submissions, and each of them underwent rigorous review by at least three reviewers, followed by in-depth discussions among reviewers and a Program Committee co-chair. The Program and General Chairs eventually selected 18 regular papers and 4 short papers for presentation during the conference and publication in the conference proceedings (around 26% acceptance rate for regular papers and 32% overall acceptance rate).

We thank the authors for submitting their inspiring work to us, and the Program Committee members and their external reviewers for providing their valuable and timely reviews. The success of ICECCS 2024 was largely due to the support of the Steering Committee and the hard work of the Organization Committee.

We hope that all the participants enjoyed ICECCS 2024!

June 2024

Guangdong Bai
Fuyuki Ishikawa
Yamine Aït-Ameur
George A. Papadopoulos

Organization

General Co-chairs

Yamine Aït-Ameur   IRIT/INPT-ENSEEIHT, France
George A. Papadopoulos   University of Cyprus, Cyprus

Program Committee Co-chairs

Guangdong Bai   University of Queensland, Australia
Fuyuki Ishikawa   National Institute of Informatics, Japan

Steering Committee

Jin Song Dong   National University of Singapore, Singapore
Mike Hinche   University of Limerick, Ireland
Xiaohong Li   Tianjin University, China
Shaoying Liu   Hiroshima University, Japan
Mauro Pezzè   University of Lugano, Switzerland
Roy Sterritt   Ulster University, UK
Jing Sun (Chair)   University of Auckland, New Zealand

Program Committee

Yamine Aït Ameur   IRIT/INPT-ENSEEIHT, France
Cyrille Valentin Artho   KTH Royal Institute of Technology, Sweden
Guangdong Bai   University of Queensland, Australia
Hadrien Bride   Griffith University, Australia
Florin Craciun   Babes-Bolyai University, Romania
Juergen Dingel   Queen's University, Canada
Guillaume Dupont   IRIT-ENSEEIHT, France
Marc Frappier   Université de Sherbrooke, Canada
Sudipto Ghosh   Colorado State University, USA
Hiroshi Hosobe   Hosei University, Japan
Zhe Hou   Griffith University, Australia
Fuyuki Ishikawa   National Institute of Informatics, Japan

Yu Jiang	Tsinghua University, China
Ferhat Khendek	Concordia University, Canada
Kung-Kiu Lau	University of Manchester, UK
Scott Uk-Jin Lee	Hanyang University, South Korea
Michael Leuschel	University of Düsseldorf, Germany
Yuekang Li	University of New South Wales, Australia
Jie Liang	Tsinghua University, China
Shang-Wei Lin	Nanyang Technological University, Singapore
Shaoying Liu	Hiroshima University, Japan
Gerald Lüttgen	University of Bamberg, Germany
Lei Ma	University of Tokyo, Japan
Kulani Mahadewa	National University of Singapore, Singapore
Weizhi Meng	Technical University of Denmark, Denmark
Dominique Mery	Université de Lorraine, France
Sadaf Mustafiz	Ryerson University, Canada
Shin Nakajima	National Institute of Informatics, Japan
Wuwei Shen	Western Michigan University, USA
Neeraj Kumar Singh	IRIT-ENSEEIHT Toulouse, France
Yulei Sui	University of New South Wales, Australia
Meng Sun	Peking University, China
Jun Sun	Singapore Management University, Singapore
Jing Sun	University of Auckland, New Zealand
Mark Utting	University of Queensland, Australia
Kailong Wang	Huazhong University of Science and Technology, China
Hironori Washizaki	Waseda University, Japan
Burkhart Wolff	Université Paris-Saclay, France
Zhilin Wu	Chinese Academy of Sciences, China
Xiao-Yi Zhang	Beijing University of Science and Technology, China
Jian Zhang	Nanyang Technological University, Singapore
Jianjun Zhao	Kyushu University, Japan
Junjun Zheng	Osaka University, Japan

Additional Reviewers

Christian Attiogbé

Hao Bu

Jianlang Chen

Siyuan Chen

Bai Jun Cheng

Wei-Yang Chiu

Mojtaba Eshghie

Harold-Nimród Földvári

J. Paul Gibson

Raju Halder

Ningke Li
Yuxi Li
Bernhard Luedtke
Deyun Lyu
Jordan Masakuna
Stephan Merz
Rosemary Monahan

Mohamed Mosbah
Tai D. Nguyen
Yang Sun
Xiaoyong Xue
Eugene Yip
Mengdi Zhang

Abstracts of the Invited Talks

Emergence and Evolution of Hierarchical Structure in Complex Systems

Constantine Dovrolis

Center for Computational Science and Technology, The Cyprus Institute, Nicosia, Cyprus
c.dovrolis@cyi.ac.cy

Abstract. It is well known that many complex systems, both in technology and in nature, exhibit hierarchical modularity: simple and general modules, each of them providing a certain function, are used within more complex modules that perform more sophisticated functions. This is the case for instance with protocol stacks in computer networking. What is not well understood however is how this hierarchical structure (which is fundamentally a network property) emerges, and how it evolves over time. We propose a modeling framework, called Evo-Lexis, that provides insight into some fundamental questions about evolving hierarchical systems. We show that deep hierarchies emerge when the population of top-layer modules evolves through tinkering and mutation. Strong selection on the cost of new top-layer modules results in reuse of more complex (longer) nodes in an optimized hierarchy. The bias towards reuse of complex nodes results in an "hourglass architecture" (i.e., few intermediate nodes that cover almost all source–target paths). With such bias, the core nodes are conserved for relatively long time periods although still being vulnerable to major transitions and punctuated equilibria. Finally, we analyze the differences in terms of cost and structure between incrementally designed hierarchies and the corresponding "clean-slate" hierarchies which result when the system is designed from scratch after a change.

Intelligent Transportation Systems: A Large-Scale and Complex Cyber-Physical System

Christos Panayiotou

Department of Electrical and Computer Engineering, University of Cyprus, Nicosia, Cyprus
christosp@ucy.ac.cy

Abstract. Transportation systems have always been a complex system with many agents (vehicles) interacting on the road network. Historically, a huge problem of these systems has been congestion with all its adverse effects: waste of time and energy, green-house emissions, accidents, and frustration. The emergence of autonomous and connected vehicles promises to solve these problems; however, they cannot solve the problem unless there is also a paradigm shift from "selfish" driving to more "social" driving. During this talk, we will address some of the problems associated with congestion and present innovative solution approaches that are based on a route reservation architecture aimed not only at alleviating congestion but also at enhancing the overall efficiency of the road network in terms of the average time vehicles spend within it.

Contents

Smart Contract

Formal Methods

Security

Program Analysis

Machine Learning and Complex Systems

DASHCHEF: A Metric Recommendation Service for Online Systems Using Graph Learning

Zilong He[1], Tao Huang[2], Pengfei Chen[1(✉)], Ruipeng Li[2], Rui Wang[2], and Zibin Zheng[3]

[1] School of Computer Science and Engineering, Sun Yat-sen University, Guangzhou, China
hezlong@mail2.sysu.edu.cn, chenpf7@mail.sysu.edu.cn
[2] Tencent, Shenzhen, China
{hidveghuang,tristonli,amurorywang}@tencent.com
[3] School of Software Engineering, Sun Yat-sen University, Guangzhou, China
zhzibin@mail.sysu.edu.cn

Abstract. To ensure the high availability of modern online systems, effective maintenance is of critical importance. Today's software maintenance techniques for online systems heavily rely on metrics, which are time series data that can describe the real-time state of a system from various perspectives. Typically, software engineers generate dashboards with metrics to aid software maintenance. Though several attempts have been devoted to metric analysis for automatic software maintenance, the primary step, i.e., dashboard generation, remains manual to a large extent. In this paper, we develop a metric recommendation service, which can automate the dashboard generation practice and greatly ease the burden in maintaining an online system. Specifically, we analyze the needs of two essential steps of online system maintenance, i.e., anomaly detection and fault diagnosis, and design metric recommendation mechanisms for them respectively. Graph learning techniques are employed in the automation of metric recommendation. Our experiments demonstrate that the proposed approach can achieve an F1-score of 0.912 in selecting metrics for anomaly detection, and an accuracy of 0.859 in retrieving metrics for faults diagnosis, which significantly outperforms the compared baselines.

Keywords: metric recommendation · graph learning · software maintenance · online systems

1 Introduction

Given that modern online systems are gradually complex and large-scale, continually maintaining these systems to ensure their high availability becomes

---

Z. He and T. Huang—Co-first authors of this work.

G. Bai et al. (Eds.): ICECCS 2024, LNCS 14784, pp. 3–22, 2025.
https://doi.org/10.1007/978-3-031-66456-4_1

increasingly important yet difficult. Maintenance primarily starts with anomaly detection and fault diagnosis. These tasks are usually developed based on metrics. Metrics, e.g., Page View Count and CPU Usage, are time series data that record the real-time state of a system. Typically, software engineers need to collect such metrics, and then generate dashboards to monitor and analyse them for daily operations. Specifically, a dashboard is a group of metrics that are organized together in a Web-hosted site to aid software maintenance. As an important part of software engineering activities, generating dashboards for software maintenance has not yet been fully investigated in prior literature.

Dashboard generation is the entrance of anomaly detection and fault diagnosis. For simplicity, this paper denotes dashboard generation for anomaly detection as metric selection, and denotes dashboard generation for fault diagnosis as metric retrieval. Metric selection is to generate a dashboard with metrics that are continuously used to trigger alerts, while metric retrieval is to generate a dashboard with metrics that are temporarily used to diagnose a specific fault.

Although various techniques have been applied to automatic anomaly detection [14,15,25,26,31] and fault diagnosis [9,19–21,30] for online systems, the common practice of dashboard generation remains manual yet. Such manual practice has several limitations: **(1) Low Scalability for Large-Scale Monitoring.** Modern online systems can be incredibly large-scale, with millions or even billions of metrics that can be collected from different components. For examples, in an online system in Tencent, the volume of metric data can be over 500 TB a day. More seriously, this number is still growing with the development of the system observability techniques [16,18,28]. Requiring software engineers to manually inspect all these data to find useful metrics is impractical. **(2) Heavy Reliance on Expert Knowledge.** There are various types of emerging techniques used in online systems. Understanding the exact meaning and usage of each metric is difficult. Therefore, it is unrealistic to assume that every engineer is experienced enough to manually generate a dashboard against actual conditions.

Metric recommendation, which automates metric selection and metric retrieval, can help a lot. If a metric recommendation service is provided, software engineers can simply trigger the service and set their dashboards according to the recommendation. As the annoying procedures are automated, the burden of software engineers can be greatly alleviated. The goal of such a task is similar to automated feature selection in the Machine Learning community, which strives to pick out useful information for a specific task. Actually, some prior work [6,8,24] has applied such automated feature selection techniques to select metrics for anomaly detection based on the correlation between metrics and historical faults. However, such feature selection techniques cannot be directly used for metric recommendation. Specifically, for metric selection, these techniques are impractical since they assume some external information, e.g., the occurrence time of all historical faults, is at hand. Continuously labelling the occurrence time of faults is labor-intensive, and hard to achieve in the maintenance of modern online systems. Moreover, for metric retrieval, such feature selection techniques is unsuitable since they cannot take the information of a specific ongoing fault into consideration.

Table 1. Number of Metrics of Some Common Components

component	Node	Container	DB (e.g., Redis)	MQ (e.g., Kafka)
#metrics	$\geq$72	$\geq$75	$\geq$186	$\geq$477

This paper proposes to automatically recommend metrics with minimal human effort. We design a data-driven metric recommendation method for dashboard generation to aid software maintenance, which can be easily applied across systems. This can pose several challenges: i) How to accommodate the need of different software maintenance tasks when performing metric recommendation. ii) How to strike a balance between the effectiveness and the reliance on human effort during the recommendation process.

To address the above challenges, we propose DASHCHEF, a metric recommendation service for online systems using graph learning. According to the different needs of software maintenance in practice, we decompose the metric recommendation service into two usage scenarios, namely metric selection for anomaly detection and unmonitored anomaly-related metric retrieval for fault diagnosis. The backgrounds and formulations of these usage scenarios will be detailed in Sect. 2.3 and 3.1, respectively. Then, graph learning techniques including random walk and graph neural network are employed to overcome the challenges discussed above. In general, our contributions are as follows,

– We propose DASHCHEF, a metric recommendation service for online systems based on graph learning, which can greatly automate metric selection for anomaly detection and metric retrieval for fault diagnosis.
– We conduct extensive experiments to validate DASHCHEF on metric data collected from Tencent®. The results confirm the effectiveness of DASHCHEF in recommending metrics for software maintenance in online systems.

2   Background

In this section, we introduce the background of metric recommendation for online systems in practice, including the basic concepts about a dashboard and their usages for software maintenance.

2.1   Metrics in Online Systems

Online systems encompass many metrics that can be collected from different angles such as workload, CPU, memory, storage and network. Table 1 shows the number of collectable metrics of some common components in online systems, including physical node [3], container [4], database (DB) [5] and message queue (MQ) [2]. Without an appropriate recommendation service, the need to handle such a large number of metrics will greatly hinder the maintenance of online systems.

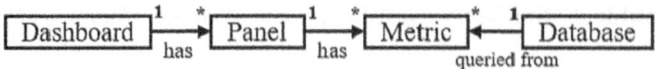

Fig. 1. Basic concepts of dashboards

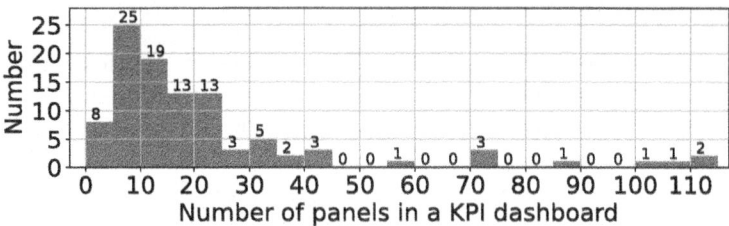

Fig. 2. The histogram of the number of panels in the collected dashboards.

2.2 Dashboard

Inside a Dashboard. Figure 1 shows the relationships of some basic concepts about a dashboard. A dashboard is the entry point for software maintenance. Software engineers usually use it to inspect the state of a system, or configure the inputs for some intelligent anomaly detection and fault diagnosis algorithms. A dashboard consist of several panels, and each panel is configured with some metrics to show specific real-time properties of a system. Typically, metrics in the same panel are metrics of the same type but with different attributes. For example, the CPU Usages (metric type) of different servers (metric attributes) are usually placed in the same panel. Figure 2 displays a histogram summarizing the number of metric panels in each dashboard collected from the 100 open-sourced dashboards with the top downloads (each downloaded by 171 thousands ∼ 18 millions users) from Grafana [1]. It can be observed that more than three quarters of dashboards have less than 25 metrics. In a nutshell, operators usually care about only some of the metrics.

Outside a Dashboard. Typically, the metrics are queried from a remote database. Since the attention of a software engineer is limited, he/she usually can only place a few metrics in a dashboard for anomaly detection. We call these metrics as *Key Performance Indicators (KPIs)* in this paper. Despite the displayed KPIs, there are still many metrics that can be queried from the database. Since these metrics are usually not used by any software engineers, we call them *unmonitored metrics* in this paper. Although they are neglected in a dashboard for anomaly detection, they may still be useful for the diagnosis of some specific faults. Therefore, in this paper, we also propose a method to recommend these metrics during the diagnosis of a specific fault.

2.3   Using Dashboards for Software Maintenance

In this section, we describe different usages of dashboards for software maintenance in practice, which can shed light on the design of our metric recommendation solution.

Monitoring and Anomaly Detection. The initial intent for creating a dashboard is to monitor some properties of a running system and build confidence on it. For example, one might need to confirm whether the system is working properly after he/she updates some configurations of the system. Apart from manual monitoring, some statistical or artificial intelligence techniques [14, 15, 25, 26, 31] can be applied to automatic anomaly detection on metrics, which greatly improves the operational efficiency. Since the effectiveness of the input metrics greatly affect the accuracy of the anomaly detection algorithms, metric recommendation in advance can help a lot.

Fault Diagnosis. If some problems occur, the fault diagnosis heavy relies on metrics too since some metrics can provide clues about the root cause of a fault. Fault diagnosis is to find these metrics. This tasks can be conducted manually, or performed by some artificial intelligence techniques developed recently [9, 19–21, 30]. However, no matter how this task is performed, selecting useful metrics on a dashboard to assess the symptom of a fault needs to be conducted first. It is noteworthy that the selected metrics here might not be KPIs since the possible faults are diverse, and the metrics for fault diagnosis should be selected according to the fault symptoms. For this usage, metric recommendation plays a vital role.

3   Motivation

3.1   Problem Formulation

For different purposes, the useful metrics varies. This paper proposes to automatically recommend useful metrics, and targets at making full use of metrics that can be collected, not limited to KPIs. The following contains the usage scenario formulations of metric recommendation for software maintenance.

Metric Selection for Anomaly Detection. In an online system, metric data are collected from enormous entities (e.g., machines, services), and then stored into a database. They can be fetched to a dashboard for monitoring, or further inputted to some machine learning models for anomaly detection. In online systems, each entity can be associated with many metrics. As a result, the primary problem is, *"For each entity, which metrics can better reveal the entity state and should be utilized for anomaly detection?"*. Specifically, for an entity, a set of metrics $\mathbf{M} = \{M_1, M_2, \cdots\}$ can be collected, with a metric $M_i = [x_i^{t-w_s}, \cdots, x_i^t]$, where x_i^t is its observation at time t and w_s is the observation window size. The objective is to gain a recommended subset of $\mathbf{M}$, and the recommended subset should comply with the fault revealing requirement, noise removal requirement and pattern diversity requirement, which will be further discussed in Sect. 3.2.

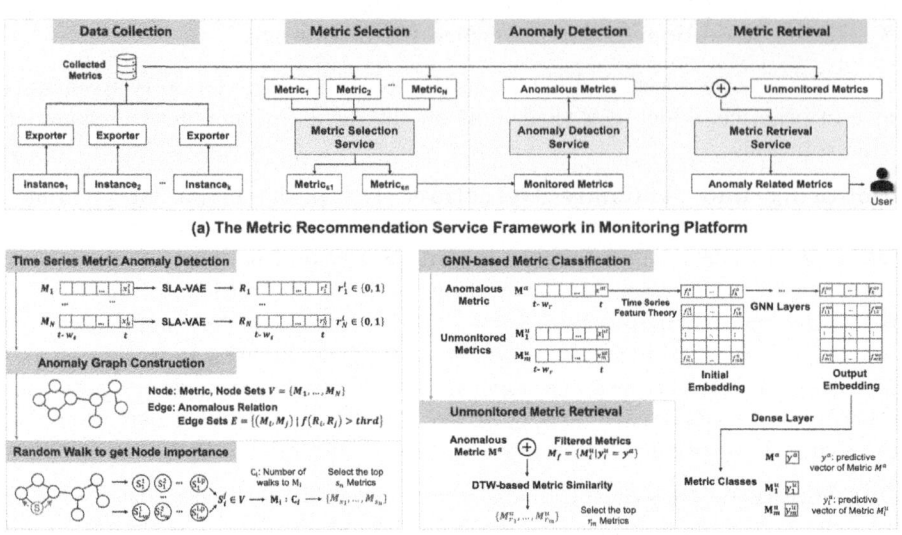

Fig. 3. The metric recommendation service framework in monitoring platform, and the metric selection and retrieval methodologies.

Unmonitored Anomaly-Related Metric Retrieval for Fault Diagnosis. Apart from the selected KPIs for anomaly detection, there may still be some unmonitored metrics which are helpful for the diagnosis of some faults. Previously, these metrics may be simply neglected by the engineers. This paper aims at making the full use of metrics that can be collected, and proposes to retrieve the unmonitored metrics on demand for fault diagnosis. Therefore, the problem is, *"Given an anomalous KPI, which unmonitored metrics may expose some clues about the anomaly and need to be retrieved for analysis?"*. Specifically, given an anomalous KPI M^a and a set of other unmonitored metric $\mathbf{M}^u = \{M_1^u, M_2^u, \cdots\}$, the objective is to calculate a recommended subset of $\mathbf{M}^u$ to retrieve for fault diagnosis. The recommended subset should comply with the fault understanding requirement, which will be discussed in Sect. 3.2.

3.2 Metric Recommendation Requirements

Based on the introduction of metric usages in Sect. 2.3 and the problem formulation described above, we summarize the requirements of metric recommendation as follows.

Overall Requirement Effort Reducing Requirement. The proposed metric recommendation method is to reduce human effort. Therefore, we should not require engineers to continuously provide some external inputs (e.g., the time of all the historical faults for a supervised learning) to gain the recommendation.

Even when such external inputs are a must, the effort should be one-time, and reusable by the same system over time as well as other systems.

Requirements for Metric Selection Fault Revealing Requirement. If a fault occurs, the selected KPIs should have a high probability to behave abnormally. Otherwise, many faults may be neglected, and thus the KPIs are actually useless. Therefore, one requirement of metric selection is to select KPIs that can successfully reveal faults in a system.

Noise Removal Requirement. Some metrics are noisy and using them for alerting might give rise to a flood of false alerts. Therefore, one requirement of metric recommendation is to exclude these noisy metrics.

Requirement for Metric Retrieval Fault Understanding Requirement. Diagnosing a fault requires an in-depth investigation into the impact of the fault. Therefore, the requirement of metric retrieval is to retrieve metrics that can provide a comprehensive view of a fault.

4 Methodology

To meet the requirements of metric recommendations, we introduce a metric recommendation service, i.e., DASHCHEF. This service encompasses offline metric selection for anomaly detection and online retrieval of unmonitored anomaly-related metrics for fault diagnosis. Figure 3 illustrates the framework of the metric recommendation service within the monitoring platform, along with the methodologies for metric selection and retrieval. In the following, we will provide a detailed explanation of the methodology design for offline metric selection and online unmonitored anomaly-related metric retrieval.

4.1 Overview

To ensure the availability of online systems, software engineers must gather metrics from multiple instances, utilize their expert knowledge to select relevant metrics for monitoring, and employ anomaly detection techniques to identify any anomalies within these metrics. When anomalies are detected, engineers need to diagnose and resolve system faults based on the anomalous metrics. However, achieving complete and accurate metric selection for anomaly detection can be challenging. Therefore, we first propose a metric selection method for anomaly monitoring and detection. Additionally, unmonitored metrics can also aid in identifying and diagnosing anomalies. In an industrial setting, the number of unmonitored metrics far surpasses the monitored ones. Relying solely on monitored metrics can impede fault diagnosis. As a result, we also design a method for retrieving anomaly-related unmonitored metrics specifically for fault diagnosis.

Figure 3(a) illustrates the framework of the proposed metric recommendation service within the monitoring platform. Initially, metrics are collected from system instances and stored in a database. Subsequently, the metric selection service recommends Key Performance Indicators (KPIs) that should be monitored. Software engineers can manually incorporate the recommendation results with their expert knowledge to create relevant monitoring tasks. Alternatively, monitoring tasks can be automatically generated based on the recommendation results. An anomaly detection service is then employed to identify anomalies. Given that the focus of this paper is not on the anomaly detection algorithm, we will not provide a detailed explanation of the technique used. For our purposes, we utilize the semi-supervised variational auto-encoder (SLA-VAE) [15] to detect anomalies in metrics. Finally, the metric retrieval service is utilized to identify unmonitored metrics that are related to anomalies. When anomalies are detected within monitored metrics, the unmonitored anomaly-related metrics are also provided to software engineers for further fault diagnosis.

4.2 Metric Selection for Anomaly Detection

The selected KPIs, are required to reveal the system faults effectively. When the system fails, the monitored KPIs should be anomalous so that the software engineers can notice and repair the fault timely. We may adopt the correlation between metric anomalies and system faults to automatic select metric in a supervised way, just as prior work [12,13] does. However, there are two issues requiring us to design a new unsupervised solution in practice. First, the sample labels about system faults are usually incomplete and even lost in industrial environments. If the samples do not contain enough information about system faults, it is difficult for the recommended KPIs to detect all system failures. Second, due to the differences between systems, we should select different KPIs which are suitable for different systems. When we adopt a supervised method to select metrics, we need to collect and generate different samples for each system, which is disadvantageous for the wide deployment of the metric selection method.

To better design the unsupervised metric selection method, we further deconstruct the target problem. If the anomaly detection results of one metric are similar to the results of others metrics, it means that this metric can also identify the anomalies of other metrics and we can select this metric as KPI. Then the correlation between metric anomalies and system faults can be transformed into the relationships between metric anomalies. Eventually, we can construct the anomaly graph of metrics based on the metric anomaly relationships and select the KPIs via graph learning.

Figure 3(b) shows the structure of the proposed unsupervised **Metric Selection** approach using **Graph** learning (MSG). Since we focus on metric anomaly relationships, we first adopt a time series metric anomaly detection method to transform the original metrics into detection results. Then we adopt the detection results to construct the graph which can represent the topological relationships between metrics. Finally, we use graph representation learning techniques to learn the metric importance and select the representative KPIs.

Time Series Metric Anomaly Detection. The objective of time series metric anomaly detection is to transform the original observations into vectors used to describe whether the observation is normal or anomalous. The metric set to be selected can be defined as $\mathbf{M} = \{M_1, M_2, \cdots, M_N\}$, where N is the number of collected metrics, $M_i = [x_i^{t-w_s}, \cdots, x_i^t]$ is the ith metric which contains collected observations from time $t - w_s$ to t, and w_s is the observation window size. Time series metric anomaly detection aims to determine whether the observations are anomalous or normal. When identifying anomalies, the anomaly detection technique should be available to all kinds of metrics. In this paper, we adopt SLA-VAE [15] to identify anomalies. SLA-VAE contains an anomaly definition based feature extraction module, allowing it to be effective in the detection of different types of anomalies. Then the detection results of the metric set can be expressed as $\mathbf{R} = \{R_1, R_2, \cdots, R_N\}$, where $R_i = [r_i^{t-w_s}, \cdots, r_i^t]$ is the detection results of the ith metric, and $r_i^k \in \{0, 1\}$ represents whether the observation of the ith metric at time k is normal or anomalous.

Anomaly Graph Construction. To characterize the relationship between metrics, we try to build an anomaly graph of time series metrics. The anomaly graph of metric set can be defined as $G = (V, E)$, where $V = \{M_1, M_2, \cdots, M_N\}$ is the metric set and $E = \{(M_i, M_j) | f(R_i, R_j) > \tau\}$ is the anomalous relation set. The function f is adopted to measure the anomalous relation between metric M_i and M_j, and the threshold τ is used to filter the unrelated metric pair (M_i, M_j). When designing the function f, we transform the measure of anomalous relation among metrics into anomaly detection performance evaluation and adopt averaged F-measure to characterize the anomalous relation. The function f can be expressed as:

$$f(R_i, R_j) = \frac{2 * p(R_i, R_j) * p(R_j, R_i)}{p(R_i, R_j) + p(R_j, R_i)} \qquad (1)$$

$$p(R_i, R_j) = \frac{\#_{k \in [t-w_s, t]}(r_i^k = 1 \ \& \ r_j^k = 1)}{\#_{k \in [t-w_s, t]}(r_i^k = 1)} \qquad (2)$$

where $\#_{k \in [t-w_s, t]}(r_i^k = 1)$ is the number of anomalies in the ith metric M_i and $\#_{k \in [t-w_s, t]}(r_i^k = 1 \ \& \ r_j^k = 1)$ is the number of anomalies in both the ith metric M_i and the jth metric M_j in the meantime. In addition, when two metrics have similar time series shapes, it means that both metrics may reveal the same thing. Thus, we prune the anomaly graph based on the similarity between metrics and remove edges between metrics with similar shapes. In this paper, we adopt dynamic time warping (DTW) [22] algorithm to compute the similarity. If the DTW distance of two metrics is lower than a relevant threshold, we will remove the edge between them. Eventually, we will obtain a pruned anomaly graph which can characterize the anomaly relations between metrics.

Random Walk to Select Metrics. Once we construct the anomaly graph, we aim to select time series metrics based on the topological relationships. Based on the anomaly graph, random walk algorithm [27], which is the classical algorithm for computing node embedding and importance in graph learning, is adopted to compute metric importance. The results of one round of random walk on the anomaly graph can be defined as

$$\{s_i^1, s_i^2, \cdots, s_i^{L_p} \mid s_i^j \in V\},$$

where L_p is the path length of random walk. We repeatedly execute L_w rounds of random walks in parallel. In more detail, during each round of random walk, we first randomly selected one metric $s_i^1 \in V$ and then sample one of its neighbor metrics based on the anomalous relation. The sampling function is defined as:

$$P(s_i^j|s_i^{j-1}) = \frac{f(s_i^j, s_i^{j-1})}{\sum_{s_i^k \in V_n(s_i^{j-1})} f(s_i^k, s_i^{j-1})} \tag{3}$$

where $P(s_i^j|s_i^{j-1})$ is the probability that metric s_i^j will be sampled and $V_n(s_i^{j-1})$ is the neighbor metric set of s_i^{j-1} based on E. Besides, we should note that the neighbor metric set $V_n(s_i^j)$ does not contain metric s_i^{j-1}. If s_i^{j-1} has no neighbor metric or L_p metrics have been sampled, we will stop the current random walk. Eventually, we compute the number of walks to each metric and select the top s_n metrics $\{M_{s_1}, M_{s_2}, \cdots, M_{s_n}\}$ as KPIs.

4.3   Anomaly-Related Metric Retrieval for Fault Diagnosis

The objective of unmonitored anomaly-related metric retrieval is to select metrics that capture the same implications as the anomalous metrics. This objective can be reformulated as a time series metric classification problem. We group metrics with similar implications into categories and employ a relevant classification approach to classify them. The unmonitored metrics that belong to the same category as the anomalous metrics can be considered as anomaly-related metrics. Therefore, we design a metric retrieval approach based on metric classification. Figure 3(c) illustrates the structure of the proposed unmonitored anomaly-related Metric Retrieval approach using Graph Neural Network (MRG). We initially utilize a Graph Neural Network (GNN) to classify metrics and subsequently apply the Dynamic Time Warping (DTW) distance to recommend the most similar metrics.

GNN-Based Metric Classification. The unmonitored metric set can be defined as $\mathbf{M}^u = \{M_1^u, M_2^u, \cdots, M_m^u\}$, where m is the number of unmonitored metrics, $M_i^u = [x_i^{u,t-w_r}, \cdots, x_i^{u,t}]$ is the ith metric which includes collected observations from time $t - w_r$ to t, and w_r is the corresponding window size. The detected anomalous metric is $M^a = [x^{a,t-t_r}, \cdots, x^{a,t}]$. The GNN-based metric classification aims to classify the metrics $\mathbf{M}^u$ and M^a into user-designed categories $\mathbf{C} = \{c_1, c_2, \ldots, c_o\}$, where o is the number of categories.

First, we introduce time series feature theory to extract features as initial embedding input of graph layers. Comparing with adopting original observations as initial embedding, extracted features can not only retain the time series information to ensure performance, but also reduce the computational complexity of the graph neural network. Hence, we adopt feature extraction approaches in this paper. Previous approaches extract time series features from three perspectives, namely *statistical domain, temporal domain, and spectral domain* [7, 10]. Here we adopt TSFEL [7] to extract features from above three domains. Eventually, the extracted features of metric M_i^u can be defined as $e_i^u = [e_{i,1}^u, e_{i,2}^u, \cdots, e_{i,d}^u]$, where d is the number of features.

Second, we adopt Graph Neural Network (GNN) to leverage and process the topology information of metrics. Comparing with directly using the extracted features, GNN utilizes both the time series information within metrics and the topology information between metrics for metric classification, which may result in better performance. In practice, there may involve tens of thousands of unmonitored metrics need to be recommended. We are required to optimize the process of aggregating neighbor metrics to ensure the efficiency of graph neural network. The aggregation process can be expressed as:

$$\hat{e}_i^{u,t+1} = \text{AGG}_{t+1}(\{e_j^{u,t}|e_j^{u,t} \in V_m(e_i^{u,t})\}) \tag{4}$$

$$e_i^{u,t+1} = \sigma(W_{t+1} \cdot [e_i^{u,t+1}\|\hat{e}_i^{u,t+1}]) \tag{5}$$

where $e_i^{u,t}$ and $e_i^{u,t+1}$ are the embedding at layer t and $t + 1$, $V_m(e_i^{u,t})$ is the embedding of neighbor metrics of metric M_i^u, AGG is the aggregation method for aggregating neighbor information, $\hat{e}_i^{u,t+1}$ is the aggregated information, and σ is the logistic sigmoid function. Our optimization for the aggregation process lies in the selection of $V_m(e_i^{u,t})$. Due to the large number of neighbor metrics, we randomly select metric samples from the whole metric set. Then we adopt euclidean distance to measure the similarity between $e_i^{u,t}$ and $e_j^{u,t}$ and select the top metrics with similar embedding as neighbor metrics. Finally, based on the neighbor metrics, the output embedding $e_i^u = [e_{i,1}^u, e_{i,2}^u, \cdots, e_{i,d}^u]$ can contain both the time series information within metrics and the topology information between metrics.

Third, we adopt another dense layer to process the output embedding of GNN and classify metrics. The output categories of unmonitored metrics can be denoted as $\hat{\mathbf{y}}^u = \{\hat{y}_1^u, \hat{y}_2^u, \cdots, \hat{y}_m^u\}$, where $\hat{y}_i^u$ is the output vector of unmonitored metric M_i^u. In addition, we train the graph layers and dense layer using cross entropy loss. The loss of metric M_i^u is denoted as:

$$\mathcal{L}_i = -\frac{1}{o}\sum_{j=1}^{o}[y_{i,j}^u \log \hat{y}_{i,j}^u + (1 - y_{i,j}^u) \log(1 - \hat{y}_{i,j}^u)] \tag{6}$$

where $\hat{y}_{i,j}^u$ is the predictive probability that M_i^u belongs to category c_j, and y_i^u is the ground truth vector. We choose the top predictive probability in the vector to determine the predictive category. Ultimately, the metric set $\mathbf{M}_f = \{M_i^u|y_i^u = y^a\}$, which are classified into the same category as M_a, is determined as target candidate set to be recommended.

Table 2. Statistics of Evaluation Samples for Metric Selection

VM Number	Window Size	Total Points	Anomalous Points
400	7 days	4032000	328/0.008%

Unmonitored Anomaly-Related Metric Retrieval. Unmonitored metric retrieval is to generate metric subset $\mathbf{M}_r^u = \{M_{q_1}^u, M_{q_2}^u, \cdots, M_{q_m}^u \mid M_{q_i}^u \in \mathbf{M}^u\}$ related to M_a, where r_m is the number of recommended metrics. Comparing with directly adopting $\mathbf{M}_f$ to reveal and localize anomalies, we still need to rank the target candidate set. In practical environment, there may exist hundreds or even thousands of metrics fall into the same category as M_a. The large number of unordered results will make it difficult for software engineers to make decisions. Hence, we further rank the candidate set and select the top q_m metrics for recommendation. In this paper, we adopt the DTW [22] algorithm to measure the similarity between M_a and metrics in the candidate set $\mathbf{M}_f$ and use this similarity to rank metrics. Eventually, we recommend the sorted anomaly-related metrics $\mathbf{M}_r^u$ to engineers for decision making.

5 Evaluation

In this section, we will respectively evaluate whether the proposed metric selection method MSG contributes to anomaly detection and whether the metric retrieval approach MRG benefits fault localization.

5.1 RQ1: How Effective is Metric Selection for Monitoring and Anomaly Detection?

Experiment Design. To assess the proposed MSG in DASHCHEF, we conduct experiments that combine metric selection and anomaly detection. In practice, generic exporters like those in Prometheus collect over fifty resource-related metrics for users to detect anomalies. We observe variations in the number and types of metrics selected across different industrial environments, highlighting the absence of a unified evaluation standard. Since directly evaluating the performance of metric selection solely at the metric level is challenging, we compare the anomaly detection performance using the metrics selected based on MSG against other baselines. Additionally, we focus on detecting anomalies in cloud virtual machines (VMs), a key concern for software engineers, to evaluate the effectiveness of the proposed method in selecting metrics.

For evaluation, we gather real-world metrics of cloud VMs from Tencent®. Table 2 presents the statistics of the evaluation samples, where we randomly select 400 cloud VMs. For each VM, we collect all metrics over a window size (w_s) of 7 days, with observations recorded at one-minute intervals. This results in a total of 4,032,000 data points to be detected. The exporter captures 56 metrics from 9 perspectives, including disk usage, I/O usage, memory usage, CPU

Table 3. The top 10 concerned metrics of engineers

Metric Perspective	Metric Name	Metric Meaning
I/O Usage	io_util	average io usage
	io_avgrq_sz	average size of requests
CPU Usage	cpu_detail	CPU usage of single VM
	cpu_summary	average CPU usage
	cpu_detail_max	max CPU usage
Memory Usage	pct_used	used physical memory
	pct_usable	unused physical memory
	psc_pct_used	used psc memory
Disk Usage	disk_in_use	average disk usage
Network	speed_packets_recv	speed of packet receive

Table 4. The top 10 selected metrics based on MSG

Metric Perspective	Metric Name	Metric Meaning
I/O Usage	io_util	average io usage
CPU Usage	cpu_detail	CPU usage of single VM
Memory Usage	pct_used	used physical memory
	psc_pct_used	used psc memory
Disk Usage	disk_in_use	average disk usage
Network	speed_packets_recv	speed of packet receive
	speed_packets_send	speed of packet send
Env	proc_blocked_current	number of blocked proc
CPU load	load1	average load in 1min
	load_per_cpu	average load per cpu

usage, etc. Taking disk usage as an example, the exporter collects three metrics: *disk_total, disk_used,* and *disk_in_use*. We utilize these multiple metrics to train anomaly detection models and determine whether each data point is anomalous or not.

To evaluate the anomaly detection performance, we employ Precision, Recall, and F1-score, where Precision $= \frac{TP}{TP+FP}$, Recall $= \frac{TP}{TP+FN}$, and $F1-score = \frac{2 \times Precision \times Recall}{Precision+Recall}$. We compare the performance difference between the experimental group (using metrics generated by the MSG) and the baseline groups, which consider expert experience and random selection. Furthermore, we uniformly adopt the semi-supervised variational auto-encoder (SLA-VAE) [15] as the anomaly detection model in this study.

Experiment Results. We first compare the proposed MSG and expert experience (EE) to evaluate performance. To better understand the expert experience,

Table 5. The performance comparison between MSG and baseline

Method	Precision	Recall	F1-score
EE	0.779	0.706	0.741
MSG	**0.909**	**0.915**	**0.912**

we counted tens of thousands of related tasks running on monitoring platform. We also surveyed the main concerns of software engineers on monitoring cloud VM. The expert users are engineers using our monitoring platform, who manage different services from different products. We rely on all these expert users to select the baseline metrics. Specifically, when an expert user concerns one metric, he/she can configure a simple strategy to monitor this metric in our monitoring platform and the monitoring configurations can be stored. We have tens of thousands of monitoring configurations built by more than a thousand of users. For confidentiality, we only show the conclusion here without relevant statistics.

Table 3 shows the top 10 concerned metrics of engineers which are selected as baseline metrics here. Besides, we adopt MSG to generate the top 10 metrics, which are presented in Table 4. Then we use the selected metrics to train model and detect anomalies respectively. The performance comparison results are shown in Table 5. We can find that the anomaly detection performance based on MSG is significantly better than expert experience. The selected metric comparison and performance comparison can illustrate that metric selection based on expert experience may ignore important metrics that contribute to anomaly detection. In contrast, based on the anomalous behavior of metrics, MSG can assist in monitoring and detecting system anomalies comprehensively.

We then focus on the impact of the number of metrics on anomaly detection performance. We randomly select $n = \{10, 20, 30, 40, 50\}$ metrics from the whole collected metrics to train model and detect anomalies, and compare the performance with MSG. Figure 4 presents the anomaly detection comparison results between MSG and randomly chosen baselines. We can find that using more metrics to detect anomalies will result in better performance when randomly selecting metrics. However, we only get relatively good results when selected metric set is close to the full set. It means that only some of the collected metrics are useful for anomaly detection. When these metrics are not selected, the detection performance is negatively affected. In addition, we also find that MSG outperforms baselines. It illustrates that MSG can correctly select the useful metrics for anomaly detection.

5.2 RQ2: How Effective is Metric Retrieval for Recommending Unmonitored Anomaly-Related Metric?

Experiment Design. When evaluating the performance of the proposed MRG, we mainly focus on the accuracy that whether the unmonitored metrics will be classified into the same category as those detected as anomalies. In practice, it is difficult to establish and label the relationship between all unmonitored

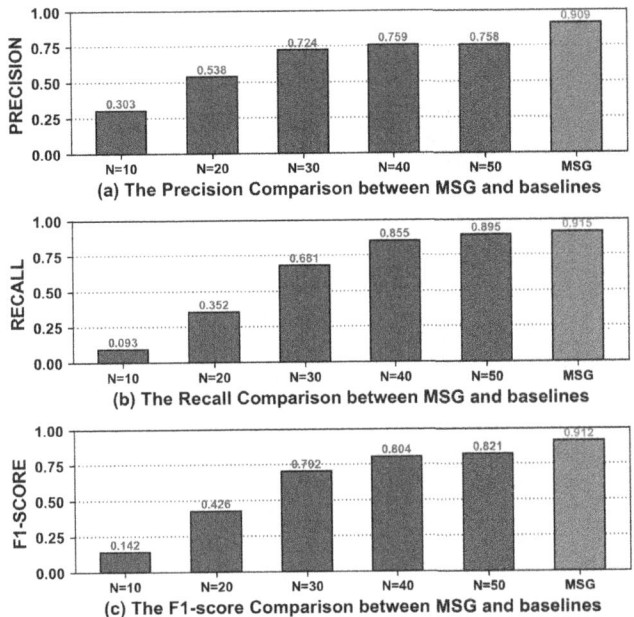

(a) The Precision Comparison between MSG and baselines

(b) The Recall Comparison between MSG and baselines

(c) The F1-score Comparison between MSG and baselines

Fig. 4. The anomaly detection comparison results between MSG and randomly chosen baselines.

metric and anomalous metrics. To solve it, we transform the research question into time series classification problem as mentioned in Sect. 4.3. Both anomalous metrics and unmonitored metrics are adopted as input to MRG synchronously. The unmonitored metrics that fall into the same category as anomalous metrics can be understood as anomaly-related metrics. Therefore, we instead evaluate the accuracy that whether the metrics will be classified into the correct category.

We adopt the metrics collected from one system in Tencent® for evaluation. Table 6 presents the statistics of evaluation sample for evaluating the proposed MRG. This evaluation sample contains all metrics collected from 400 cloud VMs, and the window size w_r of metrics is 2 days. Then we label the metrics from 8 label domains, and the details of the label domains are shown in Table 7. We focus on the normal patterns and anomalous patterns when labeling the metrics. If metrics have the same normal and anomalous patterns, we consider that they may belong to the same category. Eventually, the metrics are labeled into 15 categories.

We use accuracy score to evaluate the metric classification performance, where accuracy $= \frac{TP+TN}{TP+FP+TN+FN}$. We compare the accuracy difference between experimental group and baseline group to verify the effectiveness of metric retrieval model. For experimental group, we use MRG to classify the metrics. For baseline group, we adopt previous classification work to classify the metrics. In addition, it is difficult to evaluate the correctness of the ranking of retrieved metrics. We do not evaluate it here but will discuss it in Sect. 6.2.

Table 6. Statistics of Evaluation Samples for Metric Retrieval

Metric Num	Window Size	Label Domain Num	Label Size
22400	2 days	8	15

Table 7. The label domain for metric classification

Metric Domain	Label Domain
Normal Patterns	continuity VS discreteness
	stationarity VS non-stationarity
	periodicity VS non-periodicity
	high variance VS low variance
	up trend VS down trend VS no trend
Anomalous Patterns [30]	up VS down VS fluctuation
	sudden VS steady
	continuous VS transient

Experiment Results. When designing the baseline models, we focus on evaluating two modules in MRG, namely feature extraction and classification model. We adopt $2 * 2$ factorial design [29] to conduct experiments and evaluate these two factors. For the first factor, the baseline adopts original observations to train classification models, while the experimental group uses extracted features. For the second factor, since MRG introduces GNN layers and Dense layer to classify metrics, we introduce two baseline models to evaluate it. The first one is Random Forest (RF) classifier [23], which is the representative classification method. The second one is only Dense layer, which is used to evaluate the performance of GNN layers.

Figure 5 presents the performance comparison results between MRG and baselines. For the feature extraction factor, we can observe that the experimental group has a better performance no matter what classification models are adopted. It illustrates that the feature extraction method can extract important information in metrics so that the overall performance is significantly improved. For the classification model factor, we can find that random forest classifier performs better when trained by original observations. When using the extracted features to train model, MSG outperforms baselines and random forest classifier performs worst. The previous result is rooted in the size of model parameters of GNN layers and Dense layer. When using original observations to train the model, the input size is 1440 per day $\times$ 2 day = 2880, which may result in large size of model parameters. We need more samples to train the GNN layers and Dense layer when directly using original observations. However, the sample size in the industrial environment is difficult to meet this requirement, which also illustrates the importance of feature extraction. When using extracted features, MSG can leverage and process both the topology information between metrics

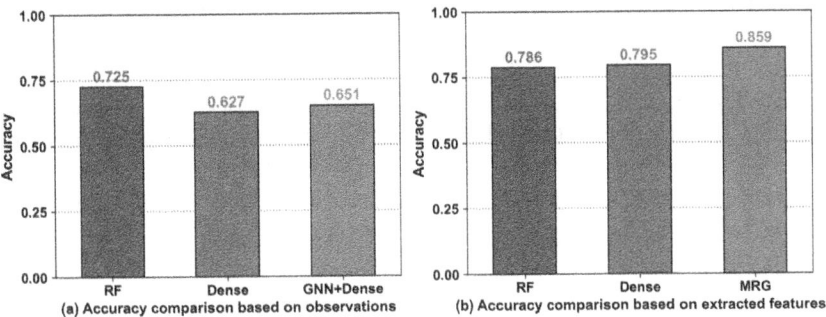

Fig. 5. The metric classification performance comparison results between MRG and baselines.

and time series information within metrics, which results in better performance. In summary, due to the better classification performance, the proposed MSG is effective for recommending unmonitored anomaly-related metric.

6   Discussion

6.1   Practical Implication

The practical implication of this paper is that graph learning is suitable for characterizing the relationship between metrics. This paper introduces graph to represent the anomalous relationship between metrics and adopt graph learning to process the topology information of metrics. Comparing with traditional approaches, graph learning can better characterize the unstructured topology of metrics. The experimental results have also shown that graph learning can effectively improve the performance of metric recommendation. In addition, in the industrial environment, metrics are not only used to detect and locate anomalies, but also for further analysis such as load balancing, hotspot discovery and analysis. Due to the capacity of characterizing the relationship between metrics, graph learning can also assist software engineers in further analysis in these new scenarios.

6.2   Limitations

The limitation of the proposed metric recommendation service lies in the lack of the ranking of the retrieved metrics for fault localization. The ranking of the retrieved metrics can further favor fault analysis and diagnosis. Currently, all of the retrieved anomaly-related metrics are recommended to engineers, who will analyze the root cause manually. In future work, we will design an automatic fault diagnosis methods to locate the root cause, and combine with it to improve the ranking of retrieved metrics.

7 Related Work

Anomaly Detection. Current anomaly detection work can mainly be categorized into traditional statistical methods [25] and deep learning based methods [14,15,26,31]. Deep learning based methods are popular recently due to their powerful capacity in modeling complex data. However, no matter which method is applied, the selected metrics matter. The proposed method in this paper can aid prior work in selecting effective metrics for anomaly detection in online systems.

Automatic Feature Selection for Anomaly Detection. There also exist some work [11–13] that provides automatic mechanism for selecting metrics for anomaly detection. For example, Fu [12] selected metrics that can maximize the mutual information [17] between the selected metrics and the historical faults. Farshchi et al. [11] leveraged the correlation between metrics and the operator's activity log to select metrics based on regression analysis. Guan et al. [13] selected the most relevant principal components of different failure types to identify anomalies in cloud infrastructures. However, all of them heavily rely on some external information such as the time and types of historical faults to perform metric selection, rendering these methods difficult to apply in practice.

Fault Diagnosis. A considerable amount of literatures [9,19–21,30] has investigated fault diagnosis in online systems. For example, MicroCause [21] designs a path condition time series algorithm to learn the relationships between metrics and use random walk to infer the root cause. PatternMiner [30] uses anomaly pattern classification to pick out important anomaly patterns set by engineers and perform root cause metric ranking. Dejavu [19] train a classifier to diagnose recurring failures in online systems. In general, fault diagnosis can be performed from different perspectives, so the exact problem formulations of different work can be quite different. The proposed method in this study can be applied together with prior methods to facilitate fault diagnosis in online systems.

8 Conclusion

This paper proposes DASHCHEF, a metric recommendation service for online systems on the basis of graph learning. After an investigation on prior dashboard practice, we design metric recommendation mechanisms for two essential usage scenarios of software maintenance in online systems, namely metric selection for anomaly detection and metric retrieval for fault diagnosis. Extensive experiments confirm the effectiveness of DASHCHEF. Moreover, DASHCHEF has been successfully merged into the software maintenance practice in Tencent® online systems.

Acknowledgments. The research is supported by the National Natural Science Foundation of China (No. 62272495) and the Guangdong Basic and Applied Basic Research Foundation (No. 2023B1515020054), and sponsored by Tencent.

References

1. Grafana dashboards (2023). https://grafana.com/grafana/dashboards/. Accessed 12 Dec 2023
2. Kafka monitoring (2023). https://kafka.apache.org/documentation
3. Node exporter (2023). https://github.com/prometheus/node_exporter
4. Prometheus monitoring for containers (2023). https://github.com/google/cadvisor/blob/master/metrics/prometheus.go
5. Redis monitoring (2023). https://redis.io/commands/info/
6. Baradari, I., Shoar, M., Nezafati, N., Motadel, M.: A new approach for KPI ranking and selection in ITIL processes: using simultaneous evaluation of criteria and alternatives (SECA). J. Ind. Eng. Manag. Stud. **8**(1), 152–179 (2021)
7. Barandas, M., et al.: TSFEL: time series feature extraction library. SoftwareX **11**, 100456 (2020)
8. Beyer, B., Jones, C., Petoff, J., Murphy, N.R.: Site Reliability Engineering: How Google Runs Production Systems. O'Reilly Media, Inc. (2016)
9. Chen, P., Qi, Y., Zheng, P., Hou, D.: CauseInfer: automatic and distributed performance diagnosis with hierarchical causality graph in large distributed systems. In: IEEE INFOCOM 2014-IEEE Conference on Computer Communications, pp. 1887–1895. IEEE (2014)
10. Christ, M., Braun, N., Neuffer, J., Kempa-Liehr, A.W.: Time series feature extraction on basis of scalable hypothesis tests (tsfresh-a python package). Neurocomputing **307**, 72–77 (2018)
11. Farshchi, M., Schneider, J.G., Weber, I., Grundy, J.: Metric selection and anomaly detection for cloud operations using log and metric correlation analysis. J. Syst. Softw. **137**, 531–549 (2018)
12. Fu, S.: Performance metric selection for autonomic anomaly detection on cloud computing systems. In: 2011 IEEE Global Telecommunications Conference-GLOBECOM 2011, pp. 1–5. IEEE (2011)
13. Guan, Q., Fu, S.: Adaptive anomaly identification by exploring metric subspace in cloud computing infrastructures. In: 2013 IEEE 32nd International Symposium on Reliable Distributed Systems, pp. 205–214. IEEE (2013)
14. He, Z., et al.: A spatiotemporal deep learning approach for unsupervised anomaly detection in cloud systems. IEEE Trans. Neural Netw. Learn. Syst. **34**(4), 1705–1719 (2020)
15. Huang, T., Chen, P., Li, R.: A semi-supervised VAE based active anomaly detection framework in multivariate time series for online systems. In: Proceedings of the ACM Web Conference 2022, pp. 1797–1806 (2022)
16. Jha, D.N., Lenton, G., Asker, J., Blundell, D., Wallom, D.: Holistic runtime performance and security-aware monitoring in public cloud environment. In: 2022 22nd IEEE International Symposium on Cluster, Cloud and Internet Computing (CCGrid), pp. 1052–1059. IEEE (2022)
17. Kraskov, A., Stögbauer, H., Grassberger, P.: Estimating mutual information. Phys. Rev. E **69**(6), 066138 (2004)
18. Levin, J., Benson, T.A.: ViperProbe: rethinking microservice observability with eBPF. In: 2020 IEEE 9th International Conference on Cloud Networking (CloudNet), pp. 1–8. IEEE (2020)
19. Li, Z., et al.: Actionable and interpretable fault localization for recurring failures in online service systems. In: Proceedings of the 2022 30th ACM Joint Meeting on European Software Engineering Conference and Symposium on the Foundations of Software Engineering. ESEC/FSE 2022 (2022)

20. Lin, J., Chen, P., Zheng, Z.: Microscope: pinpoint performance issues with causal graphs in micro-service environments. In: Pahl, C., Vukovic, M., Yin, J., Yu, Q. (eds.) ICSOC 2018. LNCS, vol. 11236, pp. 3–20. Springer, Cham (2018). https://doi.org/10.1007/978-3-030-03596-9_1

21. Meng, Y., et al.: Localizing failure root causes in a microservice through causality inference. In: 28th IEEE/ACM International Symposium on Quality of Service, IWQoS 2020, Hangzhou, China, 15–17 June 2020, pp. 1–10. IEEE (2020)

22. Müller, M.: Dynamic time warping. In: Müller, M. (ed.) Information Retrieval for Music and Motion, pp. 69–84. Springer, Heidelberg (2007). https://doi.org/10.1007/978-3-540-74048-3_4

23. Paul, A., Mukherjee, D.P., Das, P., Gangopadhyay, A., Chintha, A.R., Kundu, S.: Improved random forest for classification. IEEE Trans. Image Process. **27**(8), 4012–4024 (2018)

24. Ramadona, S., Haryadi, S., Aryanti, D.R.: Over the top call service key performance indicator. In: 2015 1st International Conference on Wireless and Telematics (ICWT), pp. 1–4. IEEE (2015)

25. Siffer, A., Fouque, P., Termier, A., Largouët, C.: Anomaly detection in streams with extreme value theory. In: Proceedings of the 23rd ACM SIGKDD International Conference on Knowledge Discovery and Data Mining, Halifax, NS, Canada, 13–17 August 2017, pp. 1067–1075. ACM (2017)

26. Su, Y., Zhao, Y., Niu, C., Liu, R., Sun, W., Pei, D.: Robust anomaly detection for multivariate time series through stochastic recurrent neural network. In: Proceedings of the 25th ACM SIGKDD International Conference on Knowledge Discovery & Data Mining, pp. 2828–2837. ACM (2019)

27. Tong, H., Faloutsos, C., Pan, J.Y.: Fast random walk with restart and its applications. In: Sixth International Conference on Data Mining (ICDM 2006), pp. 613–622. IEEE (2006)

28. Weng, T., Yang, W., Yu, G., Chen, P., Cui, J., Zhang, C.: Kmon: an in-kernel transparent monitoring system for microservice systems with eBPF. In: 2021 IEEE/ACM International Workshop on Cloud Intelligence (CloudIntelligence), pp. 25–30. IEEE (2021)

29. Wohlin, C., Runeson, P., Höst, M., Ohlsson, M.C., Regnell, B., Wesslén, A.: Experimentation in Software Engineering. Springer, Heidelberg (2012)

30. Wu, C., et al.: Identifying root-cause metrics for incident diagnosis in online service systems. In: 32nd IEEE International Symposium on Software Reliability Engineering, ISSRE 2021, Wuhan, China, 25–28 October 2021, pp. 91–102. IEEE (2021)

31. Xu, H., et al.: Unsupervised anomaly detection via variational auto-encoder for seasonal kpis in web applications. In: Proceedings of the 2018 World Wide Web Conference on World Wide Web, WWW 2018, Lyon, France, 23–27 April 2018, pp. 187–196. ACM (2018)

SC-WGAN: GAN-Based Oversampling Method for Network Intrusion Detection

Wuxia Bai, Kailong Wang, Kai Chen$^{(\boxtimes)}$, Shenghui Li, Bingqian Li, and Ning Zhang

Huazhong University of Science and Technology, Wuhan 430074, China
{wuxiabai,wangkl,kchen,lishenghui,libq2022,zn_hust}@hust.edu.cn

Abstract. The growing cyber threat landscape necessitates robust defenses, with Network Intrusion Detection Systems (NIDS) at the forefront. Leveraging Deep Neural Networks (DNN) has significantly improved detection accuracy in NIDS. Nonetheless, the inherent data imbalance between malicious and normal network traffic impairs the efficacy of DNN-based methods. Traditional approaches employ Generative Adversarial Networks (GANs) to mitigate this by generating minority class samples. However, these methods often struggle with the scarcity of specific data types during training, leading to low-quality synthetic samples and sub-optimal detection performance.

Addressing this, we introduce the Synchronous Classifier Wasserstein Generative Adversarial Network (SC-WGAN). This novel model extends the WGAN framework by integrating a classifier that processes all traffic data types. This classifier collaborates with the discriminator during the training process, guiding the generator to produce higher-quality synthetic samples. We evaluate SC-WGAN's performance on three well-known open-source benchmark datasets. Additionally, we compare it against contemporary GAN-based solutions tackling data imbalance. Our findings reveal that SC-WGAN, surpasses existing methods in generating more representative samples and enhancing NIDS detection accuracy.

Keywords: Network Intrusion Detection · Imbalanced Data · Oversampling · GAN

1 Introduction

In the era of rapid Internet technology advancement, the global community of Internet users is expanding, leading to a continuous surge in network traffic. Consequently, an increasing amount of vital, sensitive, and confidential information is intricately linked to the Internet [1]. With the existence of security vulnerabilities within systems, attackers can exploit these weaknesses to compromise the confidentiality, integrity, and availability of network assets through various network attacks [2]. To counteract such threats, the utilization of Intrusion Detection Systems (IDS) has emerged as a highly effective and widely adopted approach.

K. Wang—Co-first author.

G. Bai et al. (Eds.): ICECCS 2024, LNCS 14784, pp. 23–42, 2025.
https://doi.org/10.1007/978-3-031-66456-4_2

IDS can be categorized based on their detection approach: signature-based [3], known for low error rates but limited by an inability to detect new, zero-day attacks, and anomaly-based [4], which models normal behavior to identify unknown threats. As network environments become increasingly complex and the volume of network data surges, these traditional IDS methods struggle to maintain accuracy and efficiency. Recent research has begun to explore the integration of deep learning techniques in anomaly-based IDS, aiming to leverage these methods' ability to handle large-scale data and complex patterns.

Ideal deep learning-based classification methods operate under the assumption that the target classifications share the same distribution. In the context of network traffic classification, this assumption is invalid since normal network traffic far outweighs the malicious traffic in real-world. This implies an imbalance issue in the dataset used for model training, causing deep learning-based models to lean towards normal traffic. Consequently, there is insufficient learning of features associated with malicious traffic, making the models prone to classifying all unknown types of traffic as normal, leading to the successful execution of attacks.

To address the issue of data imbalance, oversampling methods are the most commonly used strategy, which directly increase the number of minority class samples [5]. As a generative model, Generative Adversarial Network (GAN) [6], based on the principles of adversarial game learning, demonstrate significant advantages by continually learning the distribution of original data and generating high-quality data in various domains such as image, sound, and text [7]. As a deep learning method, however, the training of GAN requires a large amount of data, and the limited volume of minority class makes it challenging to ensure the effectiveness of model training and the quality of generated data.

Therefore, this paper proposes an improved GAN model called SC-WGAN (Synchronous Classifier Wasserstein Generative Adversarial Network). SC-WGAN addresses the issue of insufficient sample quantity during the training process by effectively utilizing all types of traffic data rather than specific one. This enhancement aims to improve the quality of generated data. And through balancing dataset to improve the performance of deep learning-based IDS. The main contributions of this paper are summarized as follows:

- We have designed and developed a variant of GAN called SC-WGAN. This model effectively addresses the issue of insufficient sample quantity during the model training process.
- In comparison to other oversampling methods, the dataset balanced by our proposed method has better results in training the classifier leading to the improvement of classification.
- Experimental results conducted on the NSL-KDD, UNSW-NB15, and CIC-IDS2017 datasets confirm that our proposed method effectively tackles with the data imbalance problem for minority attack classes and improves the performance of attack detection.

The rest of the paper is organized as follows: Sect. 2 introduces the related work of oversampling methods and briefly introduces GAN and WGAN and their applications in the field of NIDS. Section 3 presents the proposed method. Section 4 describes the experimental setting. Section 5 describes results. Section 6 provides a conclusion of the paper.

2   Related Work

In this section, we will discuss related work on oversampling methods and provide a brief introduction of GAN and WGAN and their application for NIDS.

2.1   Oversampling Methods

The effectiveness of oversampling methods can largely be traced back to two key factors. First, oversampling acts as a data preprocessing technique that is applicable across various classification methods. Second, it directly tackles the fundamental issue of data imbalance by compensating for the scarcity of data in minority classes [8].

The most basic form of oversampling, Random Oversampling (ROS), increases the number of samples by simply replicating minority class samples. However, ROS can lead to overfitting since it does not add new information. To counter this, the Synthetic Minority Oversampling Technique (SMOTE) [9] synthesizes new samples through linear interpolation between two minority class samples, mitigating ROS's overfitting issues. Nevertheless, SMOTE might still produce samples that suffer from issues like blurred class boundaries and over-laps, due to its reliance on interpolation. Borderline-SMOTE [10] enhances SMOTE by focusing on minority class samples near the decision boundary to create new samples, thus refining the class sample distribution. In contrast to algorithms like SMOTE that generate the same number of samples for each minority class, Adaptive Synthetic (ADASYN) [11] algorithm determines the number of samples to be synthesized for each minority class based on the data distribution, generating more samples for the harder-to-learn minority classes.

Although these methods are straightforward to implement, they are prone to overfitting when dealing with complex feature spaces or high sample diversity. In contrast, GAN-based oversampling methods generate samples closely resembling the statistical characteristics of real data by learning data distributions. This approach not only effectively increases the number of minority class samples but also enhances sample diversity, reducing the risk of overfitting.

2.2   A Brief Introduction of GAN and WGAN

In 2014, the proposal of the GAN sparked a revolution in the field of deep learning [6]. As shown in Fig. 1, a GAN consists of two networks: a generator and a discriminator. The generator learns the distribution of the original data and transforms input noise data to generate synthetic samples, which are then fed into the discriminator. The discriminator's task is to determine whether the input comes from the real dataset or the generated samples. Both the generator and the

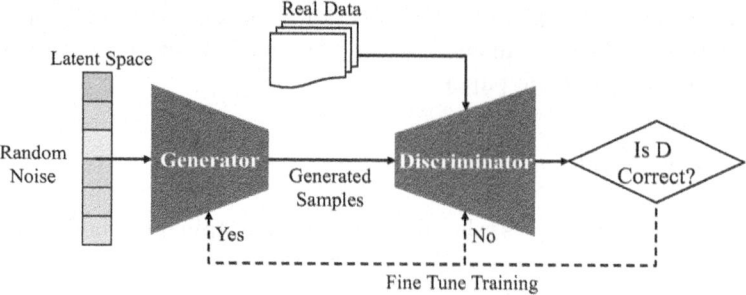

Fig. 1. Framework of GAN.

discriminator are trained simultaneously. During this process, the discriminator aims to maximize the probability of assigning the correct label to both real and generated samples, while the generator aims to maximize the probability of the discriminator assigning the wrong label to its generated samples. The ultimate goal of the GAN is to achieve a Nash equilibrium, where the generated samples are indistinguishable from real data, and the discriminator is unable to determine whether the input data is real or generated.

Wasserstein Generative Adversarial Network (WGAN) [12] replaces the Jensen-Shannon divergence used in GAN with the Wasserstein distance to evaluate the distribution difference between real and generated samples. The Wasserstein distance, known for its better smoothness properties, theoretically resolves the gradient vanishing problem faced by the generator under the approximation of an optimal discriminator. This eliminates the need to carefully balance the training of the generator and discriminator, leading to a more stable training process. Additionally, the introduction of Wasserstein distance addresses the mode collapse issue in GAN, ensuring the diversity of generated samples.

2.3   Application of GANs in NIDS

Due to the data generation capabilities of GANs, many researchers in recent years have applied them to the field of network intrusion detection to address the issue of imbalanced classification.

Some studies directly employed GANs or their variants to address the imbalance in network traffic datasets. For instance, Vu et al. [13] used the Auxiliary Classifier Generative Adversarial Network, Douzas et al. [14] adopted the Conditional Generative Adversarial Network, Lee et al. [15] utilized GANs, and Liu et al. [16] implemented WGAN with Gradient Penalty. These efforts confirmed the effectiveness of GANs in enhancing data balance in network intrusion detection systems and in improving the detection capabilities of models.

The aforementioned studies focus on applying GANs and their variants to network traffic generation. Additionally, some research emphasizes integrating domain expertise in network traffic to optimize data processing. Ring et al. [17]

utilized a WGAN-GP-based approach to generate network traffic with similar underlying characteristics. To address the challenges posed by the continuous and discrete attributes of the data, they proposed three preprocessing methods to suit the requirements of GAN processing. Hao et al. [18] enhanced the performance of GANs by preprocessing network traffic features and employing the Earth-Mover distance and an encoder structure. This allowed for better capturing of data distributions and learning of latent space representations.

Additionally, some studies have introduced innovative mechanisms into GANS to address data imbalance issues and improve classification performance. Guo et al. [5] introduced an end-to-end framework, TA-GAN, that utilizes a feedback mechanism to optimize sample generation, integrating both sample generation and classifier training to produce samples that better meet the learning requirements of classifiers, thus enhancing classification performance. Gong et al. [19] proposed the SA-ACGAN-GN model, which incorporates a self-attention mechanism and gradient normalization to address the long-tail distribution and stability issues in mobile traffic data. Ding et al. [20] developed the RVGAN-TL model that combines variational autoencoders and similarity measurements, along with roulette wheel selection and transfer learning, to effectively balance data distribution. Yuan et al. [21] introduced a method named B-GAN, which uses long short-term memory networks to generate high-quality anomaly samples, addressing the imbalanced data classification problem in industrial control system intrusion detection. Dlamini et al. [22] proposed the DGM model, based on conditional GANs introducing KL divergence for generating flow data, comparing it with four common machine learning classification methods and traditional oversampling methods. Subsequently, Kumar et al. [23] introduced the WCGAN combined with an XGBoost classifier to enhance network intrusion detection accuracy, achieving superior results compared to experiments based on the DGM model. Following this, Srivastava et al. [24] building upon the Wasserstein Conditional GAN, incorporated gradient penalty (WCGAN-GP) and employed a genetic algorithm for feature selection, proving that the WCGAN-GP model outperforms both WCGAN and DGM.

The application of GANs in NIDS marks a significant advancement in addressing data imbalance issues using deep learning techniques. The success of GANs requires a large amount of training data; however, the aforementioned studies do not account for the fact that the malicious traffic data used to train GANs for generating minority class samples often suffers from insufficient sample quantities. Therefore, designing appropriate GAN-based oversampling methods tailored to the characteristics of network traffic datasets is crucial for enhancing intrusion detection performance.

3   Methodology

3.1   Problem Definition

In the task of network traffic classification, assume that the dataset (X, Y) contains a total of n different types and N records, where X represents the network traffic data and Y represents the corresponding network traffic type. The n types of traffic are labeled from 1 to n in descending order of sample quantity. The i-th class of network traffic has N_i samples. Thus, $N_1 > \ldots > N_i > \ldots > N_n$. And then $\frac{N_1}{N_1} < \ldots < \frac{N_1}{N_i} < \ldots < \frac{N_1}{N_n}$. The Imbalance Ratio (IR) is defined using Eq. (1).

$$IR_i = \frac{N_1}{N_i} \tag{1}$$

IR defines the degree of imbalance for each type of data. The larger the value of IR, the higher the degree of imbalance in the dataset. When $IR_i > \epsilon$ (where ϵ is a predefined constant), the dataset (X, Y) is considered to be imbalanced. Types with labels less than i are designated as the majority class, while those with labels greater than i are designated as the minority class.

3.2   System Overview

To mitigate the imbalance among different types of traffic in the dataset, a GAN is employed to generate a specified number of samples for the minority class. The enriched dataset is subsequently utilized to train a deep learning-based traffic classification model.

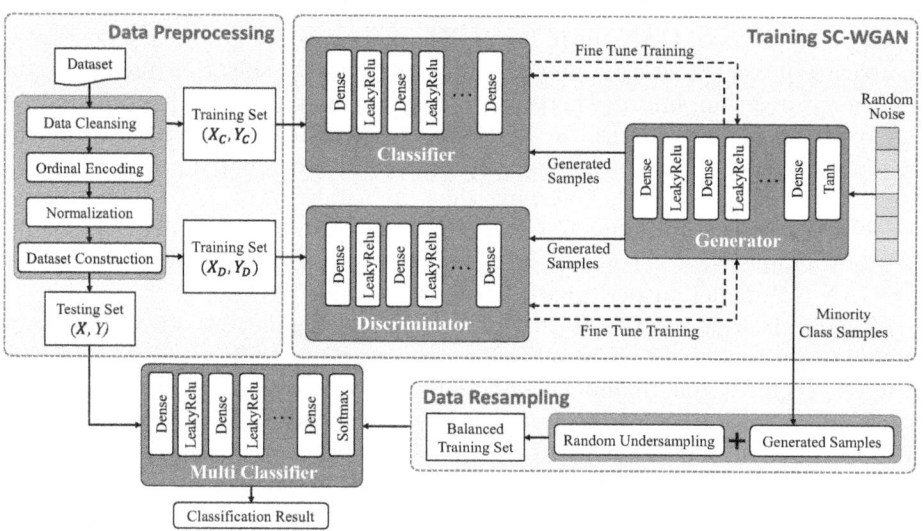

Fig. 2. Network intrusion detection system based on SC-WGAN.

To address the issue of limited samples during GAN training, we introduce a synchronous training classifier based on WGAN. The classifier operates under a similar theoretical framework as the discriminator. However, unlike the discriminator, the classifier accepts all types of traffic as input, thereby teaching the generator to recognize the common features across all real traffic. Subsequently, the discriminator fine-tunes the generator to learn the characteristics of the specific data distribution. By integrating the classifier and discriminator to jointly guide the generator, this approach not only ensures effective model training but also enhances the quality of the generated samples.

As illustrated in Fig. 2, the overall workflow consists of four parts: data preprocessing, training the SC-WGAN model, data resampling, and training the multi-class classifier to classify unknown types of traffic. To facilitate the understanding of our approach, we further provide details of the three critical processes in the rest of this section: data preprocessing, training the SC-WGAN, and data resampling.

3.3   Data Preprocessing

The data preprocessing stage is divided into four key steps. First, data cleansing is performed to eliminate errors or irrelevant information. Next, given the variety of data types, data encoding is applied to transform the data into a format suitable for machine learning models. Subsequently, data normalization, which considers the different scales of data, is conducted to enhance the efficiency of model training, accelerate model convergence, and improve model stability. Finally, dataset construction is carried out to fulfill the specific requirements of model training. A particular type of network traffic is selected from the original dataset and labeled as *True* to form the dataset (X_D, Y_D). Simultaneously, all types of traffic from the original dataset are also labeled as *True* to construct the dataset (X_C, Y_C). (X_D, Y_D) serves as the input for the discriminator, while (X_C, Y_C) is used as the input for the classifier.

3.4   Training SC-WGAN

The SC-WGAN introduces a classifier that functions as an additional discriminator, guiding the generator to learn the entire real data distribution from the dataset. Subsequently, the discriminator fine-tunes the generator to refine its output further. The model diagram of SC-WGAN is illustrated in Fig. 3. Solid lines represent inputs, and dashed lines represent the parameter training process. The classifier, discriminator, and generator undergo concurrent training. Subsequently, we will provide a detailed introduction to each of these components.

Classifier. The classifier receives input comprising all types of traffic from the training set (X_C, Y_C) along with samples generated by the generator. On one hand, the classifier processes real traffic data, adjusting its neural network parameters accordingly. On the other hand, it evaluates samples produced by the generator, performing classifications on these inputs. Functioning similarly to a discriminator, the classifier strives to discern whether the inputs originate from real data or are produced by the generator. It engages in a strategic interplay with the generator, analogous to the interaction between the discriminator and

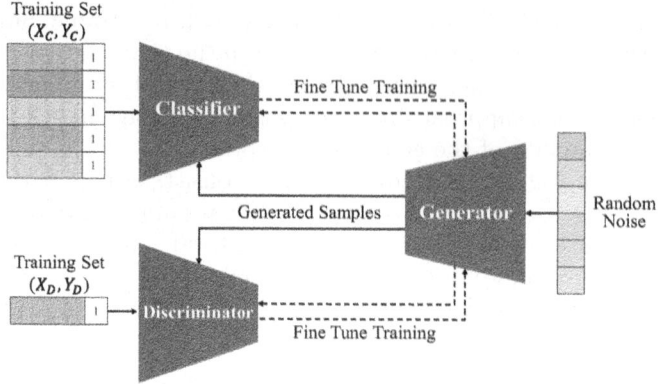

Fig. 3. Framework of SC-WGAN.

the generator. The primary role of the classifier is to guide the generator in capturing the common features of real traffic, after which the discriminator further refines the generator's output. In accordance with the principles of WGAN, the loss function for the classifier is defined as follows:

$$\mathcal{L}_C = \mathbb{E}_{z \sim P_z} \left[c_\varphi \left(g_\theta(z) \right) \right] - \mathbb{E}_{x_C \sim P_r} \left[c_\varphi(x_C) \right] \tag{2}$$

where z represents the random noise, P_z represents the probability distribution of random noise (usually Gaussian distribution). $c_\varphi(\cdot)$ represents the classifier, and $g_\theta(\cdot)$ represents the generator, φ and θ are the parameters of the classifier and generator respectively. x_C represents an instance selected from the dataset (X_C, Y_C), and P_r represents the probability distribution of the data.

Discriminator. Similar to the classifier, the discriminator receives input comprising a specific type of traffic from the training set (X_D, Y_D) and samples generated by the generator. The discriminator's objective is to determine whether the inputs come from (X_D, Y_D) or generated by the generator. A key distinction exists between the classifier and the discriminator: while the classifier encourages the generator to produce samples that are as realistic as possible without considering the type of traffic, the discriminator specifically guides the generator to produce samples that match a particular type of traffic corresponding to the training set (X_D, Y_D). This targeted approach helps refine the generator's ability to replicate specific traffic patterns accurately. The loss function of the discriminator is as follows:

$$\mathcal{L}_D = \mathbb{E}_{z \sim P_z} \left[f_\omega \left(g_\theta(z) \right) \right] - \mathbb{E}_{x_D \sim P_r} \left[f_\omega(x_D) \right] \tag{3}$$

where x_D represents an instance selected from the dataset (X_D, Y_D), $f_\omega(\cdot)$ represents the discriminator, ω is the parameters of the discriminator.

Generator. The generator updates its parameters based on feedback from both the classifier and the discriminator, aiming to minimize the Wasserstein distance between the distribution of generated samples and real data. The classifier is trained first, guiding the generator to learn common features from the entire dataset. Then, the discriminator directs the generator to generate specific types of data. While the objectives of the classifier and the discriminator are to accurately identify generated samples, the generator strives to create samples indistinguishable from real data, thereby challenging the classifier and discriminator to misclassify them as genuine. Therefore, the loss function of the generator is given by:

$$\mathcal{L}_G = -\mathbb{E}_{z \sim P_z}\left[c_\varphi\left(g_\theta(z)\right)\right] - \mathbb{E}_{z \sim P_z}\left[f_\omega\left(g_\theta(z)\right)\right] \tag{4}$$

In conclusion, SC-WGAN aims to maximize the following objective function:

$$\arg\max_{\theta, \omega, \varphi} \quad \mathbb{E}_{x_c \sim P_r}\left[c_\varphi(x_c)\right] - \mathbb{E}_{z \sim P_z}\left[c_\varphi(g_\theta(z))\right]$$
$$+ \mathbb{E}_{x_D \sim P_r}\left[f_\omega(x_D)\right] - \mathbb{E}_{z \sim P_z}\left[f_\omega(g_\theta(z))\right] \tag{5}$$

The pseudocode for training SC-WGAN is shown in Algorithm 1.

Algorithm 1. Training process of SC-WGAN

Input: The training set (X_C, Y_C), (X_D, Y_D) after preprocessing, the minibatch size m, the number of iterations T, the clipping parameter c
Output: Generator from SC-WGAN
1: **for** $t = (0, ..., T)$ **do**
2:     **for** each batch from (X_C, Y_C), (X_D, Y_D) **do**
3:       Sample m examples from X_C, with labels Y_C
4:       Sample m noise samples from noise prior P_z
5:       update the classifier, keeping G fixed:

$$\nabla_\varphi \frac{1}{m} \sum_{i=1}^{m} \left[-c_\varphi\left(x^{(i)}\right) + c_\varphi\left(g_\theta\left(z^{(i)}\right)\right) \right]$$

6:       φ weight clipping
7:       Sample m noise samples from noise prior P_z
8:       update the generator: $-\nabla_\theta \frac{1}{m} \sum_{i=1}^{m} c_\varphi\left(g_\theta\left(z^{(i)}\right)\right)$
9:       Sample m examples from X_D, with labels Y_D
10:      Sample m noise samples from noise prior P_z
11:      update the discriminator, keeping G fixed:

$$\nabla_\omega \frac{1}{m} \sum_{i=1}^{m} \left[-f_\omega\left(x^{(i)}\right) + f_\omega\left(g_\theta\left(z^{(i)}\right)\right) \right]$$

12:      ω weight clipping
13:      Sample m noise samples from noise prior P_z
14:      update the generator: $-\nabla_\theta \frac{1}{m} \sum_{i=1}^{m} f_\omega\left(g_\theta\left(z^{(i)}\right)\right)$
15:    **end for**
16: **end for**

3.5 Data Resampling

To adjust the proportion of different types of traffic and obtain a balanced training set where each type of network traffic contains m samples, this paper uses Random Under-Sampling method to sample the majority class with $N_i > m$. For the minority class with $N_i < m$, the generator is used to generate $m - N_i$ samples. Finally, the n types of network traffic are combined, resulting in a balanced training set of $n \times m$ samples, which is used to train a multi-class traffic classifier. By eliminating the impact of imbalanced data distribution on deep learning-based classification methods, the balanced dataset aims to improve the performance of the classifier.

4 Experiment Setting

This section mainly introduces the experimental settings. Comparative experiments and ablation experiments were conducted on three datasets: NSL-KDD, UNSW-NB15 and CIC-IDS2017 to verify the advancement of SC-WGAN. The proposed method was implemented using the Pytorch framework. All experiments were conducted on a workstation with Intel(R) Xeon(R) Gold 6234 CPU @ 3.30 GHz and 3 NVIDIA Quadro RTX 5000 GPUs with Ubuntu 18 64-bit operating system.

4.1 Dataset

NSL-KDD. The NSL-KDD dataset is derived from the KDD-99 dataset by Tavallaee et al. [25] through data cleaning. Each record is a description of the overall characteristics of the TCP packet sequence from start to end in a certain period of time, with a total of 41 feature attributes and 1 label attribute. The dataset is divided into five categories: Normal, DoS, Probe, R2L and U2R. Among them, the training set can be subdivided into 23 types, while the test set can be subdivided into 38 types. As a result, only 83.37% of the data in the test set corresponds to traffic types that appear in the training set. The quantity and proportion of data for each type in NSL-KDD is described in Table 1.

Table 1. The overview of NSL-KDD.

Data set	Class	Training Data			Testing Data		
		Number	Proportion	IR	Number	Proportion	IR
NSL-KDD	Normal	67343	53.46%	1	9711	43.08%	1
	DoS	45927	36.45%	1.47	7458	33.08%	1.30
	Probe	11656	9.25%	5.78	2421	10.74%	4.01
	R2L	995	0.79%	67.68	2754	12.22%	3.53
	U2R	52	0.04%	1295.06	200	0.89%	48.56
	Total	125973	–	–	22544	–	–

UNSW-NB15. The UNSW-NB15 dataset [26] is a network-based intrusion detection dataset introduced in 2015. The dataset includes traffic data consisting of normal flows and various malicious attacks collected from modern network environments. The dataset consists of 42 feature attributes (excluding serial number, IP address, and port number) and 2 label attributes, and contains 10 types of network traffic. The quantity and proportion of data for each type in UNSW-NB15 is described in Table 2.

Table 2. The overview of UNSW-NB15.

Data set	Class	Training Data			Testing Data		
		Number	Proportion	IR	Number	Proportion	IR
UNSW-NB15	Normal	56000	31.94%	1	37000	44.94%	1
	Generic	40000	22.81%	1.4	18871	22.92%	1.96
	Exploits	33393	19.04%	1.68	11132	13.52%	3.32
	Fuzzers	18184	10.37%	3.08	6062	7.36%	6.10
	DoS	12264	6.99%	4.57	4089	4.97%	9.05
	Reconnaissance	10491	5.98%	5.34	3496	4.25%	10.58
	Analysis	2000	1.14%	28	677	0.82%	54.65
	Backdoors	1746	1.00%	32.07	583	0.71%	63.46
	Shellcode	1133	0.65%	49.43	378	0.46%	97.88
	Worms	130	0.07%	430.77	44	0.05%	840.91
	ToTal	175341	–	–	82332	–	–

CIC-IDS2017. The CIC-IDS2017 dataset, released in 2017 by the Canadian Institute for Cybersecurity, comprises a comprehensive array of normal and the latest malicious attack traffic, specifically designed to mimic real-world network environments. The dataset encompasses the most common and destructive attack types of that time. It features 78 attribute variables and 1 label attribute. CIC-IDS2017 contains 13 types of traffic data, with a total of 2 813 797 pieces of data. We set the ratio of training set to test set to 9:1. Since the training process of SC-WGAN requires all types of data in the training set, this paper samples approximately 1/10 of the data for training purposes. The quantity, proportion and sampling proportion of data for each type in CIC-IDS2017 is described in Table 3.

Taking into account the imbalance in the dataset and the computational constraints, we standardized the number of instances for all types of network traffic in the NSL-KDD dataset to 4 000, in the UNSW-NB15 dataset to 10 000, and in the CIC-IDS2017 dataset to 4 000 during the experimental process.

Table 3. The overview of CIC-IDS2017.

Network Traffic Type	Total Number	Training Data				Testing Data		
		Number	Proportion	IR	Adjusted ratio	Number	Proportion	IR
BENIGN	2260360	2034324	80.331%	1	0.0875	226036	80.329%	1
DoS Hulk	229198	206279	8.146%	10	0.1	22920	8.145%	10
Port Scan	157703	141933	5.605%	14	0.1	15771	5.605%	14
DDoS	127082	114374	4.516%	18	0.1	12709	4.517%	18
DoS GoldenEye	10289	9261	0.366%	220	0.8	1029	0.366%	220
FTP-Patator	7894	7105	0.281%	286	0.8	790	0.281%	286
SSH-Patator	5861	5275	0.208%	386	0.8	587	0.209%	385
DoS Slowloris	5771	5194	0.205%	392	0.8	578	0.205%	391
DoS Slowhttptest	5485	4937	0.195%	412	0.8	549	0.195%	412
Web Attack	2166	1951	0.077%	1043	1	217	0.077%	1042
Bot	1943	1749	0.069%	1163	1	195	0.069%	1159
Infiltration	34	31	0.001%	65623	1	4	0.001%	56509
Heartbleed	11	10	0.001%	203432	1	2	0.001%	113018
Total	2813811	2532423	–	–	–	281388	–	–

4.2   Setting of SC-WGAN

In the SC-WGAN model, three MLP neural networks are employed to construct the generator, discriminator and classifier. The number of fully connected layers and nodes per layer for these three models are as shown in Table 4. LeakyReLU activation function with a negative slope of 0.2 is applied after each Fully Connected Layer. The number of nodes in the output layer of the generator, as well as the input nodes for the discriminator and classifier depends on the feature dimensions present in different datasets.

During the training process, the following settings were used: an epoch value of 100, a batch size of 1 024, the optimizer selected was RMSProp, and the learning rate was set to 1×10^{-3}.

Table 4. SC-WGAN model parameter setting.

Layer	Generator	Discriminator	Classifier
FC-1	(128, 256)	(*, 512)	(*, 512)
FC-2	(256, 512)	(512, 256)	(512, 256)
FC-3	(512, 256)	(256, 128)	(256, 128)
FC-4	(256, 128)	(128, 64)	(128, 64)
FC-5	(128, *)	(64, 1)	(64, 1)

4.3   Performance Metrics

This paper uses Precision, Recall, F1-score, and AUC to evaluate the performance of network traffic classifier. Precision, Recall, and F1-score are commonly used in binary classification scenarios. In the case of multi-class classification, they can be calculated using the one-versus-all principle. The metrics for each class are then averaged based on weights if necessary.

Precision quantifies the ratio of true positive predictions to the total number of samples predicted as positive. It is calculated using the following formula:

$$Precision = \frac{TP}{TP + FP} \tag{6}$$

In this context, TP (True Positive) denotes instances where the actual value is True and correctly predicted as True. FP (False Positive) signifies cases where the actual value is False but is incorrectly predicted as True. TN (True Negative) corresponds to situations where the actual value is False and accurately predicted as False. Lastly, FN (False Negative) refers to instances where the actual value is True but is erroneously predicted as False.

Recall indicates the fraction of actual positive instances that are accurately predicted as positive among all instances that are truly positive. The calculation formula is as follows:

$$Recall = \frac{TP}{TP + FN} \tag{7}$$

F1-score is the harmonic average of Precision and Recall, calculated by the following formula:

$$F1 - score = \frac{2 \times Precision \times Recall}{Precision + Recall} \tag{8}$$

AUC is defined as the area under the ROC curve. AUC can be understood as the probability that, given a randomly selected positive sample and a randomly selected negative sample, the classifier assigns a higher score to the positive sample than to the negative sample. Using the AUC value as performance metric for model is justified because, in many instances, the ROC curve may not distinctly indicate which classifier performs better. As a numerical representation, a classifier with a higher AUC is considered to have better performances.

5   Experiment Results and Discussion

Due to the variation in sample quality generated by different oversampling methods, this will impact the quality of the balanced dataset used to train the classifier model and ultimately affect the classification performance on the test set. Therefore, this paper demonstrates the effectiveness and advancement of SC-WGAN by comparing the training performance of the classification model on balanced dataset constructed using different oversampling methods.

5.1 Comparing with Other Oversampling Methods

In order to demonstrate the advancement of SC-WGAN as a DNN-based generative model, we conducted comparative experiments with five traditional oversampling methods: ROS, SMOTE, SMOTEENN, Borderline-SMOTE, and ADASYN. The baseline was established using the classification results on the original dataset. For the experiments, an MLP model was selected as the unified traffic classifier. The comparative experimental results on three datasets are presented in Table 5. The highest value for each metric is highlighted in bold, while the second-highest values are underlined.

Table 5. Experimental results compared with other methods, presented in %.

Methods	NSL-KDD				UNSW-NB15				CIC-IDS2017			
	Precision	Recall	F1-score	AUC	Precision	Recall	F1-score	AUC	Precision	Recall	F1-score	AUC
Baseline	72.63	75.65	70.82	87.87	76.12	**69.38**	70.53	**92.52**	93.11	92.76	91.60	76.23
ROS	81.86	80.07	77.19	90.02	82.27	55.92	63.89	80.79	96.86	94.34	95.40	87.81
SMOTE	80.83	78.13	75.33	88.89	**84.77**	63.70	69.90	76.65	95.16	94.06	94.19	87.79
SMOTE_ENN	81.07	78.52	76.10	**91.11**	81.70	60.75	66.63	73.90	95.03	95.47		73.82
SMOTE_Borerline	81.73	79.64	77.16	89.48	84.32	65.11	70.74	78.20	94.98	92.47	93.25	90.63
ADASYN	82.42	80.44	**78.93**	88.48	83.60	61.04	67.84	76.93	95.49	92.66	93.52	88.42
SC-WGAN	**82.66**	**81.12**	78.87	90.64	84.37	67.39	**72.37**	87.95	**97.64**	**95.42**	**96.33**	**95.56**

From Table 5, it is clear that compared to the baseline, all oversampling methods used to balance the dataset enhance classification results. This confirms that imbalanced data distribution in the original dataset impedes classification performance, highlighting the importance of adjusting data proportions to boost model effectiveness.

Compared to other oversampling methods, SC-WGAN consistently achieves optimal or near-optimal solutions across three datasets. On the NSL-KDD dataset, the SC-WGAN oversampling method outperformed others in Precision and Recall metrics, and was slightly behind the best in F1-Score and AUC by only 0.06% and 0.47%, respectively. On the UNSW-NB15 dataset, although the SC-WGAN method only achieved the best value in F1-Score, it obtained suboptimal values in the remaining three metrics and showed a significant advantage over other oversampling methods in Recall, F1-Score, and AUC. On the CIC-IDS2017 dataset, SC-WGAN topped all metrics, especially in AUC, scoring 95.56%. It improved by 21.74% over the lowest-performing SMOTE-ENN method and by 4.93% over the next best Borerline-SMOTE. Overall, SC-WGAN enhances classifier performance significantly better than other methods.

To visually assess the impact of the SC-WGAN method on classification models, Figs. 4 and 5 provide t-SNE visualizations that compare the original datasets with those balanced using SC-WGAN. The t-SNE algorithm, as described in [27], transforms high-dimensional data points into a lower-dimensional space, facilitating visualization in scatter plots while preserving the relative distances between original data points as much as possible. Figure 4a reveals that the NSL-KDD dataset primarily consists of three majority classes, making it difficult to distinguish the other two types. However, Fig. 4a shows a clearer separation among the five types, enhancing the classifier's ability to perform multi-class classification. Similarly, in Fig. 5a, the UNSW-NB15 dataset is predominantly comprised of Normal-type data, while Fig. 5b shows a more balanced distribution among various data types. Nonetheless, there is still some overlap in data distribution in Fig. 5b, which poses challenges for accurate classification, as evidenced by the results in Table 5. Owing to the extensive size of the CIC-IDS2017 dataset, the t-SNE visualization for this dataset is omitted in this presentation.

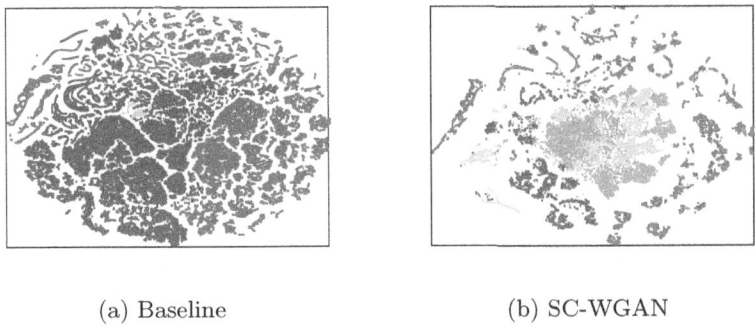

(a) Baseline   (b) SC-WGAN

Fig. 4. T-SNE visualization of NSL-KDD dataset.

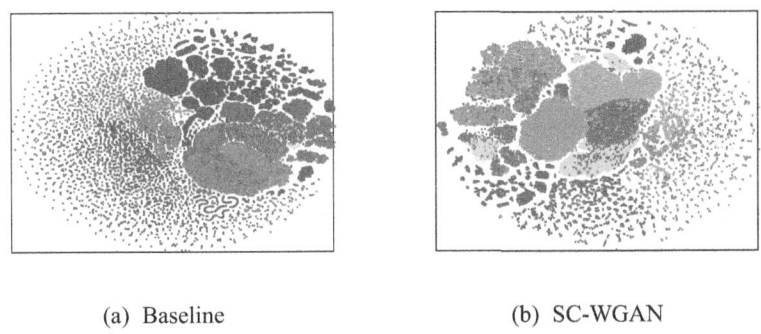

(a) Baseline   (b) SC-WGAN

Fig. 5. T-SNE visualization of UNSW-NB15 dataset.

5.2   Comparing with GAN-Based Methods

To further demonstrate the advanced nature of SC-WGAN, comparative experiments with WCGAN are conducted on two datasets. The WCGAN model was proposed by Kumar et al. in 2023 [23], which stands for Wasserstein Conditional Generative Adversarial Network. Since Kumar et al.'s work involved using machine learning methods for traffic classification, the experiments additionally employed SC-WGAN combined with Decision Tree (DT) and Support Vector

Machine (SVM) to train network traffic classification models. The results are presented in Table 6.

Table 6. Experimental results compared with WCGAN, presented in %.

Dataset	Methods	WCGAN			SC-WGAN		
		Precision	Recall	F1-score	Precision	Recall	F1-score
NSL-KDD	DT	79.55	73.40	76.35	76.72	74.71	73.59
	SVM	18.55	43.06	25.93	83.25	78.95	77.19
UNSW-NB15	DT	70.85	71.39	71.12	45.68	46.65	43.69
	SVM	72.07	53.83	61.63	81.32	66.75	70.19

Table 6 shows that on the NSL-KDD dataset, the SC-WGAN method performs slightly worse than WCGAN when using the DT. However, when using SVM, the SC-WGAN method significantly outperforms WCGAN. Similarly, in the UNSW-NB15 dataset, SC-WGAN achieves better classification results with SVM, but performs worse than WCGAN when using DT. Overall, the combination of SC-WGAN with SVM yields the best classification results on both datasets. This suggests that it is beneficial to experiment with different classification methods to select the most suitable one.

The comparative results on the two datasets mentioned above reveal that the proposed SC-WGAN model exhibits a certain level of advancement compared to existing models.

5.3   Ablation Experiment

Ablation experiments can reveal the importance of different components or features in a model and how each part affects the final experimental results. By using four methods-GAN, WGAN, SC-GAN (adding a synchronous classifier to GAN), and SC-WGAN-to construct balanced datasets for training classification models, we conduct ablation experiments for in-depth analysis. This helps verify the rationale behind the improvements in the SC-WGAN method.

Figures 6, 7 and 8 display normalized confusion matrix heatmaps for three datasets. These heatmaps are adjusted according to the classification outcomes of each type, providing a detailed analysis of the effectiveness of each oversampling method. Due to the oversampling methods specifically target the minority class, our focus is primarily on the classification of the last two types of data in NSL-KDD, the last four types in UNSW-NB15, and the last four types in CIC-IDS2017. It can be observed that with the SC-WGAN method, the R2L and U2R types of data in the NSL-KDD dataset achieve the highest accuracy of 0.19 and 0.18, respectively. By comparing the diagonal elements of the four matrices on the UNSW-NB15 dataset, it can be observed that the values for the SC-WGAN

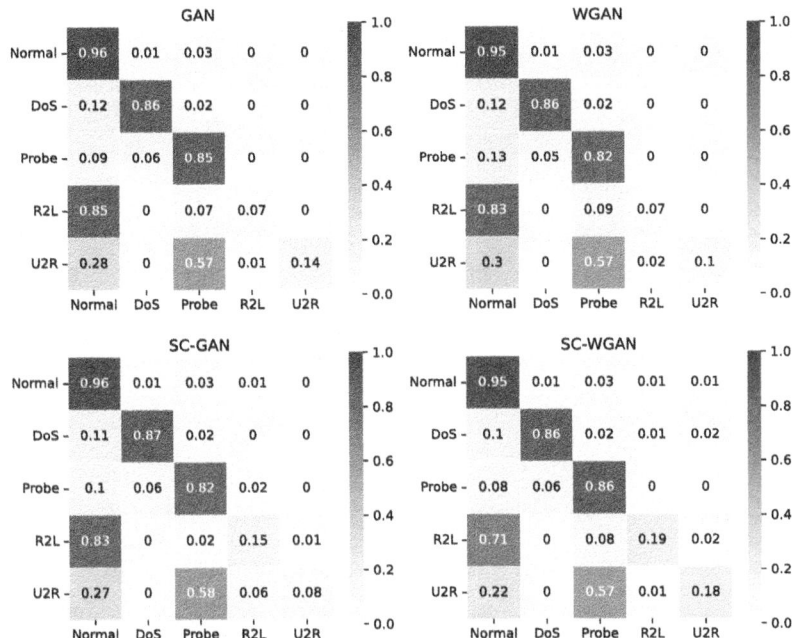

Fig. 6. Confusion matrix for ablation experiments on the NSL-KDD dataset.

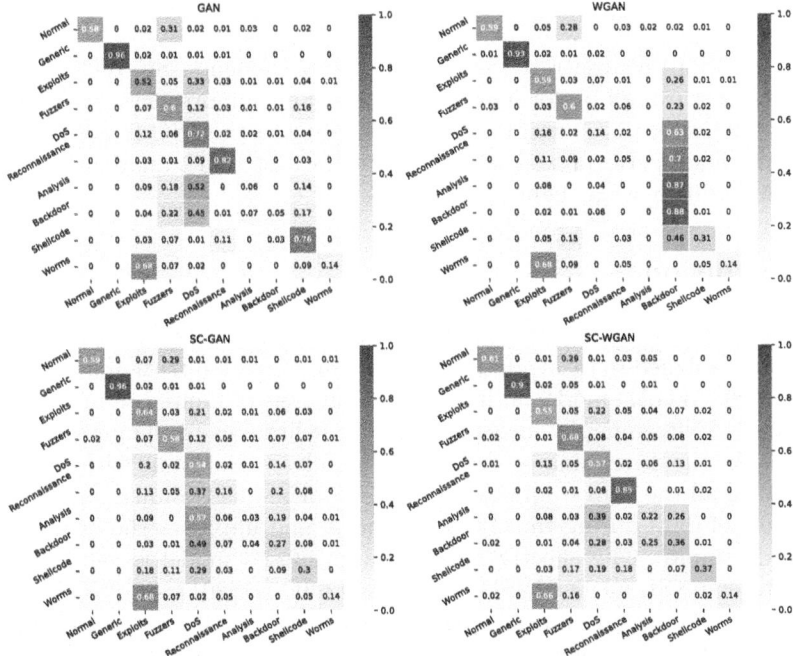

Fig. 7. Confusion matrix for ablation experiments on the UNSW-NB15 dataset.

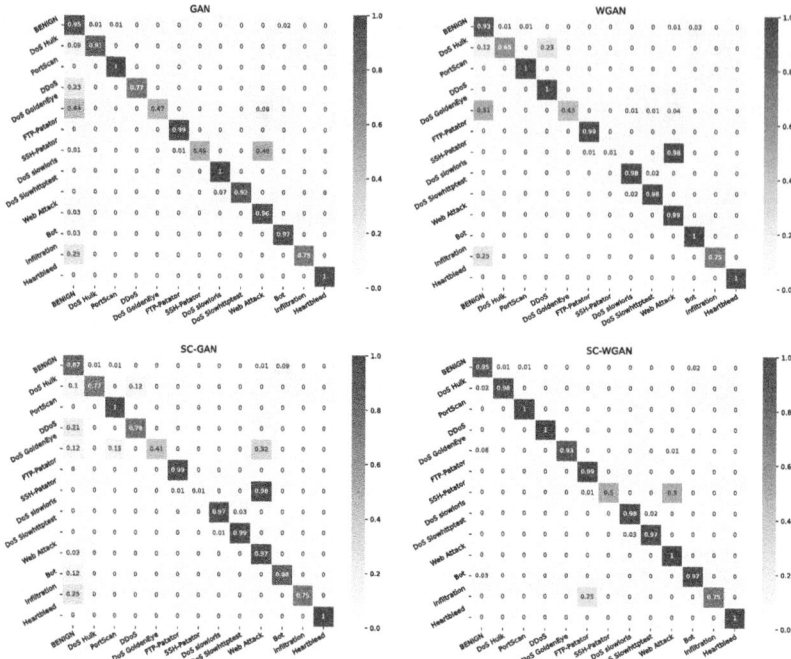

Fig. 8. Confusion matrix for ablation experiments on the CIC-IDS2017 dataset.

method are more concentrated on the diagonal elements. SC-WGAN demonstrates the best overall performance on the last four minority classes, with accuracies of 0.22, 0.36, 0.37, and 0.14, respectively. On the CIC-IDS2017 dataset, all four methods perform well in classifying the last four minority classes, but SC-WGAN still shows a slight advantage.

6   Conclusion

This paper introduces SC-WGAN, a novel method based on an enhanced GAN framework, designed to tackle data imbalance in network intrusion detection systems. The SC-WGAN model addresses the challenge of limited sample availability during GAN training by incorporating a synchronous training classifier. The classifier, like the discriminator, engages in an adversarial game with the generator, utilizing all the data from the training set for model training. This increases the volume of data during the training process and improves the quality of the generated samples. The effectiveness of the proposed model is validated using three datasets: NSL-KDD, UNSW-NB15, and CIC-IDS2017. SC-WGAN consistently achieves optimal or near-optimal results across these datasets, surpassing traditional oversampling methods. Further comparisons with the advanced WCGAN show that SC-WGAN outperforms when coupled with an SVM classifier. Additionally, ablation studies confirm SC-WGAN's superior performance on minority classes, demonstrating its ability to significantly improve the quality of generated samples for these groups.

Moreover, given that GANs are designed for generating outputs in a continuous space, while network traffic comprises discrete values, future research is encouraged to investigate the utilization of GAN models explicitly tailored for discrete value training, such as BGAN and SeqGAN. This exploration may contribute to further enhancing the quality of generated samples.

References

1. Wang, K., Zhang, J., Bai, G., Ko, R., Dong, J.S.: It's not just the site, it's the contents: intra-domain fingerprinting social media websites through CDN bursts. In: 30th The Web Conference (WWW) (2021). https://doi.org/10.1109/ICECCS2018.2018.00011
2. Abdulganiyu, O.H., Ait Tchakoucht, T., Saheed, Y.K.: A systematic literature review for network intrusion detection system (IDS). Int. J. Inf. Secur. **22**, 1–38 (2023). https://doi.org/10.1007/s10207-023-00682-2
3. Asad, H., Adhikari, S., Gashi, I.: A perspective-retrospective analysis of diversity in signature-based open-source network intrusion detection systems. Int. J. Inf. Secur. **23**, 1–16 (2023). https://doi.org/10.1007/s10207-023-00794-9
4. Louk, M.H.L., Tama, B.A.: Dual-IDS: a bagging-based gradient boosting decision tree model for network anomaly intrusion detection system. Expert Syst. Appl. **213**, 119030 (2023)
5. Guo, Y., Xiong, G., Li, Z., Shi, J., Cui, M., Gou, G.: TA-GAN: Gan based traffic augmentation for imbalanced network traffic classification. In: 2021 International Joint Conference on Neural Networks (IJCNN), pp. 1–8. IEEE (2021)
6. Goodfellow, I., Pouget-Abadie, J., Mirza, M., Xu, B., Warde-Farley, D., Ozair, S., Courville, A., Bengio, Y.: Generative adversarial networks. Commun. ACM **63**(11), 139–144 (2020)
7. Wang, Z., Wang, P., Zhou, X., Li, S., Zhang, M.: FLOWGAN: unbalanced network encrypted traffic identification method based on GAN. In: 2019 IEEE International Conference on Parallel & Distributed Processing with Applications, Big Data & Cloud Computing, Sustainable Computing & Communications, Social Computing & Networking (ISPA/BDCloud/SocialCom/SustainCom), pp. 975–983. IEEE (2019)
8. Kovács, G.: An empirical comparison and evaluation of minority oversampling techniques on a large number of imbalanced datasets. Appl. Soft Comput. **83**, 105662 (2019)
9. Chawla, N.V., Bowyer, K.W., Hall, L.O., Kegelmeyer, W.P.: SMOTE: synthetic minority over-sampling technique. J. Artif. Intell. Res. **16**, 321–357 (2002)
10. Han, H., Wang, W.-Y., Mao, B.-H.: Borderline-SMOTE: a new over-sampling method in imbalanced data sets learning. In: Huang, D.-S., Zhang, X.-P., Huang, G.-B. (eds.) ICIC 2005, Part I. LNCS, vol. 3644, pp. 878–887. Springer, Heidelberg (2005). https://doi.org/10.1007/11538059_91
11. He, H., Bai, Y., Garcia, E.A., Li, S.: ADASYN: adaptive synthetic sampling approach for imbalanced learning. In: 2008 IEEE International Joint Conference on Neural Networks (IEEE World Congress on Computational Intelligence), pp. 1322–1328. IEEE (2008)

12. Arjovsky, M., Chintala, S., Bottou, L.: Wasserstein generative adversarial networks. In: International Conference on Machine Learning, pp. 214–223. PMLR (2017)
13. Vu, L., Bui, C.T., Nguyen, Q.U.: A deep learning based method for handling imbalanced problem in network traffic classification. In: Proceedings of the 8th International Symposium on Information and Communication Technology, pp. 333–339 (2017)
14. Douzas, G., Bacao, F.: Effective data generation for imbalanced learning using conditional generative adversarial networks. Expert Syst. Appl. **91**, 464–471 (2018)
15. Lee, J., Park, K.: GAN-based imbalanced data intrusion detection system. Pers. Ubiquit. Comput. **25**, 121–128 (2021)
16. Liu, X., Li, T., Zhang, R., Wu, D., Liu, Y., Yang, Z.: A GAN and feature selection-based oversampling technique for intrusion detection. Secur. Commun. Netw. **2021**, 1–15 (2021)
17. Ring, M., Schlör, D., Landes, D., Hotho, A.: Flow-based network traffic generation using generative adversarial networks. Comput. Secur. **82**, 156–172 (2019)
18. Hao, X., et al.: Producing more with less: a GAN-based network attack detection approach for imbalanced data. In: 2021 IEEE 24th International Conference on Computer Supported Cooperative Work in Design (CSCWD), pp. 384–390. IEEE (2021)
19. Gong, X., Jia, L., Li, N.: Research on mobile traffic data augmentation methods based on SA-ACGAN-GN. Math. Biosci. Eng. **19**, 11512–11532 (2022)
20. Ding, H., et al.: RVGAN-TL: a generative adversarial networks and transfer learning-based hybrid approach for imbalanced data classification. Inf. Sci. **629**, 184–203 (2023)
21. Yuan, L., Yu, S., Yang, Z., Duan, M., Li, K.: A data balancing approach based on generative adversarial network. Futur. Gener. Comput. Syst. **141**, 768–776 (2023)
22. Dlamini, G., Fahim, M.: DGM: a data generative model to improve minority class presence in anomaly detection domain. Neural Comput. Appl. **33**, 13635–13646 (2021)
23. Kumar, V., Sinha, D.: Synthetic attack data generation model applying generative adversarial network for intrusion detection. Comput. Secur. **125**, 103054 (2023)
24. Srivastava, A., Sinha, D., Kumar, V.: WCGAN-GP based synthetic attack data generation with GAN based feature selection for IDS. Comput. Secur. **134**, 103432 (2023)
25. Tavallaee, M., Bagheri, E., Lu, W., Ghorbani, A.A.: A detailed analysis of the KDD CUP 99 data set. In: 2009 IEEE Symposium on Computational Intelligence for Security and Defense Applications, pp. 1–6. IEEE (2009)
26. Moustafa, N., Slay, J.: UNSW-NB15: a comprehensive data set for network intrusion detection systems (UNSW-NB15 network data set). In: 2015 Military Communications and Information Systems Conference (MilCIS), pp. 1–6. IEEE (2015)
27. Van der Maaten, L., Hinton, G.: Visualizing data using t-SNE. J. Mach. Learn. Res. **9**(11), 2579–2605 (2008)

Automated Parameter Determination for Enhancing the Product Configuration System of Renault: An Experience Report

Hao Xu[1,3]($\boxtimes$)(iD), Souheib Baarir[1,2](iD), Tewfik Ziadi[1](iD), Siham Essodaigui[3](iD), and Yves Bossu[3](iD)

[1] LIP6, Paris, France
{hao.xu,souheib.baarir,tewfik.ziadi}@lip6.fr
[2] Paris Nanterre University, Paris, France
[3] Renault Group, Paris, France
{hao.xu,siham.essodaigui,yves.bossu}@renault.com

Abstract. The problem of configuring the variability models is pervasive in plenty of domains. Renault, a leading automobile manufacturer, has developed an internal product configuration system to model its vehicle diversity. This system is based on the well-known knowledge compilation approach and is associated with a set of parameters. Different input parameters have a strong influence on the system's performance. The parameters actually used are determined manually. Our work aims to study and determine these parameters automatically. This paper studies Renault's variability models and product configuration system and presents a parameter prediction model for this system. The results show the predicted parameters' competitiveness compared with the parameters by default.

Keywords: Variability model · Knowledge compilation · Machine learning · Parameter tuning

1 Introduction

Variability Modeling Problems [2] are very common in real life. In the car industry, such problems are important since they are related to different business activities, including engineering design, manufacturing, etc.

As an example of Variability Model (VM) in car industry, let's consider a VM M: it has a variable *model*, which represents the vehicle model with a value range in the domain $\{m_1, m_2\}$; a variable *fuel* with the domain $\{petrol, diesel, lpg\}$. Then, variable dependencies describe activity-related constraints (business, technical, legal requirements, and many others). For instance, the constraint $model = m_1 \Rightarrow fuel = lpg$ leads to four possible combinations that form the different configurations for this vehicle. We refer to the set of possible configurations for a VM as the *configuration* (or the *solution*) space.

© The Author(s), under exclusive license to Springer Nature Switzerland AG 2025
G. Bai et al. (Eds.): ICECCS 2024, LNCS 14784, pp. 43–63, 2025.
https://doi.org/10.1007/978-3-031-66456-4_3

Renault, a world-leading automobile manufacturer [30], uses such VMs to model its vehicle range. Some ranges of its vehicles can reach 10^{32} possible configurations. With such a large configuration space, a common requirement is to be able to search for satisfying configurations based on users' queries. These queries can include consistency checks (to determine if a specified vehicle model exists) or requests for all the possible satisfied configurations, among others.

To deal with such requests, Renault has adopted a *knowledge compilation*[12] based approach. The idea of *knowledge compilation* is using symbolic structures (e.g., BDDs [1], SDDs [11], etc.) to represent the problem *configuration space*. An internal product configuration system has been developed to assist in building the configuration space for VMs using these symbolic structures.

Although this system provides an efficient method to handle large vehicle models, its performance can be limited by the memory size of the symbolic structure. The total size of such compiled structures can reach dozens of gigabytes, which puts pressure on memory usage. In Renault's configuration system, the compilation[1] of such symbolic structures is associated with several parameters that greatly influence the structure's size. Determining the best-performing parameters *manually* for each VM is difficult and tedious. This paper addresses this problem, proposes, and implements an automated parameters prediction model to obtain the best-performing parameters for each VM.

The paper is organized as follows: Sect. 2 presents related research; Sect. 3 describes the internal system with its associated parameters; Sect. 4 presents the parameter tuning and prediction process; Sect. 5 presents the obtained results; and finally, Sect. 6 concludes the paper and provides perspectives for future work.

2 Related Work

The performance of many algorithms relies heavily on carefully tuned parameter configurations based on user preferences or performance criteria [7]. Over the years, various automatic parameter tuners have been proposed, which can be categorized into two types: *local* methods «««and *model-based* methods.

In the category of local methods, notable examples include GGA [3], paramILS [19], SPOT [5], and irace [21]. These methods employ local search strategies in the configuration space, such as genetic algorithms, iterated local search, racing procedures, and more. They have demonstrated strong competitiveness when applied to solvers in diverse problem domains like mixed integer programming (MIP) [17], machine learning, and propositional satisfiability solving [20]. However, these tools suffer from a limitation of being problem feature-independent. In other words, the parameter tuning results are static and cannot be adjusted based on the characteristics of the input problem instance [14]. Theoretical and empirical studies on various algorithms and problems have shown that algorithm parameters are highly dependent on specific instance features of

---

[1] Here, "compilation" refers to the process of building the configuration space in knowledge compilation terminology.

the target problem [13]. Indeed, the optimal parameter values can vary significantly with different input instance features, such as problem size [28].

On the other hand, model-based tuners consider problem features during the
tuning process by leveraging machine learning techniques to build a model. Two
notable examples in this category are SMAC and PIAC.

SMAC, sequential model-based algorithm configuration [18], is an algorithm
that automates the process of finding the optimal parameter set for algorithms.
It is often used for parameter tuning in machine learning models. SMAC iteratively runs the model with different parameter combinations and uses the results
to learn which parameters are likely to yield good performance. It employs a
Bayesian optimization strategy that takes into account both the model's performance and the uncertainty in estimated performances. This enables efficient
search in the parameter space, achieving good solutions with fewer iterations
compared to other methods.

PIAC (per instance algorithm configuration) relies on learning an empirical
performance model (EPM) that can predict algorithm performance based on the
instance and specified parameter settings [8]. The empirical performance model
captures the relationship between instance features, parameters, and algorithm
performance, enabling performance prediction with given features and parameters. However, PIAC faces challenges when it comes to predicting or searching
for the best parameter setting in cases where the parameter space is large.

In our case, we aim to develop a feature-dependent, model-based automated
algorithm tuner for each instance. In the following sections, we describe the
problem and present our contributions.

3   Background and Problem Statement

The product configuration system used by Renault involves the generation of
VMs and their associated configuration spaces. In this section, we provide a
brief overview of Renault's variability model and its configuration system. We
also introduce the system's parameters and their usage. Finally, we outline the
objectives and analyze the challenges associated with the tuning process.

3.1   Variability Model: Definitions

A variability model represents variables (also known as features) and their
options, along with the relationships and dependencies between them. Formally,
the VM can be defined as follows [29]:

A Variability Model is a triple (V, O, C), where V is a set of variables, O is
a set of options for the variables, and C is a set of constraints. Each variable
$v \in V$ is associated with a domain $Domain(v) \in O$. A *constraint* $c \in C$ can be
intentional or extensional:

- *Extension constraint:* Also known as a table constraint, it is defined by
  enumerating a list of allowed or forbidden value tuples. It has the form
  $extension(V, S)$, where $V = \langle v_1, .., v_n \rangle$ and S is a set of supported/forbidden
  value tuples, $S = \langle \langle d_1, ..., d_n \rangle, ... \rangle$ (with $d_i \in Domain(v_i)$).

- *Intension constraint:* It is a constraint of the form $intension(V, P)$, where $V = \langle x_1, ..., x_n \rangle$ is a sequence of n variables (the scope of the constraint), and P is a predicate expression with n formal parameters on the variables of V.

A *literal* is a statement of the form $v = d$, where $d \in Domain(v)$. An *assignment* is a set of literals covering all the variables in V. A *partial assignment* is a set of literals covering a subset of V. A *solution* is an assignment consistent with all the constraints in C.

Here we give an example of a VM M:

- $Variables = \{model, fuel, airconditioning, dustfilter\}$
- $Options = \{\{m_1, m_2\}, \{petrol, diesel, lpg\},$
  $\{manual, auto\}, \{with, none\}\}$
  with, $Domain(model) = \{m_1, m_2\}$,
  $Domain(fuel) = \{petrol, diesel, lpg\}$,
  $Domain(airconditioning)$ $=$ $\{manual, auto, none\}$ and
  $Domain(dustfilter) = \{with, none\}$
- $Constraints$:

$$c_1 : extension(\langle model, fuel \rangle, \langle \langle m_1, lpg \rangle, \langle m_2, petrol \rangle,$$
$$\langle m_2, diesel \rangle, \langle m_2, lpg \rangle \rangle)$$

$$c_2 : intension(\langle model, fuel, airconditioning \rangle,$$
$$\langle (((model = m_1) \lor (model = m_2)) \land$$
$$((fuel = petrol) \lor (fuel = diesel))) \Rightarrow airconditioning = auto \rangle)$$

$$c_3 : intension(\langle airconditioning, dustfilter \rangle,$$
$$\langle airconditioning = manual \Rightarrow dustfilter = with \rangle)$$

3.2 Variability Model as Undirected Graphs

We introduce two undirected graphs that are used to encode a VM. These graphs are defined as follow:

- *variable-constraint graph*: each constraint is represented by a node; Each variable is represented by a node; An arc exists between a constraint node and a variable node if the variable is involved in the constraint.
- *variable graph*: each variable is represented by a node; An arc exists between two variable nodes if the two variables are involved in the same constraint.

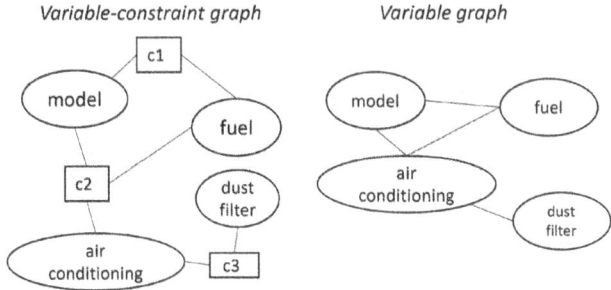

Fig. 1. *Variable-constraint graph* and *variable graph* of *M*.

3.3   The Product Configuration System of Renault: Overview and Challenges

Renault's product configuration system can be divided into two parts: the offline part generates and saves the configuration space, while the online part deals with different requests. The parameters are taken as inputs for the offline part and are then used to help control and make decisions during the compilation process. The online process searches for solutions in the compiled structure generated by the offline phase. Figure 2 presents the system architecture. As mentioned before, the configuration space is represented in the form of a symbolic structure, which implicitly saves all the possible configurations. More specifically, it is based on a private compiled representation of vehicle diversity in the form of a cluster tree, which has been used in various applications at Renault since 1995 [23]. Here, we detail the offline process of the construction of the cluster tree (more details can be found in [9,23]).

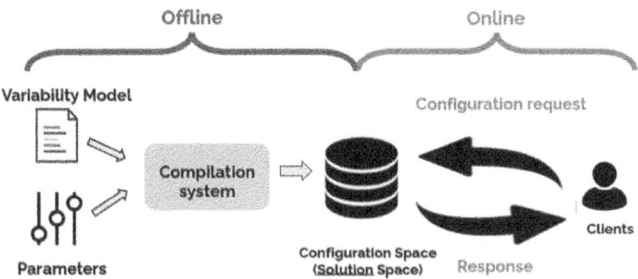

Fig. 2. Renault's product configuration system.

– A *cluster* is a group of variables associated with a set of partial solutions of a set of constraints. These constraints should only involve the variables in the cluster. The compilation process encodes literals as Boolean variables, where a literal represents a variable value assignment. Each partial solution is encoded as a vector of bits.

For M, consider that $cluster_1$ refers to variables $model$ and $fuel$, and $cluster_2$ refers to variables $airconditioning$ and $dustfilter$. c_1 contains two variables, $model$ and $fuel$, so c_1 is associated with $cluster_1$. c_3 is associated with $cluster_2$. Boolean variables encode literals: a for $model = m_1$, b for $model = m_2$, c for $fuel = petrol$, d for $fuel = diesel$, and e for $fuel = lpg$. f, g, h encode the choices $manual, auto, none$ for $airconditioning$. i, j encode the choices $with, none$ for $dustfilter$. The partial solutions for each cluster, $cluster_1$ and $cluster_2$, are presented in Fig. 3.

– **A cluster tree** is a tree where each node represents a cluster. An arc between two clusters indicates a dependency between them in terms of constraints. These constraints involve variables from both clusters. The arc between the clusters contains a *Matrix* which evaluates whether the partial solutions within the linked clusters are consistent with the constraints. The *Matrix* enables the restoration of complete configurations from the partial solutions in each cluster. Figure 4 presents an example of a cluster tree and the *Matrix* is contained in the red arrow.

Multiple cluster trees can be derived for the same problem, depending on how the variables are arranged within clusters. To optimize the size of the cluster tree, the current system constructs it using heuristic analysis of the *variable-constraint graph* [6]. The system follows a process of *variable-constraint graph-based partitioning* to assign variables to clusters. Two important parameters are considered during this process to achieve an optimized cluster tree size:

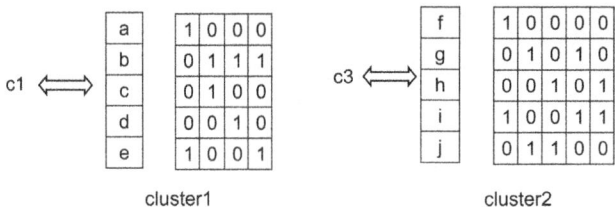

Fig. 3. $cluster_1$ and $cluster_2$.

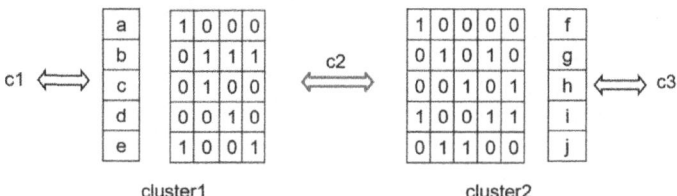

Fig. 4. Cluster tree example.

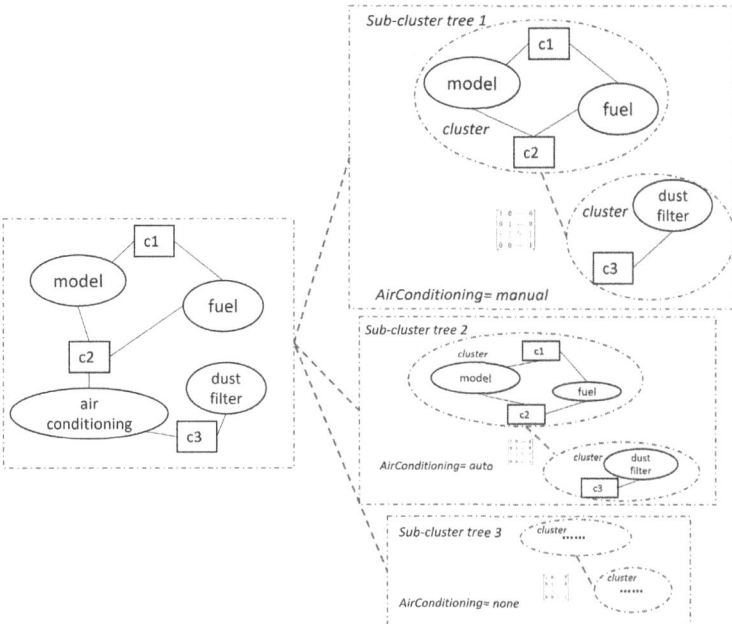

Fig. 5. Example of graph partition

- **PartitionVariables:** This parameter represents a list of variables in the variability model. The order of variables in the list is significant. The system uses these variables to partition the *variable-constraint graph* and construct the cluster tree. Each graph partition results in several sub-graphs (or sub-cluster trees). In each sub-cluster tree, the partitioned variable is assigned a value from its domain to remove it. For example, in Fig. 1, if we partition the graph using the node *airconditioning*, we obtain three sub-cluster trees as shown in Fig. 5. Each sub-cluster tree consists of two clusters linked by an arc, with each arc containing a matrix that represents the consistency information of partial solutions.
- **MatrixSize:** This parameter represents the maximum size threshold of the matrices in the cluster tree. A matrix between clusters reflects the consistency of partial solutions within the linked clusters. The value of this parameter has a significant impact on the final size of the cluster tree.

  A large **MatrixSize** allows for variables to be dispatched with large matrices between clusters, resulting in a larger compiled structure. However, a small **MatrixSize** can lead to a smaller overall size of the compiled structure. Nevertheless, it also leads to more graph partitions. As a consequence, more sub-cluster trees are created in the compiled structure, which in turn increases the time required to iterate through all the sub-cluster trees when responding to requests.

  In summary, the choice of **MatrixSize** affects both the size of the compiled structure and the computational efficiency of processing requests. It is crucial to strike a balance between the size of the compiled structure and the runtime performance.

The primary objective of the tuning process is to propose a parameter prediction model that accurately predicts the appropriate values of **MatrixSize** and **PartitionVariables** for each input instance. This prediction model aims to reduce the size of the cluster tree compared to the default parameters. The next section presents the process of tuning and predicting these parameters.

4   Parameters Tuning and Predicting for Renault's Product Configuration System

Our objective is to develop an automated prediction model that is capable of generating feature-dependent parameter predictions. In this section, we outline the process of training such a model and integrating it into the configuration system.

4.1   Parameter Prediction Model (PPM) Training

Training a parameter prediction model involves two main processes: parameter *tuning* and *learning*. The tuning process is used to search for the best-performing parameters for each instance, while the learning process focuses on training a machine learning model that maps the features of instances to the corresponding best-performing parameters.

Parameters Tuning Process. The tuning process is a prep-processing work to prepare data for training set. It aims to identify the best-performing parameters that optimize a given objective. It is important to note that the best-performing parameter setting we search for is not necessarily the theoretically best parameter setting, as that would be computationally expensive and challenging to determine. Instead, we aim to find the parameter setting that optimizes the target algorithm the most within a given computational resource and time limit, based on specific evaluation metrics. In our case, the evaluation metric is the compilation result size. Figure 6 illustrates the steps involved in this process.

Parameter Settings Space Construction. The initial step involves defining the parameter domains to create the parameter settings space. For the **MatrixSize** parameter, we have constrained the domain to integers ranging from 10 to 300. This choice is informed by practical experience, encompassing all suitable values for **MatrixSize**. Through observation, we have determined that values exceeding 300 result in a single, unpartitioned large sub-cluster tree, whereas values that are too small lead to excessive graph partitions. This, in turn, increases the response time, as previously mentioned.

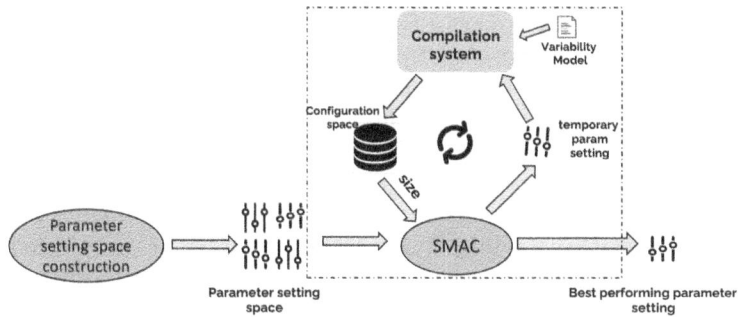

Fig. 6. Parameters tuning process

To specify the domain for 'PartitionVariables,' we employ the concept of 'betweenness centrality' as introduced in [4]. Betweenness centrality is a measure of centrality in a graph based on shortest paths [10]. We create a 'variable graph' for each instance and compute the 'betweenness centrality' for each variable node. Subsequently, we rank the variables based on their centrality values and select the top five.

All subsets of this list are considered as potential domains for **PartitionVariables**, with different orders of the same variables treated as separate parameters. It's worth mentioning that the actual number of variables used in the daily parameters[2] ranges from 0 to 10. From extensive tests with multiple instances, we observed that the first five variables used in **PartitionVariables** often matter much more than the later variables over the compilation result. The alteration of the variables at the back position slightly change the compilation result size. So, to facilitate the experiments, we fix the number of used variables in **PartitionVariables** to five.

SMAC Activity. Once the parameter domains are defined, we initiate the tuning phase to search for the best-performing parameter settings within the parameter settings space for each instance. To accomplish this, we employ SMAC, an automated algorithm configuration tool. As discussed in Sect. 2, SMAC optimizes the performance of an algorithm by executing it with different parameter settings. It employs various strategies, such as random forest [25] and Bayesian optimization [15], to guide the search process. Notably, SMAC excels in handling categorical parameters.

We configure SMAC to run a maximum of 60 iterations for each instance. The choice of 60 iterations is based on extensive testing of multiple instances, where we observed that the compilation size does not significantly decrease beyond approximately 50 iterations. The target for optimization is the compilation size. We record the best-performing parameter settings found by SMAC and also measure the time consumed during this process.

---

[2] The daily parameter is also called the production parameter, which is manually determined and updated by the system developers.

Parameters Learning Process. Before delving into the learning process, it is important to address the issue of choosing an appropriate machine learning model for the two parameters. Predicting **MatrixSize** is a classical regression task that involves mapping problem features to the optimal **MatrixSize** for each instance. On the other hand, predicting **PartitionVariables** is a classification task rather than regression. The model for predicting **PartitionVariables** should be capable of selecting variables from the instance and determining their position in the **PartitionVariables** list. To tackle this issue, we decided to train two separate models: one regression model for predicting **MatrixSize**, and one classification model for **PartitionVariables**. The choice of input instance features, models, and training processes for each model will be explained separately in this section. Figure 7 illustrates this process.

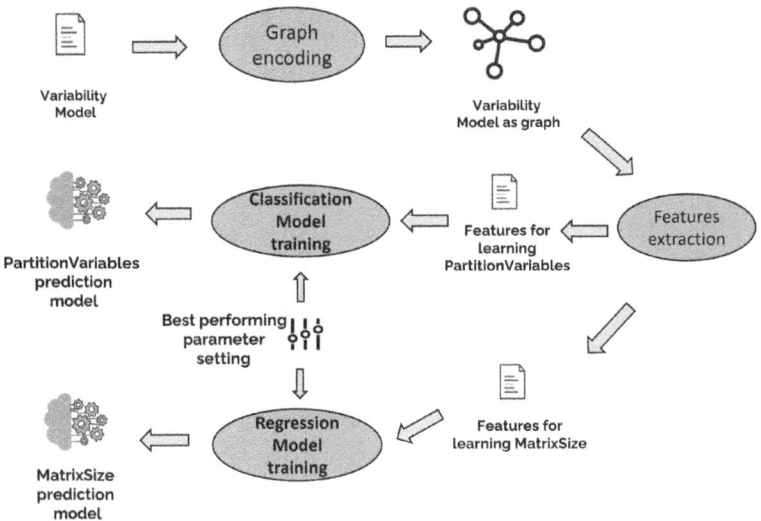

Fig. 7. Parameters learning process

Graph Encoding Activity. The first step involves encoding the VM as a *variable-constraint graph* and a *variable graph* to facilitate feature extraction.

Features Extraction Activity.

– *Features for learning **MatrixSize**:* To train the model for predicting **Matrix-Size**, we draw inspiration from a similar study called SATZILLA [31], which extracts 48 features to analyze SAT instances by encoding them as *variable graphs* and *variable-constraint graphs*. In our case, we also rely on graph-based features and select 14 specific features:

- *Problem size features:*
  * Number of variables in the VM
  * Number of constraints in the VM
- *Variable-constraint graph features:*
  * Minimum, average, and maximum *degree* values[3] of constraint nodes
  * Minimum, average, and maximum *degree* values of variable nodes
  * Minimum, average, and maximum *betweenness centrality* values of constraint nodes
  * Minimum, average, and maximum *betweenness centrality* values of variable nodes

  Since the construction of the cluster tree is based on the partitioning of the *variable-constraint graph*, it is reasonable to select features from such graphs as inputs for the learning process.
- *Features for learning **PartitionVariables***: **PartitionVariables** is a parameter that is related to each variable of the instance. In order to build a model for the classification task, we need to extract features for each variable in the instance. For each variable, we calculate its node degree and betweenness centrality value separately in the *variable-constraint graph* and the *variable graph*. We use a four-bit vector to represent these pieces of information as features for each variable.

Regression Model Training Activity. With the target **MatrixSize** and available features for each instance, we proceed to the learning phase to train the regression model. Various standard machine learning models, such as Linear Regression [27], Support Vector Machines (SVM) [22], Gaussian Regression [32], etc., can be utilized. After careful evaluation, we select **Random Forest Regression** as our choice. This decision is based on its ability to provide reasonable predictions without requiring extensive hyper-parameter tuning. Furthermore, it effectively addresses the problem of overfitting that can occur with decision trees [25].

Classification Model Training Activity. The process of selecting a variable partition for the compilation process involves evaluating the suitability of all variables in the instance. To accomplish this, we develop a scoring model based on the **SVM classification model** [22]. This model aims to assign scores to all variables in an instance, with a higher score indicating a more favorable position in the **PartitionVariables** list. To train this model, we begin by labeling the variables in the **PartitionVariables** list identified by SMAC. Each variable is assigned a score ranging from 0 to 1, with the leading variable receiving the highest score and the last obtaining the lowest score. Variables not included in the list are assigned a score of 0. Subsequently, we train a model to learn the relationship between the features of these variables and their corresponding scores. This trained model is capable of scoring variables for unseen instances.

---

[3] The degree of a vertex in an undirected graph is the number of edges incident with (meeting at or ending at) itself [16].

By ordering the variables based on their scores, we can dynamically define the number of required variables for each instance. The resulting ordered list serves as the predicted **PartitionVariables**.

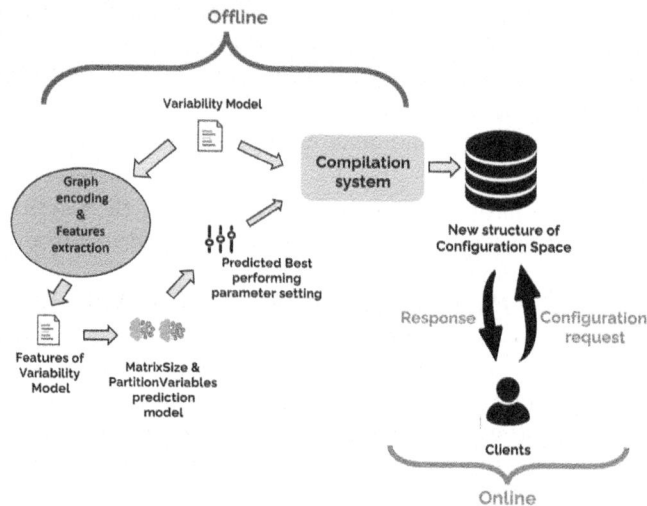

Fig. 8. Integration of the parameters prediction model into the configuration system

4.2   Integrating the Prediction Models into Renault's Product Configuration System

Now that we have outlined all the necessary steps for the tuning and learning process, we proceed to integrate the prediction models into the configuration system. Figure 8 illustrates the enhanced system with the inclusion of the prediction models. When a VM is processed, it is initially passed through the "graph encoding and features extraction" activity to extract its features. This step is relatively quick and does not significantly impact the overall compilation time. Subsequently, the extracted features are fed into the prediction models, which generate the predicted values for **MatrixSize** and **PartitionVariables**. By applying these predicted parameters to the configuration system, we obtain a new data structure representing the configuration space. Figure 8 gives a global overview of the newly developed system.

In the next section, we will evaluate the performance of the parameters discovered by SMAC. Additionally, we will conduct separate and combined tests on the predicted **MatrixSize** and **PartitionVariables** to assess their effectiveness.

5   Experiments

In this section, we present the results of the tuning and prediction processes described earlier. Firstly, we obtain the best-performing parameters using

SMAC. Then, with the parameter settings discovered by SMAC, we train two separate models to predict **MatrixSize** and **PartitionVariables**. The compilation results for different parameter configurations and the response time of the requests are analyzed. All experiments are conducted on the same execution environment: a CPU i7 with 3.00 GHz and 32 GB RAM.

5.1 Datasets

We collected a dataset consisting of 600 instances, each representing a variable model of Renault. These instances are divided into a training set containing 200 instances and a validation set containing 400 instances.

To evaluate the influence of parameter tuning, we classify the instances into three classes based on a custom indicator provided by Renault. This indicator is associated with the cluster tree compilation process discussed in Sect. 3.3 and approximately measures the solving difficulty of each instance within the configuration system. It is calculated by analyzing the *variable-constraint graph* and determining the number of cycles present. Specifically, we compute the number of connected components (Ncc) in the graph and utilize Tarjan's cycle enumeration algorithm [26] to calculate the number of cycles (Nci) for each connected component. Additionally, we determine the number of nodes ($Nnode$) in each connected component. Finally, we use the sum of the product of Nci and $Nnode$ ($\sigma = \sum_{i=1}^{Ncc}(Nci_i \times Nnode_i)$) as an indicator for classifying the models. It's worth mentioning that we chose to use the number of cycles to calculate this indicator because the cycle is more difficult to deal with during graph partitioning. It takes more partitioning for the cycle parts in the graph to obtain the separate sub-cluster trees. Generally, we detect more cycles in the graph of large instances compared to the small instances, which proves the reliability of this indicator.

We order the 600 instances based on the σ value and assign them to three classes, each consisting of 200 instances. Within each class, one-third of the instances are allocated to the training set, while the remaining two-thirds form the validation set. The details of the dataset subdivision are presented in Table 1. The column $\#VM$ indicates the number of VMs in each class, while $Avg(\sigma)$ represents the average σ value for each class. Furthermore, $Min(\sigma)$ (resp. $Max(\sigma)$) denotes the minimum (resp. maximum) σ value within each class.

Table 1. Subdivision of 600 instances in three classes using the calculated σ.

Class	#VM	$Avg(\sigma)$	$Min(\sigma)$	$Max(\sigma)$
C_1	200	210,517,750	1,593,181	2,462,269,412
C_2	200	138,079	235	1,593,181
C_3	200	61	0	235

5.2 Results

In this section, we present the results of the parameter tuning and prediction processes. We begin by introducing the **Production parameters**, which are currently used by Renault's configuration system as a reference. Then, we compare the performance of the parameters found by SMAC with the production parameters.

Production Parameters. The production parameters are manually determined and updated by the system developers. They are obtained through an analysis of the input VM's *variable-constraints graph* and extensive simulation tests. All the experimental results are compared to the compilation results achieved using the production parameters.

Parameters Found by SMAC. In this part, we evaluate the performance of the parameters discovered by SMAC on the *training set* instances, in comparison with the production parameters. The results are presented for each class. We calculate the sum of the instances' compilation sizes for each class and present the comparison in Table 2. The column $size_{pp}$ represents the compilation size with the production parameters, while $size_{smac}$ represents the compilation size with the parameters found by SMAC. The column Δ indicates the percentage difference between the two sizes ($\Delta = (size_{pp} - size_{smac})/size_{pp}$). Additionally, the time taken by SMAC to find these parameters is presented in Table 3.

Table 2. Compilation result comparison between *production parameters* and *parameters found by SMAC*. (Unit:MB)

Class	#VM	$size_{pp}$	$size_{smac}$	Δ
C_1	67	44,948	22,030	51%
C_2	67	5,286	1,682	68%
C_3	67	17.31	17.19	0.12%

Table 3. Time usage of SMAC. (Unit:hour)

Class	#VM	time
C_1	67	68.73
C_2	67	2.61
C_3	67	0.17

We observe that for all classes of instances, the parameters found by SMAC outperform the production parameters. In the case of the complex classes C_1 and C_2, the reduction in compilation size is up to 68%. However, it is worth noting that SMAC requires a significant amount of time to find these high-performance parameters.

MatrixSize Prediction Model. This experiment evaluates the performance of the model for predicting **MatrixSize**. For the instances in the *validation set*, we compile them using the **PartitionVariables** of the production parameters along with the predicted **MatrixSize**. The compilation results are presented in Table 4, where $size_{pp}$ represents the total compilation size of the class with production parameters, $size_{matrixmodel}$ represents the compilation size with the predicted **MatrixSize**, and Δ indicates the difference ($\Delta = (size_{pp} - size_{matrixmodel})/size_{pp}$).

Table 4. Compilation result comparison between production parameters and parameters predicted by MatrixSize prediction Model. (Unit:MB)

Class	#VM	$size_{pp}$	$size_{matrixmodel}$	Δ
C_1	133	155,845	101,030	35%
C_2	133	5,441	4,806	11.6%
C_3	133	54.13	64.91	−19.9%

From Table 4, we observe that the **MatrixSize** prediction model performs well for the complex classes C_1 and C_2. The parameters with the predicted **MatrixSize** can reduce the total compilation size by up to 35%. However, for the simple class C_3, it results in an increase in the compilation size. Upon further investigation, we found that for most instances in C_3, the compilation size remains the same, except for one instance where the compilation size increases.

PartitionVariables Prediction Model. This experiment evaluates the performance of the model for predicting **PartitionVariables**. For the instances in the *validation set*, we compile them using the **MatrixSize** of the production parameters and the predicted **PartitionVariables**. We trained a scoring model to assist in selecting the partition variables. Specifically, for each instance, we rank the variables based on the model's score and choose the top 5 variables as the **PartitionVariables**. The compilation results are presented in Table 5, where $size_{varmodel}$ represents the total compilation size with the predicted parameters.

Table 5. Compilation result comparison between production parameters and parameters predicted by PartitionVariables Prediction Model (Unit:MB)

Class	#VM	$size_{pp}$	$size_{varmodel}$	Δ
C_1	133	155,845	313,958	−101.45%
C_2	133	5,441	5,609	−2.9%
C_3	133	54.13	54.13	0.00%

From Table 5, we observe that for all the classes, the predicted parameters perform the same or worse compared to the production parameters. Specifically, for the complex class C_1, it results in a twofold increase in the compilation size. We suspect that this is due to the lack of controlled **MatrixSize**. Apparently, the default **MatrixSize** of the production parameters is not suitable for the predicted **PartitionVariables**. As mentioned in Sect. 3.3, improper variable dispatch without appropriate **MatrixSize** control can lead to a large cluster tree. The results of this experiment validate this observation, especially for the complex instances. Additionally, we observe that the compilation size remains unchanged for the 133 instances of the simple class C_3 with the predicted parameters.

Combination of MatrixSize and PartitionVariables Prediction Models.
This experiment combines the parameters predicted by the MatrixSize and PartitionVariables models. For the instances in the *validation set*, we compile them using the predicted **MatrixSize** and **PartitionVariables** parameters. Table 6 presents the compilation results, comparing them with the production parameters, where $size_{both}$ represents the total compilation size with the combined predicted parameters by the two models above.

Table 6. Compilation result comparison between production parameters and parameters predicted sequentially by Models for PartitionVariables and for MatrixSize. (Unit:MB)

Class	#VM	$size_{pp}$	$size_{both}$	Δ
C_1	133	155,845	216,827	−39.12%
C_2	133	5,441	5,022	7.7%
C_3	133	54.13	64.91	−19.9%

From Table 6, we observe that for the most complex class C_1, the compilation size still increases with the predicted parameters. However, the increment is smaller compared to the results in Table 5. The performance of the predicted parameters improves when the predicted **MatrixSize** is considered.

For class C_2, the predicted parameters perform positively, reducing the compilation size by 7.7%. In the case of C_3, the increment is the same as in Table 4, indicating that it is caused by the same instance. The compilation size remains unchanged for all the other instances when using the predicted parameters compared to the production parameters.

Response Time for Requests. The online part of the compilation system is responsible for responding to requests with the compiled structure. In this section, we compare the response times of requests using different compilation results obtained with the parameters mentioned above.

We begin by specifying the type of request used for this test. For each instance, we generate 2000 requests for satisfiability checks [24]. These requests consist of partial configurations where certain variables are assigned specific values to check their satisfiability. Out of these 2000 requests, 1000 are satisfiable and 1000 are not. This choice is motivated by the actual number and type of requests used by Renault to verify the consistency of their product offer [29].

Next, for each instance in the validation set, we compile it separately using the production parameters, parameters predicted by the **MatrixSize** model, parameters predicted by the **PartitionVariables** model, and parameters predicted by both models. We measure the response times of the 2000 requests for each compiled structure. The results are presented in Table 7. RT_{pp} represents the total response time using the compiled structure with production parameters. $RT_{matrixmodel}$ represents the time using the predicted **MatrixSize** parameters. $RT_{varmodel}$ represents the time using the predicted **PartitionVariables** parameters. RT_{both} represents the time using parameters predicted by both models.

Table 7. Comparison of response times between compilation results with production parameters and predicted parameters (Unit: Seconds)

Class	RT_{pp}	$RT_{matrixmodel}$	$RT_{varmodel}$	RT_{both}
C_1	694	1133	2030	3349
C_2	22	22	28	27
C_3	3	3	3	3

From Table 7, we observe that for all compilations using predicted parameters, the response time is longer compared to using the production parameters. There are two reasons:

- For compilations using predicted **PartitionVariables** parameters, the size of the compiled structures increases (Table 5 and Table 6), which reasonably leads to longer response times.
- For compilations using predicted **MatrixSize** parameters and default **PartitionVariables** parameters, although the compiled size is smaller than the default parameters (Table 4), the response time still increases. This is because more sub-cluster trees are generated during compilation with predicted MatrixSize values. As mentioned in Sect. 3.3, a smaller **MatrixSize** can result in more sub-cluster trees, reducing the size of the global structure but increasing the response time for requests. When we verify the predicted **MatrixSize** values, we confirm that they are generally smaller than the **MatrixSize** of the production parameters.

In fact, the remarkable reduction in size achieved by the predicted **MatrixSize** outweighs the slight increase in response time, making it a highly favorable trade-off. It adds only 3.3 s ((1133-694)/133) to the response time per instance when handling 2000 requests, resulting in a mere 1.6 ms (3.3/2000) longer response time per request. Considering the substantial benefits gained from the reduced size, this minor impact on response time can be deemed negligible.

5.3   Summary

Based on the five experiments conducted, we draw the following conclusions:

- SMAC demonstrates excellent performance in searching for the best parameters for each instance, resulting in a reduction of up to 68% in compilation size. However, the drawback is its time-consuming nature.
- The **MatrixSize** prediction model proves to be successful, achieving a significant reduction of up to 35% in compilation size for the complex class C_1. The trade-off is a mere 1.6ms increase in response time per request. The operational impact of this reduction is substantial.
- For simple instances, applying the predicted **MatrixSize** and **Partition-Variables** separately or together has minimal impact on the compilation result. In most cases, the compilation size remains unchanged, with only a few instances experiencing a slight increase.
- The automated tuning and prediction of **PartitionVariables** require further exploration. The related experiments yielded negative results thus far.

Overall, the parameters found by SMAC perform the best among the parameters above, but with the drawback of time-consuming. After discussing with the stakeholders, we discovered the compilation process can happen frequently. So, applying SMAC each time before the compilation process is impractical. On the other hand, with the **MatrixSize** model obtained, we gain a 35% reduction in the compilation size with the parameters predicted within seconds. So considering both the compilation size reduction and the rapid parameters prediction, integrating the **MatrixSize** model into the product configuration system of Renault seems promising.

To facilitate the replication of the experiment process, we have provided a file accessible at: https://github.com/chiyanfly/Automated-Parameters-Tuning. The file includes:

- Source code for building the "variable-constraint graph" and "variable graph" and calculating "betweenness centrality".
- Source code demonstrating the usage of SMAC.
- Examples of **MatrixSize** and **PartitionVariables** prediction.

It is important to note that the verification of the experiment results related to the configuration system is confidential, as it is an internal proprietary system of Renault.

6   Conclusion

This paper presents an automated parameters tuning and predicting process to optimise Renault's product configuration system. The aim is to reduce the variability model's compilation size. We have shown the strong competitiveness of the **MatrixSize** parameter predicted by our model, and the reduction of compilation size is up to 35%.

Our first perspective is to continue exploiting the tuning process for **PartitionVariables**, since the current experiments show the negative results with its prediction model. We also intend to enlarge the data sets to improve the performance of models. In addition, since the **MatrixSize** prediction model has brought very promising results, this model is expected to be put into real usage for the operational system of Renault.

References

1. Akers, S.B.: Binary decision diagrams. IEEE Trans. Comput. **27**(06), 509–516 (1978)
2. Berger, T., et al.: A survey of variability modeling in industrial practice. In: Gnesi, S., Collet, P., Schmid, K. (eds.) The Seventh International Workshop on Variability Modelling of Software-intensive Systems, VaMoS 2013, Pisa, Italy, 23–25 January 2013, pp. 7:1–7:8. ACM (2013)
3. Ansótegui, C., Sellmann, M., Tierney, K.: A gender-based genetic algorithm for the automatic configuration of algorithms. In: Gent, I.P. (ed.) CP 2009. LNCS, vol. 5732, pp. 142–157. Springer, Heidelberg (2009). https://doi.org/10.1007/978-3-642-04244-7_14
4. Barthelemy, M.: Betweenness centrality in large complex networks. Eur. Phys. J. B **38**(2), 163–168 (2004)
5. Bartz-Beielstein, T., Flasch, O., Koch, P., Konen, W., et al.: SPOT: a toolbox for interactive and automatic tuning in the R environment. In: Proceedings, vol. 20, pp. 264–273 (2010)
6. Becker, A., Geiger, D.: A sufficiently fast algorithm for finding close to optimal clique trees. Artif. Intell. **1–2**, 3–17 (2001)
7. Belkhir, N., Dréo, J., Savéant, P., Schoenauer, M.: Feature based algorithm configuration: a case study with differential evolution. In: Handl, J., Hart, E., Lewis, P.R., López-Ibáñez, M., Ochoa, G., Paechter, B. (eds.) PPSN 2016. LNCS, vol. 9921, pp. 156–166. Springer, Cham (2016). https://doi.org/10.1007/978-3-319-45823-6_15
8. Belkhir, N., Dréo, J., Savéant, P., Schoenauer, M.: Per instance algorithm configuration of CMA-ES with limited budget. In: Proceedings of the Genetic and Evolutionary Computation Conference, pp. 681–688 (2017)
9. Bernard, P.: Extending cluster tree compilation with non-Boolean variables in product configuration: a tractable approach to preference-based configuration. In: Proceedings of the IJCAI, vol. 3. Citeseer (2003)
10. Brandes, U.: A faster algorithm for betweenness centrality. J. Math. Sociol. **25**(2), 163–177 (2001)
11. Choi, A., Darwiche, A.: Dynamic minimization of sentential decision diagrams. In: desJardins, M., Littman, M.L. (eds.) Proceedings of the Twenty-Seventh AAAI Conference on Artificial Intelligence, 14–18 July 2013, Bellevue, Washington, USA. AAAI Press (2013)

12. Darwiche, A., Marquis, P.: A knowledge compilation map. J. Artif. Intell. Res. **17**, 229–264 (2002)
13. Doerr, B., Le, H.P., Makhmara, R., Nguyen, T.D.: Fast genetic algorithms. In: Proceedings of the Genetic and Evolutionary Computation Conference, pp. 777–784 (2017)
14. El Yafrani, M., Scoczynski, M., Sung, I., Wagner, M., Doerr, C., Nielsen, P.: MATE: a model-based algorithm tuning engine: a proof of concept towards transparent feature-dependent parameter tuning using symbolic regression. In: Zarges, C., Verel, S. (eds.) EvoCOP 2021. LNCS, vol. 12692, pp. 51–67. Springer, Cham (2021). https://doi.org/10.1007/978-3-030-72904-2_4
15. Frazier, P.I.: Bayesian optimization. In: Recent Advances in Optimization and Modeling of Contemporary Problems, pp. 255–278. Informs (2018)
16. Grami, A.: Chapter 18 - graphs. In: Grami, A. (ed.) Discrete Mathematics, pp. 327–350. Academic Press (2023). https://doi.org/10.1016/B978-0-12-820656-0.00018-6, https://www.sciencedirect.com/science/article/pii/B9780128206560000186
17. Hutter, F., Hoos, H.H., Leyton-Brown, K.: Automated configuration of mixed integer programming solvers. In: Lodi, A., Milano, M., Toth, P. (eds.) CPAIOR 2010. LNCS, vol. 6140, pp. 186–202. Springer, Heidelberg (2010). https://doi.org/10.1007/978-3-642-13520-0_23
18. Hutter, F., Hoos, H.H., Leyton-Brown, K.: Sequential model-based optimization for general algorithm configuration. In: Coello, C.A.C. (ed.) LION 2011. LNCS, vol. 6683, pp. 507–523. Springer, Heidelberg (2011). https://doi.org/10.1007/978-3-642-25566-3_40
19. Hutter, F., Hoos, H.H., Leyton-Brown, K., Stützle, T.: ParamILS: an automatic algorithm configuration framework. J. Artif. Intell. Res. **36**, 267–306 (2009)
20. Hutter, F., Lindauer, M., Balint, A., Bayless, S., Hoos, H., Leyton-Brown, K.: The configurable sat solver challenge (CSSC). Artif. Intell. **243**, 1–25 (2017)
21. López-Ibáñez, M., Dubois-Lacoste, J., Cáceres, L.P., Birattari, M., Stützle, T.: The irace package: iterated racing for automatic algorithm configuration. Oper. Res. Perspect. **3**, 43–58 (2016)
22. Noble, W.S.: What is a support vector machine? Nat. Biotechnol. **24**(12), 1565–1567 (2006)
23. Pargamin, B.: Vehicle sales configuration: the cluster tree approach. In: ECAI 2002 Configuration Workshop, pp. 35–40 (2002)
24. Pohl, R., Lauenroth, K., Pohl, K.: A performance comparison of contemporary algorithmic approaches for automated analysis operations on feature models. In: 2011 26th IEEE/ACM International Conference on Automated Software Engineering (ASE 2011), pp. 313–322. IEEE (2011)
25. Rodriguez-Galiano, V., Sanchez-Castillo, M., Chica-Olmo, M., Chica-Rivas, M.: Machine learning predictive models for mineral prospectivity: an evaluation of neural networks, random forest, regression trees and support vector machines. Ore Geol. Rev. **71**, 804–818 (2015)
26. Tarjan, R.: Enumeration of the elementary circuits of a directed graph. SIAM J. Comput. **2**(3), 211–216 (1973)
27. Weisberg, S.: Applied Linear Regression, vol. 528. Wiley, Hoboken (2005)
28. Witt, C.: Tight bounds on the optimization time of a randomized search heuristic on linear functions. Comb. Probab. Comput. **22**(2), 294–318 (2013)
29. Xu, H., Baarir, S., Ziadi, T., Essodaigui, S., Bossu, Y., Messan Hillah, L.: An experience report on the optimization of the product configuration system of Renault. In: 2023 27th International Conference on Engineering of Complex Computer Systems (ICECCS), pp. 197–206. IEEE Computer Society (2023)

30. Xu, H., Baarir, S., Ziadi, T., Essodaigui, S., Bossu, Y., Messan Hillah, L.: Optimization of the product configuration system of Renault. In: Proceedings of the 38th ACM/SIGAPP Symposium on Applied Computing, pp. 1486–1489 (2023)
31. Xu, L., Hutter, F., Hoos, H.H., Leyton-Brown, K.: SATzilla: portfolio-based algorithm selection for sat. J. Artif. Intell. Res. **32**, 565–606 (2008)
32. Zhu, H., Williams, C.K., Rohwer, R., Morciniec, M.: Gaussian regression and optimal finite dimensional linear models (1997)

Neural Network Verification

Optimal Solution Guided Branching Strategy for Neural Network Branch and Bound Verification

Xiaoyong Xue[ID] and Meng Sun[(✉)][ID]

School of Mathematical Sciences, Peking University, Beijing, China
{xuexy,sunm}@pku.edu.cn

Abstract. Adversarial examples reveal the vulnerability of neural networks, thereby increasing the demand for formal verification of their robustness. Verification methods employing branch and bound technique have shown excellent performance for this task and are widely adopted to provide robustness assurance. A key component in branch and bound is the branching strategy, which determines how to split the feasible region. A good branching strategy can reduce the number of branches that need to be explored, and thus improve the verification efficiency. In this paper, we present a novel branching strategy. For each sub-problem in the branch and bound process, we use its optimal solution to compute a score for each branching choice. This score approximates the potential improvement brought by branching at that choice, and a higher score indicates a better branching decision. Our branching strategy also includes out-of-bound compensation and score truncation. These techniques can help better assess the quality of each branching choice. In addition, we demonstrate that some sub-problems generated by certain branching choice can be directly resolved, which reduces the number of calls to the bounding algorithm. We implement our branching strategy as OptGBS and conduct experiments on widely used open source benchmarks to evaluate its performance. Experimental results show that our branching strategy achieves a roughly 25% time cost reduction compared to the state-of-the-art branching strategies, and verification methods using our branching strategy can achieve higher verified accuracy.

Keywords: Neural Network · Robustness Verification · Branch and Bound · Branching Strategy

1 Introduction

In recent years, deep neural networks have made inspiring breakthroughs and are widely adopted in security-related fields [4,9]. However, numerous studies reveal that deep neural networks are vulnerable to adversarial attack [8,11], which states that subtle and imperceptible perturbations can cause drastic changes in the prediction results. Faced with such challenges, verification techniques are

© The Author(s), under exclusive license to Springer Nature Switzerland AG 2025
G. Bai et al. (Eds.): ICECCS 2024, LNCS 14784, pp. 67–87, 2025.
https://doi.org/10.1007/978-3-031-66456-4_4

employed to provide rigorous assurances on the robustness of neural networks. Many different techniques are used in the verification of neural networks, such as satisfiability modulo theory [5,10,13], abstract interpretation [1,20], and linear approximation [23,25]. These methods face two primary challenges, *completeness* and *efficiency*. Complete methods tend to have lower efficiency, while efficient methods may fail to verify certain properties. Method based on branch and bound [17,22,24] achieves a good balance between these two aspects. The branching process ensures the completeness of verification, and the efficient bounding technique improves efficiency. The branch and bound verification methods have shown excellent performance and are becoming mainstream.

Branching strategy is a critical component in the branch and bound framework and has significant impact on verification efficiency. Early branch and bound verification methods performed branching on the input region. The branching strategy in these methods considered which input dimension to split [21]. Later research found that verification methods branching on hidden neurons have better performance [2]. Therefore, the mainstream branching strategies today focus on selecting which hidden neuron to split. BaBSR [2] and FSB [17] are the mainstream branching strategies, both of which use the dual problem of the verification task to estimate the improvement brought by branching. However, their improvement estimation is relatively coarse and neglects the impact of neuron splitting constraints. ACS [6] is a branching strategy specifically designed for multi-neuron constraints, which can not be applied to regular neuron splitting constraints. GNN [14] uses graph neural networks to make branching decisions, which requires additional training process.

In this paper, we propose a novel branching strategy to improve the efficiency of branch and bound verification methods. It follows a greedy principle, which is selecting the neuron that can bring the most improvement. The improvement refers to the difference between the optimal value of the sub-problem produced by branching and the optimal value of the original problem. Since there are a large quantity of neurons in a neural network, it is impractical to compute the exact improvement for each neuron. Therefore, we need to estimate the branching improvement for each neuron without actually solving the sub-problems. We use the optimal solution of a problem to make the improvement estimation for the upcoming branching. The optimal solution of the input layer is propagated to each hidden layer, which leads to more accurate estimation. We also propose out-of-bounds compensation and score truncation techniques to enhance our branching strategy. For sub-problems where the original optimal solution is out of constraints boundary, we need to compensate their scores. And if the score of a sub-problem is too high, we need to truncate it to reduce its interference in the final decision. Using the estimated improvement, we assign a score to each neuron and choose the neuron with the highest score as the branching decision. Our branching strategy can be completed in one forward propagation and is therefore computationally efficient. Moreover, we also prove that some sub-problems generated in branching processes can be directly solved without invoking the bounding algorithm. This further improves the efficiency of the branch and bound verification method.

The rest of the paper is organized as follows. We first provide the preliminaries in Sect. 2. Section 3 makes a thorough inquiry into the bounding process and uses the optimal solution given by the bounding algorithm to estimate the improvement of a sub-problem. The design of our branching strategy is presented in Sect. 4. In Sect. 5, we evaluate our branching strategy against the mainstream strategies and show the experiment results. We conclude this paper in Sect. 6.

2 Preliminary

2.1 Neural Network Verification

The properties involved in neural network verification can be divided into two parts, input constraints P_{in} and output constraints P_{out}. A neural network satisfies this property if for any input that satisfies P_{in}, its corresponding output satisfies P_{out}. A verification problem of this property can be transformed into an optimization problem. For example, the widely studied local robustness property [26], which states that the neural network's predication does not change when a small perturbation is added to the given input, can be transformed into the following optimization problem.

$$
\begin{aligned}
\min \ & y_t - y_o \\
s.t. \quad & y = W^L z_{L-1} + b^L \\
& \hat{z}_k = W^k z_{k-1} + b^k \quad k \in [1, L-1], \\
& z_k = \sigma(\hat{z}_k) \quad\quad\quad k \in [1, L-1], \\
& ||z_0 - x_0||_p \leq \epsilon.
\end{aligned} \tag{1}
$$

The W and b represent weights and biases, and L is the number of layers. The hidden neurons are divided into pre-activation neurons and post-activation neurons, denoted as $\hat{z}$ and z respectively. The input constraint is $||z_0 - x_0||_p \leq \epsilon$, which indicate that the perturbation is within a ℓ_p-norm ball of radius ϵ. The optimization objective $\min y_t - y_o$ corresponds to the output constraint $y_t - y_o \geq 0$, which means that the input is always classified into class t.

If the optimal value of this optimization problem is greater than zero, the robustness property is satisfied. In the context below, when we talk about a verification problem, it refers to its corresponding optimization problem.

2.2 Activation Function Relaxation

The main obstacle in solving (1) is the constraints on activation functions. Due to the non-convexity of the activation function, it is difficult to find the exact minimum value. To address this problem, linear relaxations [20, 25] are introduced to provide upper and lower linear relaxations for the activation function. The commonly used relaxtions for ReLU activation function are:

$$
z_{k,i} \leq \frac{u(\hat{z}_{k,i} - l)}{u - l}, \ z_{k,i} \geq 0, \ z_{k,i} \geq \hat{z}_{k,i}
$$

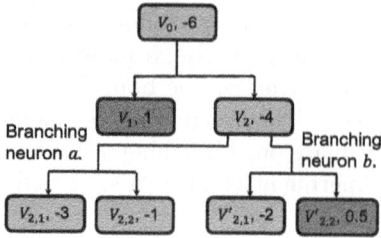

Fig. 1. Branch and bound verification process.

where u and l are the upper and lower bounds of pre-activation neuron $\hat{z}_{k,i}$, respectively. There are two alternative lower relaxations, and the choice of which one to use depends on the values of u and l. If $u + l \leq 0$, the lower relaxation is $z_{k,i} \geq 0$. Otherwise, the lower relaxation is $z_{k,i} \geq \hat{z}_{k,i}$.

With such relaxations, we can get a relaxed optimization problem, which can be easily solved. The minimum value of the relaxed problem is a lower bound for the minimum value of problem (1). Therefore, if the minimum value of the relaxed problem is larger than 0, we can also say that the property is satisfied.

2.3   Branch and Bound Verification

While methods based on linear relaxation can efficiently solve verification problems, they inevitably introduce errors to the verification results. This results in the inability to verify some properties that is actually satisfied. To overcome this issue, the branch and bound technique is introduced to obtain a complete verification method [17, 22].

The branch and bound verification process is illustrated in Fig. 1. The original verification problems are segmented into a series of sub-problems, each of which is evaluated with efficient relaxation based method. Sub-problems with minimum values larger than 0 will be pruned (green blocks in Fig. 1). The rest sub-problems (yellow blocks in Fig. 1) will be branched into smaller sub-problems until no more sub-problems need branching or a counterexample is found. The branching of a problem is accomplished through the splitting of a neuron. A relaxed neuron $z_{k,i}$ can be split into positive and negative parts, which creates two neuron splitting constraints:

$$\hat{z}_{k,i} \geq 0, \quad \hat{z}_{k,i} \leq 0$$

Each constraint corresponds to a sub-problem, and this constraint will be placed into the sub-problem. In each sub-problem, the constraints on neuron $z_{k,i}$ are accurate and free from errors. After splitting all neurons, the network becomes a linear network and we can get accurate verification results.

The worst-case time complexity of the branch and bound verification method is exponential. If a neural network has n hidden neurons, then the time required in the worst-case scenario is $O(2^n)$. Therefore, we need a good branching strategy to prune out more sub-problems and improve the efficiency of the branch and bound verification method.

3 Branch Improvement Estimation

From an intuitive perspective, the quality of the branching decision is captured by the minimum value improvement of the sub-problem. If a branching decision generates sub-problems with larger improvement, then it is considered a better decision. In Fig. 1, the improvement of sub-problem $V_{2,1}$ is the difference between its own minimum and the minimum of the parent problem V_2, which is $-3 - (-4) = 1$. Sub-problems $V'_{2,1}$ and $V'_{2,2}$ have larger improvements compared to $V_{2,1}$ and $V_{2,2}$. Therefore, neuron b is a better branching decision. However, there are numerous neurons to choose in each branch. It is impractical to compute the minimum value of all sub-problems generated by each possible neuron.

In this section, we first delve into the process of obtaining the minimum value for each sub-problem. Then, we show how to estimate the minimum value improvement of each sub-problem.

3.1 Symbolic Bound Propagation

Within the branch and bound verification process, the sub-problems are typically solved with the symbolic bound propagation methods [22,24]. The initial step involves the use of symbolic interval abstraction [19,23] to compute both the lower and upper bounds for each hidden neuron. Subsequently, the linear lower and upper relaxations for each hidden neuron are computed as outlined in Sect. 2.2

Utilizing the calculated linear relaxations, symbolic bound propagation derives a linear lower bound for each layer from the output to the input. In layer k, we use P_k and Q_k to denote the coefficients and constant term of the linear lower bound for the post-activation layer, respectively. Similarly, $\hat{P}_k$ and $\hat{Q}_k$ stand for the coefficients and constant term of the linear lower bound of the k-th pre-activation layer. The process of bound propagation can be described as follows:

$$\min y_t - y_o \geq \min P_{L-1} \cdot z_{L-1} + Q_{L-1}$$
$$\geq \min \hat{P}_{L-1} \cdot \hat{z}_{L-1} + \hat{Q}_{L-1}$$
$$\geq \cdots$$
$$\geq \min P_0 \cdot z_0 + Q_0$$

In the output layer, $\hat{P}_L$ and $\hat{Q}_L$ take the coefficient and constant of the sub-problem's optimization objective. The pre-activation of the $(k+1)$-th layer and the post-activation of the k-th layer are connected through affine layers. The propagation of P and Q between them is

$$P_k = \hat{P}_{k+1} \cdot W^k$$
$$Q_k = \hat{P}_{k+1} \cdot b^k + \hat{Q}_{k+1}$$

We use α^L and β^L to represent the slop vector and intercept vector of the linear lower relaxations for layer k. Similarly, α^U and β^U represent those of the

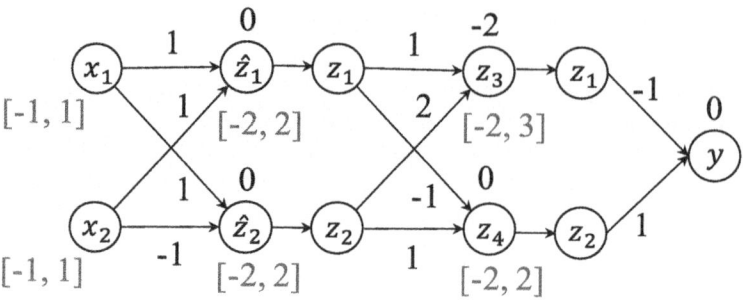

Fig. 2. Illustrative neural network

linear upper relaxations for layer k. The propagation of P and Q between the k-th pre-activation and post-activation layer is

$$\hat{P}_k = [P_k]_+ \odot \alpha_k^L + [P_k]_- \odot \alpha_k^U$$
$$\hat{Q}_k = [P_k]_+ \odot \beta_k^L + [P_k]_- \odot \beta_k^U + Q_k \tag{2}$$

where $[a]_+$ and $[a]_-$ represent the positive and negative part of a, and $\odot$ is the element-wise product. Neurons with positive $P_{k,i}$ are relaxed with the lower linear relaxation, and neurons with negative $P_{k,i}$ are relaxed with upper linear relaxation.

The result of symbolic bound propagation is obtaining a linear lower bound for the input layer variables, which is $P_0 \cdot z_0 + Q_0$. Then we can directly use Hölder Inequality [25] to find the minimum value and optimal solution $x*$ with respect to the input constraint $||z_0 - x_0||_p \le \epsilon$.

$$\min y_t - y_o = -||P_0||_q \epsilon + P_0 \cdot x_0 + Q_0$$
$$x* = \frac{1}{||P_0||_p} \cdot (\sqrt[p]{P_{0,1}^q}, \dots, \sqrt[p]{P_{0,i}^q}, \dots, \sqrt[p]{P_{0,n}^q})$$

where $1/q + 1/p = 1$ and n is the number of input neurons.

Example 1. Consider a neural network with two hidden layers. The weights and biases of each layer are shown in Fig. 2. Neurons z_1, z_2 are in the first hidden layer and z_3, z_4 are in the second hidden layer. The input neurons are x_1, x_2 and the output neuron is y. The input constraint is an infinite norm ball centered at $(0, 0)$ with a radius of 1. From this we can get the value ranges of the input neurons are $x_1 \in [-1, 1]$, $x_2 \in [-1, 1]$. Our objective is to find the lower bound of y. The bound of each pre-activation neuron is shown as the red intervals in Fig. 2. With these bounds, we can compute the linear relaxations of each neuron, and symbolic bound propagation process is as follows:

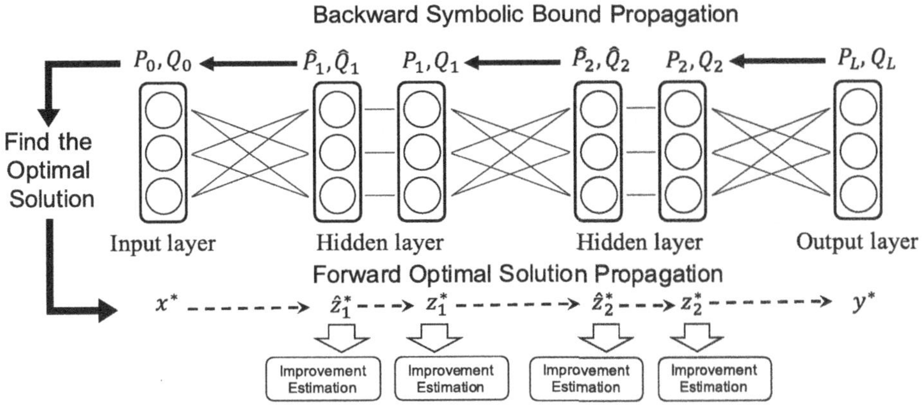

Backward Symbolic Bound Propagation

Fig. 3. Optimal solution propagation and sub-problem improvement estimation.

$$\min y \geq \min z_4 - z_3$$

$$\geq \min -\frac{3}{5}\hat{z}_3 - \frac{6}{5}$$

$$\geq \min -\frac{3}{5}z_1 - \frac{6}{5}z_2 \qquad (3)$$

$$\geq \min -\frac{3}{10}\hat{z}_1 - \frac{6}{10}\hat{z}_2 - \frac{9}{5}$$

$$\geq \min -\frac{9}{10}x_1 + \frac{3}{10}x_2 - \frac{9}{5}$$

After bound propagation, we get the linear lower bound on the input layer, that is $y \geq -\frac{9}{10}x_1 + \frac{3}{10}x_2 - \frac{9}{5}$. Combined with the value range of input neurons, we can find that the minimum value is -3 and the optimal solution is $(1, -1)$.

3.2 Optimal Solution Propagation

Estimating the improvement of a sub-problem requires estimating the minimum value of the sub-problem. We can estimate this by comparing the symbolic bound propagation process of the sub-problem and that of the parent problem. The minimum value is obtained on the input layer, but the neuron chosen for branching can be in any hidden layer. We want to incorporate the computation of the minimum value within each layer.

In the symbolic bound propagation process, we get a linear lower bound for each layer and the value range for each hidden neuron. The direct idea is to use these two to find the minimum value of the problem, but this may yield incorrect results. For example, the linear lower bound for the pre-activation neuron in the first hidden layer is $-\frac{3}{10}\hat{z}_1 - \frac{6}{10}\hat{z}_2 - \frac{9}{5}$ in Example 1. We also notice that the pre-activation neurons $\hat{z}_1$ and $\hat{z}_2$ both take values from $[-2, 2]$. With these linear lower bound and value range, the minimum value is -3.6 and the

---

Algorithm 1: Optimal Solution Propagation

Input: Optimal solution x^*,
Coefficients of linear bounds $[P_1, P_2, \ldots, P_L]$,
Linear relaxations of each hidden neuron.
Output: Optimal solution at each layer $[\hat{z}_1^*, z_1^*, \ldots, \hat{z}_{L-1}^*, \hat{z}_{L-1}^*]$

1 $z_0^* \leftarrow x^*$ for $k \leftarrow 1$ to $L - 1$ do
2 $\quad$ $\hat{z}_k^* \leftarrow W^k \cdot z_{k-1}^* + b^k$;
3 $\quad$ $n_k \leftarrow$ the number of neurons in layer k;
4 $\quad$ for $j \leftarrow 1$ to n_k do
5 $\quad\quad$ if $P_{k,j} > 0$ then
6 $\quad\quad\quad$ $z_{k,j}^* \leftarrow \alpha_{k,j}^L \cdot \hat{z}_{k,j}^* + \beta_{k,j}^L$;
7 $\quad\quad$ else if $P_{k,j} < 0$ then
8 $\quad\quad\quad$ $z_{k,j}^* \leftarrow \alpha_{k,j}^U \cdot \hat{z}_{k,j}^* + \beta_{k,j}^U$;
9 $\quad\quad$ else
10 $\quad\quad\quad$ $z_{k,j}^* \leftarrow 0$;

11 **return** $[\hat{z}_1^*, z_1^*, \ldots, \hat{z}_{L-1}^*, \hat{z}_{L-1}^*]$

---

optimal solution for this layer is $(\hat{z}_1, \hat{z}_2) = (2, 2)$. This minimum value is smaller than that calculated with symbolic bound propagation, which indicates it is an incorrect result. We also notice that the corresponding values of x_1 and x_2 are 2 and 0 respectively, which are not in the input region. Because of the error introduced by hidden neuron relaxations, the value range of the hidden neuron is larger than the actual feasible region. Therefore, direct computing using the value range of each neuron may not lead to the correct result.

We now introduce how to find the corresponding optimal solutions in each layer. Denote the optimal solution of the input layer as x^*, and the corresponding optimal solutions of the pre-activation and post-activation in the k-th hidden layer as $\hat{z}_k^*$ and z_k^*, respectively.

The optimal solution for each layer is computed sequentially from the input layer to the output layer as shown by the dashed lines in Fig. 3. For the optimal solutions of the pre-activation layer $\hat{z}_k^*$, they can be derived from the optimal solutions of the preceding post-activation layer z_{k-1}^* with the constraints one the affine layer. For optimal solutions of the post-activation layer z_k^*, they are computed according to $\hat{z}_k^*$ and the inequalities constraints. Since only boundary conditions of the inequalities constraints are used in bound propagation, we only need to determine whether to use the upper or lower bound for each neuron. If a neuron takes the lower bound, we compute z_k^* according to the lower linear relaxation, and if a neuron takes the upper bound, we compute z_k^* according to the upper linear relaxation. The computation process is carried out in a propagation manner as shown in Algorithm 1.

In Algorithm 1, $P_{k,i}$ represents the i-th element of P_k, which is the coefficient of the i-th post-activation neuron in layer k. We use $P_{k,i}$ to determine whether the upper or lower linear relaxation is used for the i-th neuron in layer k during bound propagation. If $P_{k,i}$ is positive, it indicates that the lower linear relaxation is used in bound propagation, and $z_{k,i}^*$ is computed using the lower linear relaxation (line 6). Conversely, if $P_{k,i}$ is negative, $z_{k,i}^*$ is calculated using the upper linear relaxation (line 8). In the final case, that is when $P_{k,i}$ is 0, the value of $z_{k,i}^*$ does not affect the results. For simplicity, we assign 0 to $z_{k,i}^*$ (line 10).

In this way, we can obtain the optimal solution for each layer with only a single forward propagation. The linear lower bound for each layer reaches its minimum value at this point, and this value equals to the minimum value of the input layer. Using the optimal solution for each layer, we can estimate the improvements brought by branching more precisely.

3.3   Sub-problem Improvement Estimation

To get exact the exact minimum value of the sub-problem, we need to go through the complete boundary propagation process. Considering the large number of neurons, it is not feasible to perform symbolic bound propagation on all sub-problems generated by each possible neuron. However, the symbolic bound propagation processes of the sub-problem and the parent problem have many similarities. We can use this similarity to estimate the solutions to the sub-problem.

Consider branching at the i-th neuron in the k-th layer. The original upper and lower linear relaxations for this neuron are denoted as $\alpha_{k,i}^U \hat{z}_{k,i} + \beta_{k,i}^U$ and $\alpha_{k,i}^L \hat{z}_{k,i} + \beta_{k,i}^L$, respectively. In the parent problem, the constraint corresponding to this neuron is

$$\alpha_{k,i}^U \hat{z}_{k,i} + \beta_{k,i}^U \leq z_{k,i} \leq \alpha_{k,i}^L \hat{z}_{k,i} + \beta_{k,i}^L.$$

After branching, the constraints corresponding to this neuron in the two sub-problems become $z_{k,i} = \hat{z}_{k,i}$ and $z_{k,i} = 0$, respectively.

In boundary propagation process of the parent problem, the coefficients and constant term of the linear lower bound are denoted as P and Q as shown in (2). For the sub-problem, the propagation process from the output layer to the k-th layer is same as the propagation process for its parent problem, which means P_k and Q_k in these two processes are the same. When the bound propagation process goes through the k-th layer, the linear relaxations of i-th neuron are changed, which results in the $\hat{P}_k$ and $\hat{Q}_k$ of the sub-problem being different from the parent problem. Because the split is only performed at the i-th neuron, $\hat{P}_k$ in the two bound propagation only differs at $\hat{P}_{k,i}$, and the other elements are still the same. The difference in $\hat{P}_{k,i}$ affects all neurons in the $k-1$ layer after propagating through the affine layer. The remaining propagation processes of the sub-problem is quite different from the original problem. The P_j and Q_j ($j < k$) can not be directly derived from the parent problem. From the above analysis, it can be concluded that using the linear lower bound of the k-th pre-activation layer to estimate the lower bound of the sub-problem is the most effective and efficient choice.

The linear lower bound of the k-th post-activation layer in the parent problem is

$$P_{k,1} z_{k,1} + \cdots + P_{k,i} z_{k,i} + \cdots + P_{k,n_k} z_{k,n_k} + Q_k.$$

This is also the linear lower bound of the k-th post-activation layer in the sub-problems. And we use this to compute the linear lower bound of the k-th pre-activation layer.

For the sub-problem with neuron split constraint $z_{k,i} = \hat{z}_{k,i}$, we denote the coefficient and constant term of the k-th pre-activation layer as $\hat{P}'_k$ and $\hat{Q}'_k$. The linear lower bound for the k-th pre-activation layer in the sub-problem is

$$\hat{P}'_{k,1}\hat{z}_{k,1} + \cdots + \hat{P}'_{k,i}\hat{z}_{k,i} + \cdots + \hat{P}'_{k,n_k}\hat{z}_{k,n_k} + \hat{Q}'_k.$$

According to the neuron split constraint, the i-th element of $\hat{P}'_k$ is $\hat{P}'_{k,i} = P_{k,i}$. The other elements of $\hat{P}'_k$ are the same as those of $\hat{P}_k$, which is $\hat{P}'_{k,j} = P_{k,j}, \forall j \neq i$. Moreover, $\hat{Q}'_k$ also changes because of the change in the constraint of neuron $z_{k,i}$. The relationship between $\hat{Q}'_k$ and $\hat{Q}_k$ is

$$\hat{Q}'_k = \hat{Q}_k - [P_{k,i}]_+ \cdot \beta^L_{k,i} - [P_{k,i}]_- \cdot \beta^U_{k,i}.$$

For the parent problem, we get the optimal solution at the k-th pre-activation layer with the optimal solution propagation. The linear lower bound of the k-th pre-activation layer attains its minimum value at this point. For the sub-problem, we estimate the minimum value of the sub-problem with the value of the k-th pre-activation layer's linear lower bound at this point. Suppose the optimal solution is z^*, then the estimated minimum value for the sub-problem is

$$\hat{P}'_{k,1}\hat{z}^*_{k,1} + \cdots + \hat{P}'_{k,i}\hat{z}^*_{k,i} + \cdots + \hat{P}'_{k,n_k}\hat{z}^*_{k,n_k} + \hat{Q}'_k.$$

Notice that the minimum value of the parent problem is

$$\hat{P}_{k,1}\hat{z}^*_{k,1} + \cdots + \hat{P}_{k,i}\hat{z}^*_{k,i} + \cdots + \hat{P}_{k,n_k}\hat{z}^*_{k,n_k} + \hat{Q}_k,$$

The difference between these two formulas is the improvement estimation of the sub-problem relative to the parent problem. We use Imp^P to denote such improvement, and we can get

$$
\begin{aligned}
Imp^P_{k,i} &= (\hat{P}'_{k,i} - \hat{P}_{k,i})\hat{z}^*_{k,i} + \hat{Q}'_k - \hat{Q}_k \\
&= \begin{cases} (P_{k,i} - P_{k,i}\alpha^L_{k,i})\hat{z}^*_{k,i} - P_{k,i}\beta^L_{k,i}, & P_{k,i} \geq 0 \\ (P_{k,i} - P_{k,i}\alpha^U_{k,i})\hat{z}^*_{k,i} - P_{k,i}\beta^U_{k,i}, & P_{k,i} < 0 \end{cases}
\end{aligned}
\tag{4}
$$

For the sub-problem with $z_{k,i} = 0$ constraint, We can use the same way to compute the improvement estimation, which is denoted as Imp^N.

$$
\begin{aligned}
Imp^N_{k,i} &= (\hat{P}'_{k,i} - \hat{P}_{k,i})\hat{z}^*_{k,i} + \hat{Q}'_k - \hat{Q}_k \\
&= \begin{cases} -P_{k,i}\alpha^L_{k,i}\hat{z}^*_{k,i} - P_{k,i}\beta^L_{k,i}, & P_{k,i} \geq 0 \\ -P_{k,i}\alpha^U_{k,i}\hat{z}^*_{k,i} - P_{k,i}\beta^U_{k,i}, & P_{k,i} < 0 \end{cases}
\end{aligned}
\tag{5}
$$

Example 2. We use the neural network in Example 1 to show how to compute the estimated improvement. We choose the neuron z_1 to perform branching.

In the symbolic bound propagation process, the coefficient of z_1 is $-\frac{3}{5}$, and the upper linear relaxation of z_1 is $\frac{1}{2}\hat{z}_1 + 1$. With optimal solution propagation, we have $\hat{z}_1^* = 0$.

For the sub-problem with constraint $z_1 = \hat{z}_1$, we can compute the estimated improvement with (4).

$$Imp^P = (-\frac{3}{5} - (-\frac{3}{5}) \cdot \frac{1}{2}) \cdot 0 - (-\frac{3}{5}) \cdot 1 = \frac{3}{5}$$

The estimated improvement of this sub-problem relative to the original problem is $\frac{3}{5}$, which means the estimated minimum value of this problem is $-3 + \frac{3}{5} = -\frac{12}{5}$. The exact minimum value of this sub-problem can be obtained with bound propagation. Using the constraint $z_1 = \hat{z}_1$, we can compute the lower bound of the input layer $y \geq -\frac{6}{5}x_1 - \frac{6}{5} \geq -\frac{12}{5}$. The exact minimum value is $-\frac{12}{5}$, which is same as the estimated minimum value.

For the sub-problem with constraint $z_1 = 0$, we can compute the estimated improvement with (5).

$$Imp^N = -(-\frac{3}{5}) \cdot \frac{1}{2} \cdot 0 - (-\frac{3}{5}) \cdot 1 = \frac{3}{5}$$

The estimated improvement of this sub-problem is also $\frac{3}{5}$, and the estimated minimum value is also $-\frac{12}{5}$. With bound propagation, we can find the exact minimum value is also $-\frac{12}{5}$. This demonstrates the accuracy of our estimation.

The sub-problem improvement estimation for all layers is shown in Fig. 3. Notice that our estimation method only uses information related to the neuron that is about to branch. This allows improvement estimations of multiple neurons to be performed in parallel, thereby increasing the computational efficiency.

4 Branching Strategy Design

In this section, we introduce how to use the sub-problem improvement estimation to design our branching strategy.

4.1 Branching Strategy

One neuron branching in ReLU neural network verification produces two sub-problems. We already know how to estimate the improvements of these two sub-problems. What we need to do is to evaluate the impact of branching a neuron based on these two estimated improvements.

Suppose the two improvement estimation are Imp^P and Imp^N. We evaluate the neuron with some integration of these two values, such as maximum value or minimum value. Taking the maximum value is an aggressive strategy. This can result in choosing a neuron where the improvement for one sub-problem is large, while the improvement for the other sub-problem might be small. Taking the minimum value is a conservative strategy. It focuses on the joint improvement of both sub-problems. However, it does not pay enough attention to situations where there is a large improvement in one sub-problem. These two strategy are easily influenced by the improvement of one sub-problem. In order to balance the two sub-problems, we use the average value of the improvements of these two sub-problems. Denote the impact score of a neuron as S, and we have

$$S = \frac{Imp^P + Imp^N}{2}.$$

We can directly use the score S to design a branching strategy. Each time a branching occurs, we can take the neuron with the highest score S as the branching decision. However, the design of this split score is still relatively coarse. In order to better estimate the impact of neuron branching, we propose other two techniques, out-of-bound compensation and score truncation.

Out-of-Bound Compensation. When estimating the improvement of one sub-problem, we use the value of the linear lower bound at the optimal solution for each layer. However, this optimal solution does not necessarily satisfy the neuron split constraints for that sub-problem. For example, $(0, 2)$ is the optimal solution of the first pre-activation layer in Example 1. If a branching occurs on neuron z_2, then one of the produced sub-problems contains the constraint $\hat{z}_2 \leq 0$. The component of the optimal solution in the second dimension is 2, which does not satisfy this constraint. The optimal solution lies outside the boundary. This means that we underestimate the minimum value of this sub-problem, which also means we underestimate the improvement of this sub-problem. We need to make a compensation for such sub-problems.

Assume that the neuron we want to split is $z_{k,i}$. The contribution of neuron $z_{k,i}$ to the minimum value of the sub-problem is $P_{k,i} \cdot \hat{z}_{k,i}$. And we use $\hat{z}_{k,i} = \hat{z}_{k,i}^*$ to compute the estimated minimum value. Because the optimal solution is outside the boundary of the constraint, such estimation is inaccurate. We can replace $\hat{z}_{k,i}^*$ with the closest point within the boundary, which is $\hat{z}_{k,i} = 0$. The difference between the contributions of these two points to the minimum value is

$$|P_{k,i} \cdot \hat{z}_{k,i}^* - P_{k,i} \cdot 0| = |P_{k,i} \cdot \hat{z}_{k,i}^*|.$$

This is the compensation for the improvement estimation of the sub-problem.

Notice that only the sub-problem where its neuron splitting constraint is not satisfied needs compensation. We use an indicator function to denote which sub-problem needs this compensation. The compensation terms for sub-problems with constraint $z_{k,i} = \hat{z}_{k,i}$ and $z_{k,i} = 0$ are shown in the following:

$$CP_{k,i}^P = \mathbb{1}_{\hat{z}_{k,i}^* \leq 0} \cdot |P_{k,i} \cdot \hat{z}_{k,i}^*|$$
$$CP_{k,i}^N = \mathbb{1}_{\hat{z}_{k,i}^* \geq 0} \cdot |P_{k,i} \cdot \hat{z}_{k,i}^*|.$$

Score Truncation. In the branch and bound verification framework, if the minimum value of a sub-problem is greater than 0, this sub-problem does not need further branching. Therefore, we only need to raise the minimum value of a sub-problem to above 0. It is not necessary to raise it to a higher value, and higher sub-problem improvements are not necessarily better.

For example, assume that the minimum value of the parent problem is - 5. Splitting neuron a produces two sub-problems with improvements of 1 and 8 respectively. Splitting neuron b produces two sub-problems with improvements of 2 and 6 respectively. If the average improvement is directly used as the neuron's score, then we have $S_a = 4.5$ and $S_b = 4$. Neuron a has a higher score, so it is the better choice. However, notice that the improvements of the second sub-problems are both greater than 5, which means the minimum values of them are greater than 0. These two sub-problems will be discarded and will not affect subsequent branch boundary verification process. We only need to consider the improvements of the first sub-problems produced by branching these two neurons. The improvement of the first sub-problem produced by branching neuron a, which is 1, is less than that of neuron b, which is 2. So neuron b is actually a better choice.

This example shows that the improvements of sub-problems should have an upper limit, which is the absolute value of the parent problem's minimum value. Sub-problems with improvement exceeding this limit have same effect in the following branch and bound process. Therefore, when computing the neuron's score, we need to truncate the part of the sub-problem improvement that exceeds the upper limit.

Suppose that the minimum value of the parent problem is r. The score $S_{k,i}$ of the neuron $z_{k,i}$ in our branching strategy is computed as follows:

$$S_{k,i} = \frac{\min\{Imp_{k,i}^P + CP_{k,i}^P, -r\} + \min\{Imp_{k,i}^N + CP_{k,i}^N, -r\}}{2}$$

where the minimization operator performs score truncation and CP^P, CP^N are the compensation terms.

Our branch strategy is shown in Algorithm 2, where Imp, CP and S are all vectors and $\odot$ is the element-wise product. we can also compute scores of neurons in the same layer in parallel because the improvement estimation and compensation only involves terms related to neuron $z_{k,i}$. This allows us to take full advantage of GPU acceleration. At the end of Algorithm 2, the neuron with the largest score is chosen as the branching decision.

The neuron score computation is completed in one forward propagation, and the optimal solution propagation is also completed in one forward propagation. We can merge these two processes together. Thus, the entire branching strategy only requires one forward propagation, and parallel computation with GPUs can be used for acceleration.

4.2   Sub-problem Omission

Finding the minimum value of each problem requires gradient based optimization [22], which will take a lot of time. If we know the optimal solution to the problem, we can directly obtain the minimum value of the problem.

Algorithm 2: Branching Strategy

Input: Optimal solution for each layer z_k^*,
Coefficients of linear bounds $[P_1, P_2, \ldots, P_L]$,
The minimum value of the parent problem r,
Linear relaxations of each hidden neuron.
Output: Branching decision BD.

1   $BD \leftarrow None$;
2   $Score \leftarrow 0$;
3   **for** $k \leftarrow 1$ **to** $L-1$ **do**
4      $Imp^P \leftarrow (P_k - [P_k]_+ \odot \alpha_k^L - [P_k]_- \odot \alpha_k^U) \odot \hat{z}_k^*$
5          $-[P_k]_+ \odot \beta_k^L - [P_k]_- \odot \beta_k^U$;
6      $Imp^N \leftarrow -([P_k]_+ \odot \alpha_k^L + [P_k]_- \odot \alpha_k^U) \odot \hat{z}_k^*$
7          $-[P_k]_+ \odot \beta_k^L - [P_k]_- \odot \beta_k^U$;
8      $CP_k^P \leftarrow \mathbb{1}_{\hat{z}_k^* \leq 0} \odot |P_k \odot \hat{z}_k^*|$;
9      $CP_k^N \leftarrow \mathbb{1}_{\hat{z}_k^* \geq 0} \odot |P_k \odot \hat{z}_k^*|$;
10     $S_k \leftarrow \frac{\min\{Imp_k^P + CP_k^P, -r\} + \min\{Imp_k^N + CP_k^N, -r\}}{2}$;
11     $LBest \leftarrow \arg\max S_k$;
12     $LScore \leftarrow \max S_k$;
13   **if** $BD == None$ or $Score < LScore$ **then**
14        $BD \leftarrow LBest$;

15 **return** BD

For an unsolved sub-problem, we have computed the optimal solution to its parent problem in the branching strategy. We want to check whether this optimal solution is the optimal solution to this sub-problem. Compared to the parent problem, the constraints of the sub-problem only change on the split neuron. Denote the split neuron as $z_{k,i}$. The additional constraints for the two sub-problems are $z_{k,i} = \hat{z}_{k,i}$, $\hat{z}_{k,i} \geq 0$ and $z_{k,i} = 0$, $\hat{z}_{k,i} \leq 0$, respectively. We check optimal solution propagation to obtain the optimal solution for each layer. If the optimal solution of the parent problem satisfies the new condition of the sub-problem, then it is also the optimal solution of the sub-problem.

For example, the branching is performed on neuron z_2 in the neural network in Example 1. With optimal solution propagation, we have $\hat{z}_2 = 2$ and $z_2 = 2$, which satisfy $\hat{z}_2 = z_2$ and $\hat{z}_2 \geq 0$. This indicates that the sub-problem with constraint $\hat{z}_2 = z_2$ has the same optimal solution as its parent problem. We can get that the minimum value of this sub-problem is 3 without calling the bounding algorithm.

The omission condition of the sub-problem is described by the following theorem. We also give a proof of this theorem to demonstrate the correctness of sub-problem omission.

Theorem 1. *Denote the optimal solution of the parent problem as z^*. The optimal solution of the sub-problem with neuron split constraint $z_{k,i} = \hat{z}_{k,i}$ is the* *same as that of the parent problem if $z^*_{k,i} = \hat{z}^*_{k,i}$ and $\hat{z}^*_{k,i} \geq 0$. The minimum value of this sub-problem is the same as that of the parent problem.*

Proof 1. The constraint on neuron $z_{k,i}$ in the parent problem is

$$\alpha^U_{k,i}\hat{z}_{k,i} + \beta^U_{k,i} \leq z_{k,i} \leq \alpha^L_{k,i}\hat{z}_{k,i} + \beta^L_{k,i}. \tag{6}$$

The constraints on that neuron in the sub-problem are

$$z_{k,i} = \hat{z}_{k,i} \text{ and } \hat{z}_{k,i} \geq 0. \tag{7}$$

The linear relaxations of the ReLU activation function is as introduced in Sect. 2.2. From the linear relaxation design, it can be obtained that if a feasible solution z satisfies constraint (7), then it must satisfy constraint (6). That is, a feasible solution to the sub-problem is also a feasible solution to the parent problem. From this we can get that the minimum value of the sub-problem is larger than the minimum value of the parent problem.

On the other hand, the optimal solution z^* of the parent problem satisfies constraint (7). Because the constraints of the sub-problem and the parent problem are the same on neurons other than neuron $z_{k,i}$, z^* is also a feasible solution to the sub-problem. From this we can get that the minimum value of the sub-problem is smaller than the minimum value of the parent problem. Combining the conclusion in the previous paragraph, we can get that the minimum value of the parent problem and the sub-problem are the same, and z^* is also optimal solution for the sub-problem. ∎

The omission condition of another sub-problem is described by the following theorem. The proof of this theorem is similar to the proof of Theorem 1, so we don't need to go into details.

Theorem 2. *Denote the optimal solution of the parent problem as z^*. The optimal solution of the sub-problem with neuron split constraint $z_{k,i} = 0$ is the same as that of the parent problem if $z^*_{k,i} = 0$ and $\hat{z}^*_{k,i} \leq 0$. The minimum value of this sub-problem is the same as that of the parent problem.*

For sub-problems that satisfy the omission conditions, we can directly obtain the minimum value and optimal solution. They can use these optimal solutions to find the next branching decisions. The sub-problem omission reduces the number of calls to the bounding algorithm and reduces time cost of the branch-and-bound verification process.

5   Experiments

In this section, we set up experiments to demonstrate the effectiveness of our branching strategy. For fair comparison, the bounding algorithm in the branch and bound verification framework is fixed as β-CROWN [22], which is the champion method in the International Verification of Neural Networks Competition [15]. We implement our branching strategy as OPTGBS. To show the effectiveness of the techniques proposed in Sect. 4, we also implemented the OPTGBS-C strategy, which only uses improvement estimation in Sect. 3. Our implementation is available at https://github.com/xue-xy/OptGBS.

Table 1. Time cost and percentage of verified properties on ERAN benchmark.

Model	ϵ	Metric	Methods				
			Rand	BaBSR	FSB	OptGBS-C	OptGBS
M-FNN6×100	0.02	Time (s)	129.04	114.89	112.45	**109.35**	110.70
		Verified (%)	58.1	62.9	63.0	**65.2**	64.3
M-FNN9×200	0.013	Time (s)	165.86	157.53	156.29	**154.33**	154.37
		Verified (%)	47.3	49.5	49.5	49.8	**50.0**
M-CNNSmall	0.13	Time (s)	108.09	57.61	42.32	33.31	**27.54**
		Verified (%)	64.7	92.3	94.1	**96.7**	**96.7**
M-CNNBig	0.13	Time (s)	148.36	56.62	50.71	42.70	**40.65**
		Verified (%)	51.4	85.7	87.1	89.5	**90.2**
C-FNN6×100	0.002	Time (s)	218.20	191.15	187.20	180.05	**175.80**
		Verified (%)	29.5	38.3	40.6	44.4	**46.3**
C-CNNSmall	0.008	Time (s)	116.70	61.38	52.18	46.17	**44.57**
		Verified (%)	62.1	83.1	86.1	87.7	**88.0**
C-CNNBig	0.008	Time (s)	84.82	59.25	54.39	50.69	**47.70**
		Verified (%)	72.3	81.9	83.2	84.9	**86.0**

5.1 Experimental Setup

Datasets and Networks. Our experiments are conducted on publicly available and widely adopted ERAN [7] and OVAL [14] benchmarks. We take three fully connected neural network (FNN) and four convolutional neural networks (CNN) from ERAN benchmark. They are trained on MNIST [12] or CIFAR10 [3] datasets. Similar to other experiments on ERAN benchmark [16, 18, 22], the properties on the same neural network the same ϵ value. The OVAL benchmark is consisted of three convolutional neural networks which are trained on CIFAR10 dataset. Each network is associated with 100 properties, and each property has a different ϵ value.

Metric. We use the time cost and the percentage of verified properties to measure the effectiveness of the branching strategy. A good branching strategy consumes less time and can successfully verify more properties. We also consider the number of explored sub-problems. Exploring less sub-problems indicates a better branching strategy. For neural networks in ERAN benchmark, we use the first 100 images from the corresponding test set and filter out the misclassified images. For OVAL benchmark, the experiments are conducted on all included properties.

Competitors. We compare our branching strategy against three other strategies. The trivial baseline strategy is random choice (RAND), which randomly selects a neuron to perform branching. The other two are state-of-the-art branching strategies, BABSR [2] and FSB [17]. There is also a learning based branching

Table 2. Experiment results on OVAL benchmark.

Model	Strategy	Time (s)	Sub-problems	Verified (%)
Base	Rand	992.86	10843.96	47
	BaBSR	706.39	9878.43	63
	FSB	663.15	9320.76	66
	OptGBS-C	370.3	4802.18	84
	OptGBS	**348.16**	**4672.21**	**85**
Deep	Rand	724.47	8334.96	52
	BaBSR	494.07	6242.1	73
	FSB	444.62	4728.25	76
	OptGBS-C	189.33	1932.4	90
	OptGBS	**149.79**	**1521.75**	**92**
Wide	Rand	740.21	10707.56	61
	BaBSR	490.24	9695.26	73
	FSB	494.98	9224.48	73
	OptGBS-C	367.01	6068.29	**80**
	OptGBS	**347.73**	**5430.43**	**80**

strategy GNN [14], which exploits graph neural networks to make branching decisions. Different from the above strategies, GNN needs to train the network before verification process, which leads to additional time consumption. And it was reported in [17] that the effect of GNN is not as good as FSB, so we did not compare against it.

5.2   Experiment Results on ERAN Benchmark

Table 1 shows time cost and percentage of verified properties on ERAN benchmarks, where ϵ stands for the perturbation radius. Models whose names begin with M are trained on MNIST dataset, and models whose names begin with C are trained on CIFAR10 dataset. The time limit for each property in this benchmark is 5 min. It can be seen from the table that our branching strategies OptGBS and OptGBS-C achieved the best performance on all networks. Verification using our strategy consumes the least time and verifies the most properties within the time limit.

It is noticed that our strategy exhibits varying degrees of superiority on FNN and CNN. There is not a performance gap between different strategies on FNN. Compared with the worst-performing Rand method, our best-performing strategy only achieves an average time reduction of 13.8% on the three FNN networks. We also find that there is no obvious difference between OptGBS and OptGBS-C. OptGBS only has a clear lead on the FNN6×100 network of the CIFAR10 dataset. We think that the possible reason is the dense interconnection among neurons in FNN, which indicates that individual neuron has

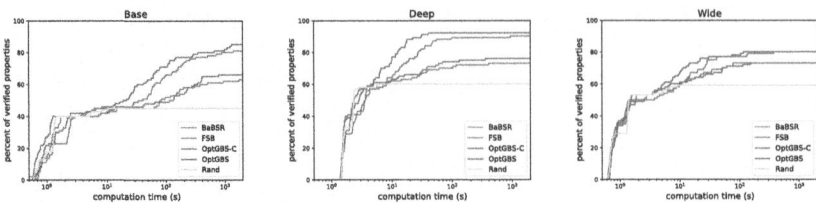

Fig. 4. Percentage of solved properties in OVAL benchmark as a function of time cost.

a relatively small impact on the minimum value. As a result, the differences between different branching decisions are not substantial. One the other hand, our method demonstrates a very obvious superiority on CNN. The OPTGBS reduces the average time consumption by 63.2% against RAND, 31.8% against BABSR, and 20.4% against FSB. Compared with OPTGBS-C, OPTGBS also reduces time usage by 7.8%. Furthermore, OPTGBS shows a significantly higher percentage of verified properties compared to other strategies. This illustrates the effectiveness of our proposed techniques.

5.3   Experiment Results on OVAL Benchmark

Table 2 shows experiment results on ERAN benchmarks. We show the time cost, the number of explored sub-problems, and the percentage of verified properties in the table. The time limit for each property in this benchmark is 30 min.

Our strategy shows significant superiority on this benchmark. In terms of time consumption, OPTGBS shows an average reduction of 64.3% compared to RAND, 49.8% compared to BABSR, 47.1% compared to FSB, and 10.7% compared to OPTGBS-C. The difference between OPTGBS and OPTGBS-C demonstrates the effectiveness of out-of-bound compensation and score truncation. The number of sub-problems explored by OPTGBS is also significantly reduced. This shows that we did make better branching decisions.

We also plot the percentage of verified properties as a function of computation time in Fig. 4. The horizontal axis represents the logarithmic scale of time cost. In the beginning, there is no obvious difference in performance among various branching strategies because only a few sub-problems are explored at this time. The properties that are verified during this time period are relatively easy, and several branches will have a great impact on the results. As the number of explored sub-problems increases, our strategy gradually shows its superior performance. OPTGBS requires less time to verify the same number of properties and successfully verifies more properties within the same time consumption. This indicates that our strategy is capable of making better branching decisions.

6   Conclusion

In this paper, we delves into the neural network branch and bound verification process and come up with a new branching strategy to improve the verification efficiency. Our branching strategy is consisted of improvement estimation, out-of-bound compensation and score truncation. We use the optimal solution of a problem to estimate improvements of the sub-problem produced by branching, which is both accurate and computationally inexpensive. The out-of-bound compensation and score truncation techniques allow us to better estimate the impact of branching a neuron. We implement our branching strategy and evaluate it on a set of neural networks with different architectures. Experiment results show that our branching strategy exhibits superior performance than other state-of-the-art strategies. In the future, we will focus on conducting research to reduce the computational burden of the bounding algorithm, thereby enhancing the branch and bound verification efficiency.

Acknowledgment. This work was sponsored by the National Natural Science Foundation of China under grant No. 62172019, and the National Key R&D Program of China (2022YFB2702200).

References

1. Bak, S., Dohmen, T., Subramani, K., Trivedi, A., Velasquez, A., Wojciechowski, P.: The octatope abstract domain for verification of neural networks. In: Chechik, M., Katoen, J.P., Leucker, M. (eds.) FM 2023. LNCS, vol. 14000, pp. 454–472. Springer, Cham (2023). https://doi.org/10.1007/978-3-031-27481-7_26
2. Bunel, R., Lu, J., Turkaslan, I., Torr, P.H.S., Kohli, P., Kumar, M.P.: Branch and bound for piecewise linear neural network verification. J. Mach. Learn. Res. **21**, 42:1–42:39 (2020)
3. Carlini, N., Wagner, D.: Towards evaluating the robustness of neural networks. In: 2017 IEEE Symposium on Security and Privacy (SP), pp. 39–57. IEEE (2017)
4. Chen, Z., Huang, X.: End-to-end learning for lane keeping of self-driving cars. In: 2017 IEEE Intelligent Vehicles Symposium (IV), pp. 1856–1860. IEEE (2017)
5. Ehlers, R.: Formal verification of piece-wise linear feed-forward neural networks. In: D'Souza, D., Narayan Kumar, K. (eds.) ATVA 2017. LNCS, vol. 10482, pp. 269–286. Springer, Cham (2017). https://doi.org/10.1007/978-3-319-68167-2_19
6. Ferrari, C., Müller, M.N., Jovanovic, N., Vechev, M.T.: Complete verification via multi-neuron relaxation guided branch-and-bound. In: The Tenth International Conference on Learning Representations, ICLR 2022, Virtual Event, 25–29 April 2022. OpenReview.net (2022)
7. Gagandeep, S., et al.: ERAN verification dataset. https://github.com/eth-sri/eran
8. Goodfellow, I.J., Shlens, J., Szegedy, C.: Explaining and harnessing adversarial examples. In: 3rd International Conference on Learning Representations (ICLR), Conference Track Proceedings (2015)
9. Julian, K.D., Lopez, J., Brush, J.S., Owen, M.P., Kochenderfer, M.J.: Policy compression for aircraft collision avoidance systems. In: 2016 IEEE/AIAA 35th Digital Avionics Systems Conference (DASC), pp. 1–10. IEEE (2016)

10. Katz, G., Barrett, C., Dill, D.L., Julian, K., Kochenderfer, M.J.: Reluplex: an efficient SMT solver for verifying deep neural networks. In: Majumdar, R., Kunčak, V. (eds.) CAV 2017. LNCS, vol. 10426, pp. 97–117. Springer, Cham (2017). https://doi.org/10.1007/978-3-319-63387-9_5

11. Kos, J., Fischer, I., Song, D.: Adversarial examples for generative models. In: 2018 IEEE Security and Privacy Workshops (SPW), pp. 36–42. IEEE (2018)

12. LeCun, Y., Bottou, L., Bengio, Y., Haffner, P.: Gradient-based learning applied to document recognition. Proc. IEEE **86**(11), 2278–2324 (1998)

13. Liu, H., Liu, S., Xu, G., Liu, A., Fang, D.: NNTBFV: simplifying and verifying neural networks using testing-based formal verification. Int. J. Softw. Eng. Knowl. Eng. **34**(02), 273–300 (2024)

14. Lu, J., Kumar, M.P.: Neural network branching for neural network verification. In: 8th International Conference on Learning Representations, ICLR 2020, Addis Ababa, Ethiopia, 26–30 April 2020. OpenReview.net (2020)

15. Müller, M.N., Brix, C., Bak, S., Liu, C., Johnson, T.T.: The third international verification of neural networks competition (VNN-COMP 2022): summary and results. CoRR abs/2212.10376 (2022)

16. Müller, M.N., Makarchuk, G., Singh, G., Püschel, M., Vechev, M.: PRIMA: general and precise neural network certification via scalable convex hull approximations. Proc. ACM Program. Lang. **6**(POPL), 1–33 (2022)

17. Palma, A.D., et al.: Improved branch and bound for neural network verification via Lagrangian decomposition (2021), https://arxiv.org/abs/2104.06718

18. Singh, G., Ganvir, R., Püschel, M., Vechev, M.: Beyond the single neuron convex barrier for neural network certification. In: Advances in Neural Information Processing Systems (NeurIPS), vol. 32, pp. 15072–15083 (2019)

19. Singh, G., Gehr, T., Mirman, M., Püschel, M., Vechev, M.: Fast and effective robustness certification. In: Advances in Neural Information Processing Systems (NeurIPS), vol. 31, pp. 10825–10836 (2018)

20. Singh, G., Gehr, T., Püschel, M., Vechev, M.: An abstract domain for certifying neural networks. Proc. ACM Program. Lang. **3**(POPL), 1–30 (2019)

21. Wang, S., Pei, K., Whitehouse, J., Yang, J., Jana, S.: Formal security analysis of neural networks using symbolic intervals. In: Proceedings of the 27th USENIX Security Symposium, pp. 1599–1614. USENIX Association (2018). https://www.usenix.org/conference/usenixsecurity18/presentation/wang-shiqi

22. Wang, S., et al.: Beta-CROWN: efficient bound propagation with per-neuron split constraints for neural network robustness verification. In: Advances in Neural Information Processing Systems (NeurIPS), vol. 34, pp. 29909–29921 (2021)

23. Wong, E., Kolter, J.Z.: Provable defenses against adversarial examples via the convex outer adversarial polytope. In: Proceedings of the 35th International Conference on Machine Learning, ICML 2018, Stockholmsmässan, Stockholm, Sweden, 10–15 July 2018. Proceedings of Machine Learning Research, vol. 80, pp. 5283–5292. PMLR (2018)

24. Xu, K., et al.: Fast and complete: Enabling complete neural network verification with rapid and massively parallel incomplete verifiers. arXiv preprint arXiv:2011.13824 (2020)

25. Zhang, H., Weng, T.W., Chen, P.Y., Hsieh, C.J., Daniel, L.: Efficient neural network robustness certification with general activation functions. In: Advances in Neural Information Processing Systems (NeurIPS), vol. 31, pp. 4944–4953 (2018)
26. Zhang, Z., Wu, Y., Liu, S., Liu, J., Zhang, M.: Provably tightest linear approximation for robustness verification of sigmoid-like neural networks. In: 37th IEEE/ACM International Conference on Automated Software Engineering, ASE 2022, pp. 80:1–80:13. ACM (2022)

AccMILP: An Approach for Accelerating Neural Network Verification Based on Neuron Importance

Fei Zheng[1,2], Qingguo Xu[1(✉)], Zhou Lei[1], and Huaikou Miao[1]

[1] School of Computer Engineering and Science, Shanghai University,
Shanghai, China
{zheng_fei,leiz,hkmiao}@shu.edu.cn, qgxu@t.shu.edu.cn
[2] Shanghai Key Laboratory of Computer Software Evaluating and Testing,
Shanghai, China

Abstract. Deep Neural Networks (DNNs) have found successful applications in various non-safety-critical domains. However, given the inherent lack of interpretability in DNNs, ensuring their prediction accuracy through robustness verification becomes imperative before deploying them in safety-critical applications. Neural Network Verification (NNV) approaches can broadly be categorized into exact and approximate solutions. Exact solutions are complete but time-consuming, making them unsuitable for large network architectures. In contrast, approximate solutions, aided by abstraction techniques, can handle larger networks, although they may be incomplete. This paper introduces AccMILP, an approach that leverages abstraction to transform NNV problems into Mixed Integer Linear Programming (MILP) problems. AccMILP considers the impact of individual neurons on target labels in DNNs and combines various relaxation methods to reduce the size of NNV models while ensuring verification accuracy. The experimental results indicate that AccMILP can reduce the size of the verification model by approximately 30% and decrease the solution time by at least 80% while maintaining performance equal to or greater than 60% of MIPVerify. In other words, AccMILP is well-suited for the verification of large-scale DNNs.

Keywords: Neural network · Formal verification · Mixed integer linear programming · Neuron importance

1 Introduction

Deep Neural Networks (DNNs) have initiated the third wave of artificial intelligence research [15] and have shown rapid development in recent years, successfully addressing complex and challenging prediction problems across various domains [19,22,29]. Nonetheless, DNNs remain inherently complex and suffer from a lack of interpretability. Most current studies on interpretability focus on passive interpretability, which falls short of Recent research has revealed that

G. Bai et al. (Eds.): ICECCS 2024, LNCS 14784, pp. 88–107, 2025.
https://doi.org/10.1007/978-3-031-66456-4_5

introducing minor perturbations to the input data of DNNs can lead to erroneous predictions [6]. Currently, a subset of researchers is actively exploring methods to enhance the resilience of DNNs against such interference by modifying the training algorithms [20].

In situations where the behavior of neural networks remains inexplicable, their accuracy is typically assessed through neural network testing. While it's essential to note that this approach does not offer an absolute guarantee of the network's decision correctness. In many non-safety critical domains, the network's behavior can be deemed acceptable if its accuracy on testing data reaches a sufficiently high level. However, in safety-critical domains, like autonomous driving [23], medical diagnosis, aircraft anti-collision systems [17], and more, traditional neural network testing fails to meet the stringent system safety requirements. In these safety-critical areas, there is a pressing need for more robust testing and verification methods to ensure the network's resilience to a defined range of disturbances.

Presently, complete Neural Network Verification (NNV) approaches tackle the challenge of NNV by reformulating it as a mathematical problem. NNV can be represented as a Mixed Integer Linear Programming (MILP) problem [1,28], a Satisfiability Modulo Theory (SMT) problem, or a Satisfiability (SAT) problem, which is then resolved using specialized solvers. Researchers subsequently assess the network's robustness based on the solver's output.

Illustrating the approach with the MILP problem, the accuracy of this verification method hinges largely on the degree of constraint relaxation applied to model the neural network's activation function. The degree of constraint relaxation is inversely proportional to the accuracy of the solution and the number of constraints.

After transforming the NNV problem into a MILP problem, it needs to be solved using a solver. However, when a model exhibits an abundance of constraints, solvers may require extended time for resolution, and even fail to provide solutions. Notably, in the context of DNNs, the problem is exacerbated as an increased number of neurons necessitates the encoding of more constraints, leading to frequent occurrences of the aforementioned issues. Consequently, many current NNV methods are primarily tailored for smaller networks and are ill-suited for application to DNNs, rendering them inadequate for verifying the robustness of DNNs across diverse fields. Hence, several researchers have initiated efforts to explore the scalability of NNV approaches. The objective of scaling NNV approaches is to enhance the speed and efficiency of verification, enabling the application of conventional, smaller-scale NNV methods to larger networks in safety-critical fields.

This paper introduces the AccMILP approach, a simplified coding technique designed for Rectified Linear Units (ReLU) neural network models. During the neural network prediction process, not all neurons exert equal influence on each label within the output layer. By assessing the influence of network neurons on the target label, a dynamic encoding scheme is proposed. This approach employs strict or relaxed constraints based on the influence level of individual neurons. The AccMILP approach can reduce the size of the NNV problem model while ensuring verification accuracy, thus facilitating the scalability of existing NNV methods to DNNs. The specific contributions of this paper are outlined below:

(i) A concept of neuron importance and a neuron ablation approach to isolate and analyze the impact of individual neurons on the overall prediction accuracy of the neural network.
(ii) An encoding approach of neurons to ensure verification accuracy while simultaneously reducing the model size based on neuron importance.
(iii) A novel approach called AccMILP, which can speed up NNV process and reduce the size of verification model. The approach is evaluated on neural networks with diverse architectures and compared with MIPVerify [28] to assess the performance and effectiveness of the AccMILP approach.

2   Related Work

This paper categorized NNV methods into two groups: exact solution and approximate solution approaches. This classification is based on the capability of an approach to yield precise results. In principle, exact solution approaches can accurately verify properties and derive entirely correct conclusions, ensuring soundness and completeness. Conversely, approximate solution approaches, while considered soundness, may yield incorrect conclusions in some cases, potentially misclassifying certain robust neural networks as non-robust.

2.1   Approaches of Exact Solution

Lomuscio et al. [1] introduced a method that employs Mixed Integer Linear Programming (MILP) to transform the Neural Network Verification (NNV) problem into a reachability problem. This innovative approach allows for verifying neural network robustness using MILP solvers. Subsequently, more and more researchers begin to improve the verification method based on MILP above using constraints [4,5], parallelization [8], dividing and conquering [10,33], or pre-processing using adversarial attacks [36].

Verification problems can also be reformulated into SMT or SAT problems. Katz et al. [16] and Guo et al. [33] translate Neural Network Verification (NNV) problems into SMT problems, leveraging SMT solvers to determine the robustness of a network. In a similar vein, Cheng et al. [7] and Narodytska et al. [24,25] utilize Boolean coding for neural networks, converting NNV problems into SAT problems for resolution. These solving approaches are typically applicable to binary neural networks, where connection weights and activation values are restricted to +1 or −1. However, we won't delve into the details of these methods here.

In addition, some researchers focus on verification for QNNs [35] because quantized neural networks (QNNs) can reduce the size of DNNs with the same accuracy and they can verify the robustness of DNNs by transforming the DNNs into QNNs.

2.2   Approaches of Approximate Solution

Approximate solution approaches are commonly used to approximate the verification problem by abstract interpretation [14,27], semidefinite programming [13,26], interval symbolic analysis [30,34], or duality [11,31].

Tjeng et al. [28] employ a combination of nonlinear formulas and pre-solving algorithms which can improve the verification speed by several orders of magnitude. DeepAbstract [2], on the other hand, adopts an abstract approach to assess the similarity between two or more neurons within the network and simplifying the network be verified by merging the similar neurons. Venus [3] classifies neurons into stable active, stable inactive and unstable categories. The interplay between different neuron types within and between layers enables the removal of redundant neuron constraints, reducing the constraints within the NNV model, and subsequently expediting the verification process.

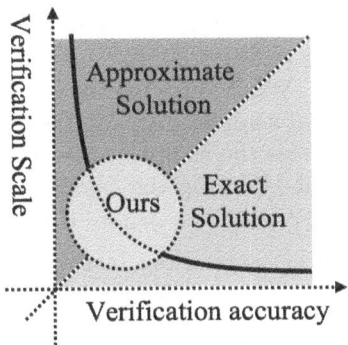

Fig. 1. The relationship between verification accuracy and verification scale in NNV

At present, the development of exact and approximate solution verification technology is separated. The relationship between verification accuracy and verification scale in NNV is depicted in Fig. 1. Scalability research in NNV has predominantly focused on the aforementioned two techniques. In this paper, we combine both exact and approximate verification techniques to strike a balance between solution accuracy and scalability.

3   Preliminaries

In this section, we illustrate the notation used in this paper and introduce the approach of transforming the neural network verification problem into the MILP problem.

3.1   Neural Network

A neural network is structured as a directed acyclic graph with multiple layers, where each layer contains several neurons. In the case of a fully connected neural network with $k+1$ layers, the first layer (layer 0) serves as the input layer, the last layer (layer k) is the output layer, and any intermediate layers between layer 0 and layer k are referred to as hidden layers, denoted as layer i, where $1 \leq i < k$. In layers except for the output layer, every neuron in one layer is connected to every neuron in the subsequent layer, and the connections are represented by edge weights. The weights and biases of neurons are iteratively updated during the training phase to optimize the network's performance. When making predictions, each neuron's input is the weighted sum of the outputs from all neurons in the preceding layer, with the neuron's bias added. The output of each neuron results from the application of an activation function to the input. Throughout this paper, the activation function used in all discussed networks is the ReLU function, which denotes the maximum between input and zero.

In this paper, $s^{(i)}$ denotes the number of neurons in layer i of the network, $neu_j^{(i)}$ denotes the neuron j in layer i, $W^{(i)}$ ($b^{(i)}$) denotes the weight matrix (bias vector) of layer i, where $i \geq 1$. When verifying the robustness of the network, the input of the network is a set of samples instead of a single sample. At this time, we use some intervals to express the sample set, so each neuron in the network has the upper and lower bounds during the calculation of the network. In this paper, we use $\hat{l}_j^{(i)}$ ($l_j^{(i)}$) and $\hat{u}_j^{(i)}$ ($u_j^{(i)}$) to denote the upper and lower bounds of the neuron $neu_j^{(i)}$ before (after) activation. In the following section, we use ub (lb) to denote the upper (lower) bound. Figure 2 is an illustration of a network with corresponding symbols for each part.

3.2   Formal Verification of DNN Using MILP

Let Net be a DNN representing a function $f(\cdot)$, whose input (output) domain is R^m (R^n). Transforming the NNV problem into a MILP problem is a matter of determining whether all the outputs of the neural network on the bounded input domain $C \in R^m$ satisfy the properties to be verified. When transforming the NNV problems into MILP problems, the neural network model, the perturbation domain, and the property to be verified must be encoded in the appropriate MILP constraints.

Neural Network Model Encoding. When transforming the NNV problem, it is essential to create variables for neurons and associate each neuron with its corresponding variable. The neural network computation between each layer is constrained by the Eq. 1.

$$n_j^{(i+1)} = \sum W_{k,j}^{(i+1)} neu_k^{(i)} + b_j^{(i+1)}, 0 \leq k \leq s^i \tag{1}$$

The accuracy of the solution can differ depending on whether constraints for the ReLU activation function are created using the big-M method or abstract

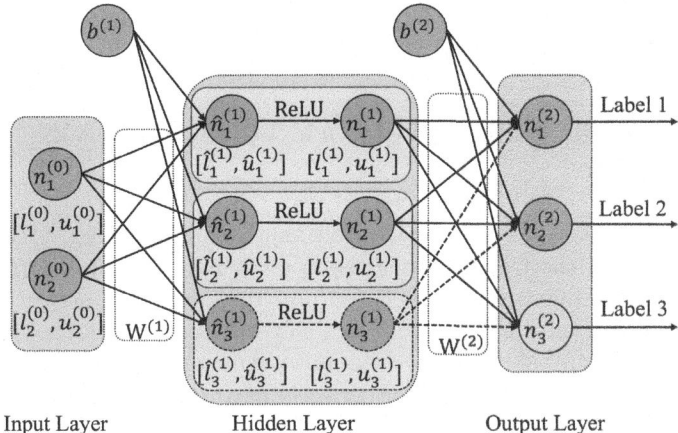

Fig. 2. An illustration of a network with corresponding symbols for each part. The dashed line denotes the neuron after ablation, the red (green) neuron denotes the decrease (increase) in the prediction accuracy of the label corresponding to that neuron after ablation. (Color figure online)

approaches. The detailed process of creating activation function constraints is elaborated in Sect. 4.

Perturbation Encoding. Taking the network used for image classification as an example. The perturbation is associated with the bounded input domain C of NNV. When encoding the perturbation as a MILP constraint, it is necessary to account for both the verification image pic and the perturbation radius ϵ and encode the perturbation using the following inequality constraints: $pic - \epsilon \leq C \leq pic + \epsilon$.

Alternatively, utilize the equation constraint $C = pic + perturbation$ directly. In this case, an additional variable $perturbation$ and a constraint $-\epsilon \leq perturbation \leq \epsilon$ need to be introduced to ensure that the domain to be verified falls within the perturbation range.

Robustness Encoding. For a network and its associated function $f(\cdot)$, Definition 1 is a definition of robustness with input pic, true label λ and perturbation ϵ.

Definition 1 (Robustness). *Let Net be a DNN representing a function $f(\cdot)$, pic be the input with label λ and perturbation radius ϵ. If there are no adversarial samples in perturbation range $[pic - \epsilon, pic + \epsilon]$, the network is robust concerning the input pic with perturbation radius ϵ.*

Hence, the robustness of neural networks can be expressed directly through a set of inequality constraints: $f_{label=\lambda}(pic') - f_{label=\mu}(pic') > 0$, where pic' represents the input after perturbation, $f_{label=\lambda}(\cdot)$ denotes the predicted probability of category λ by $Net\ f(\cdot)$, and $\mu \in [0, s^{(k)}]\backslash\{\lambda\} \in \mathbb{N}$.

4   Accelerating Neural Network Verification via Neuron Importance

This section presents an NNV acceleration approach named AccMILP based on neuron importance, the framework of the approach is shown in Fig. 3.

The AccMILP approach is structured into three primary steps: NNV problem transformation (into MILP problem), solver-based problem resolution, and robustness evaluation based on solver results. Among these steps, the first one is of paramount importance. In the process of transforming the NNV problem, AccMILP start by evaluating the importance of neurons within the neural network using a testing dataset. Subsequently, AccMILP separately encode the perturbation range, the network to be verified, and the properties to be verified (robustness). Notably, the assessment of neuron importance significantly influences the network encoding process, leading to a reduction in the number of constraints within the NNV model and, in turn, expediting the verification process. This approach utilizes the Gurobi solver to tackle the NNV model, and if the NNV model proves to be infeasible, it signifies that the neural network is robust against the input *pic* with a perturbation radius of ϵ.

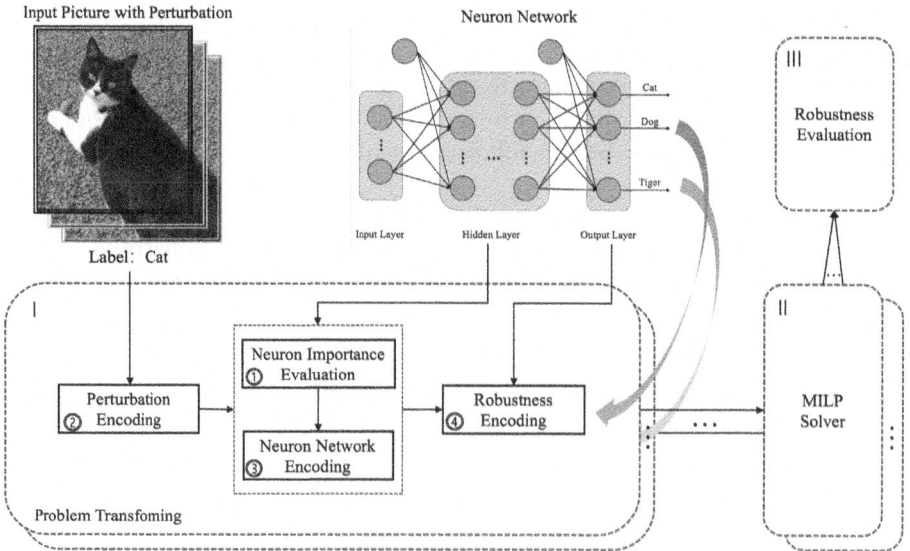

Fig. 3. Framework of the AccMILP Approach

4.1   Neuron Importance Evaluation

To understand the network behavior and discover the necessity of some modules in the network better, researchers apply the ablation experiment in the field of experimental psychology to the neural network research process. Ablation experiments for DNNs typically involve a three-step process. First, certain network modules are removed, and the network is subsequently retrained. Second, a comparison is made between the outputs of DNNs before and after the removal of these modules. Lastly, the results are analyzed to explain the significance of the ablated components.

Similar to the role of network modules, individual neurons in DNNs may also influence the network's output, with different neurons having varying impacts on the network's output. As a result, we have put forward Conjecture 1 and Conjecture 2 as follows.

Conjecture 1. The prediction accuracy of DNNs will be affected after the ablation of neuron $neu_j^{(i)}$.

Conjecture 2. The ablation of neuron $neu_j^{(i)}$ will affect the prediction accuracy of each predicted label in DNNs.

Based on the above Conjecture 1 and Conjecture 2, the concept of neuron importance with a formal definition in Definition 2 is proposed.

Definition 2 (Neuron Importance Index). *Let Net be a DNN representing a function $f(\cdot)$. The network's prediction accuracy on the testing dataset T is denoted as $acc(f(T))$, and the associated function of the network after ablating neuron $neu_j^{(i)}$ is $f_j^{(i)}(\cdot)$. The importance index $imp_j^{(i)}$ of neuron $neu_j^{(i)}$ is determined by the importance index calculation outlined in Eq. 2.*

$$imp_j^{(i)}(T) = acc(f(T)) - acc(f_j^{(i)}(T)) \tag{2}$$

Neuron $neu_j^{(i)}$ is an important neuron if $imp_j^{(i)} > 0$, and neuron $neu_j^{(i)}$ is an unimportant one if $imp_j^{(i)} \leq 0$.

The number of neurons in DNNs far exceeds that of divisible modules, resulting in prohibitively high time costs for using traditional ablation experiments. Furthermore, traditional ablation experiments necessitate retraining the network, resulting in different parameters in the retrained model compared to the original one. This disparity in parameters makes it impractical to use the comparison results as a basis for assessing the importance of neurons in the network.

In this paper, the neuron ablation process used to evaluate neuron importance is based on the approach proposed by Richard [21]. After training the network, the weights of the input edges $W_{:,j}^{(i)}$ and the corresponding bias $b_j^{(i)}$ of neuron $neu_j^{(i)}$ are manually set to zero, effectively preventing neuron $neu_j^{(i)}$ from outputting any information. This ablation process achieves the goal of ablating the target neurons without altering the structure and parameters of the original network. The procedure for the neuron ablation process is outlined in Algorithm 1.

Algorithm 1: Neuron Ablation

Input: Network Net; i, j from neuron neu_j^i;
Output: Importance index acc
Data: Testing Dataset T
/* Getting weights and biases of the layer i in network Net   */
1 $W^{(i)} \leftarrow getWeightssofLayer(Net, i)$
2 $b^{(i)} \leftarrow getBiasesofLayer(Net, i)$
/* Ablation of neuron neu_j^i */
3 $W^{(i)}[:, j] \leftarrow 0$
4 $b^{(i)}[j] \leftarrow 0$
/* Using $W^{(i)}$ and $b^{(i)}$ computed above replace $W^{(i)}$ and $b^{(i)}$ of Net   */
5 $Net \leftarrow replaceWeightsandBias(Net, W^{(i)}, b^{(i)})$
/* Evaluating network Net using datasets T */
6 $acc \leftarrow evaluate(Net, T)$
7 **return** acc

4.2   Abstract Approaches of ReLU Activation Function

The AccMILP approach, combining the exact solution and the approximate solution verification techniques through neuron importance, can balance the accuracy of the solution and the scale of verification by encoding activation functions in different abstract approaches.

For the ReLU function $y = max(0, x), x \in [l, u]$, if $u \leq 0$, $y \equiv 0$, then the neuron is in a stable inactive state, if $l \geq 0$, $y \equiv x$, then the neuron is in a stable active state. Both the above cases can be coded by equality constraints directly.

This paper just consider the case where $l < 0 < u$. Table 1 illustrates the four coding methods employed in AccMILP to encode the activation functions in neurons with varying levels of importance.

The above four neuron coding methods require neuron boundary values, and the accuracy of the boundary values has a very important impact on the accuracy of the NNV model, AccMILP use two approaches to calculate the neuron boundary values, Interval Arithmetic (IA) and Affine Arithmetic (AA), whose intersection upper and lower bounds are the neuron boundary values used in AccMILP.

Big-M Encoding. The big-M method is a technique for handling piecewise linear constraints in the encoding of linear programming problems. It involves introducing a binary variable $\delta \in 0, 1$ and manually selecting a large number, denoted as M. In this paper, when applying the big-M method to encode the ReLU function for neuron $neu_j^{(i)}$, the upper bound $\hat{l}_j^{(i)}$ and the lower bound $\hat{u}_j^{(i)}$ before the activation of neuron $neu_j^{(i)}$ are used as the values for the big number M.

Table 1. Relaxation of ReLU Activation Function.

Abstract Approaches	Scope	Constraints
Big-M	$\hat{l}_q^{(i)}$ ⋯ $\hat{u}_q^{(i)}$	$y \le x - \hat{l}_j^{(i)} * (1 - \alpha)$ $y \ge x$ $y \le \hat{u}_j^{(i)} * \alpha$ $y \ge 0$ $\alpha \in \{0, 1\}$
Triangle Relaxation	$\hat{l}_q^{(i)}$ ⋯ $\hat{u}_q^{(i)}$	$y \ge x$ $(\hat{u}_j^{(i)} - \hat{l}_j^{(i)}) * y \le \hat{u}_j^{(i)} * (x - \hat{l}_j^{(i)})$ $y \ge 0$
Linear Relaxation $(-\hat{l}_j^{(i)} \ge \hat{u}_j^{(i)})$	$\hat{l}_q^{(i)}$ ⋯ $\hat{u}_q^{(i)}$	$(\hat{u}_j^{(i)} - \hat{l}_j^{(i)}) * y \le \hat{u}_j^{(i)} * (x - \hat{l}_j^{(i)})$ $y \ge 0$
Linear Relaxation $(-\hat{l}_j^{(i)} < \hat{u}_j^{(i)})$	$\hat{l}_q^{(i)}$ ⋯ $\hat{u}_q^{(i)}$	$(\hat{u}_j^{(i)} - \hat{l}_j^{(i)}) * y \le \hat{u}_j^{(i)} * (x - \hat{l}_j^{(i)})$ $y \ge x$

Triangle Relaxation Encoding. The triangle relaxation method is an abstract approach for the ReLU activation function, as proposed by Ehlers et al. [12]. Similar to the Big-M method, the upper bound $\hat{l}_j^{(i)}$ and the lower bound $\hat{u}_j^{(i)}$ before activation are used to encode the activation function of neuron $neu_j^{(i)}$.

Linear Relaxation Encoding. The linear relaxation method is less restrictive than the triangle relaxation method, utilizing only two lines as the upper and lower bounds of the ReLU activation function. The space between these two straight lines represents the abstraction region of the ReLU function, and a smaller area within this region results in a more accurate abstraction. Therefore, we have explored various scenarios related to neuron boundary values and developed optimal upper and lower bound computation approaches tailored to these scenarios, with the aim of minimizing the abstraction region by adjusting the lower bound.

4.3 Neural Network Verification Based on Neuron Importance

The previous sections have introduced neuron importance and various abstract approaches to ReLU. In this section, we combine neuron importance with different abstract approaches to ReLU in a heuristic manner and introduce an NNV acceleration approach called AccMILP for scalability.

The number of constraints in the MILP model can significantly impact solution efficiency and verification accuracy. The degree of relaxation used to encode neurons of varying importance also influences the model's verification accuracy. The AccMILP approach employs the exact coding method, i.e., the Big-M method, to encode neurons with higher importance, while using relaxation coding methods for neurons with lower importance. This approach aims to enhance solution efficiency while maintaining verification accuracy. Algorithm 2 outlines the process of creating an NNV model for DNNs using AccMILP.

Algorithm 2: Algorithm of Network Encoding

Input: Network Net; Importance index acc; Input pic; Perturbations ϵ;
Output: MILP model m
/* The importance index acc is computed through Algorithm 1   */
/* Creating MILP model and initialise model variables   */
1 $m \leftarrow createMILPModel()$
2 $m.initModel(Net, pic, \epsilon)$
/* Choosing abstract approaches of ReLU function by neuron
   importance   */
3 **for** *neuron in Net* **do**
4   **if** *isImportant(neuron.acc)* **then**
5        $m.addConstr(exactEncoding(neuron))$
6   **else**
7        $m.addConstr(relaxationEncoding(neuron))$

8 **return** m

Taking Fig. 2 as an example: in the prediction process for label 1 (label 3), neuron $n_3^{(1)}$ is considered a neuron of high (low) importance. When conducting verifications of DNNs to determine the presence of adversarial samples that might be classified as label 1 (label 3) within the perturbation range, it becomes necessary to encode the ReLU activation function in neuron $n_3^{(1)}$ using the exact (relaxation) encoding method.

The process of verifying the robustness of DNNs is outlined in Algorithm 3.

5   Experiments

In this section, we evaluate the results of AccMILP. In Sect. 5.3, we provide evidence to support the conjectures regarding neuron importance proposed earlier. In Sect. 5.4, we assess the performance of AccMILP, considering model verification accuracy, model size, and solution time.

Algorithm 3: Algorithm of Robustness Verification

Input: Network Net; Importance index acc; Input pic; Perturbations ϵ;
Output: Boolean $isRobust$

1 **for** $targetIndex$ in $Net.outputClassNum\backslash\{pic.label\}$ **do**
       /* Creating MILP model through Algorithm 2   */
2       $m \leftarrow networkEncoding(Net, acc, pic, \epsilon)$
3       $m.optimize()$
4   **if** $m.status! = infeasible$ **then**
5   **return** False

6 **return** True

5.1 Experimental Setting

We conducted experiments using multiple fully connected neural network architectures on the following three commonly used datasets:

- MNIST [9]: A dataset of handwritten digits 0–9, with images formatted as a $28 \times 28 \times 1$-pixel grayscale image.
- Fashion-MNIST [32]: A dataset of clothing, with images also formatted as $28 \times 28 \times 1$-pixel grayscale images.
- CIFAR-10 [18]: A dataset for recognizing common objects, with images formatted as $32 \times 32 \times 3$-pixel color images.

Details of Feedforward Neural Newtworks (FNNs) used in our experiments are given in Table 2.

Table 2. Details of the FNNs used in our experiments.

Model	ReLUs	Network Architecture	Datasets
FNN_1	48	$\langle 784, 32, 16, 10 \rangle$	MNIST
FNN_2	1024	$\langle 784, 512, 512, 10 \rangle$	MNIST
FNN_3	896	$\langle 784, 512, 256, 128, 10 \rangle$	MNIST
FNN_4	60	$\langle 784, 40, 20, 10 \rangle$	Fashion-MNIST
FNN_5	1024	$\langle 784, 512, 512, 10 \rangle$	Fashion-MNIST
FNN_6	896	$\langle 784, 512, 256, 128, 10 \rangle$	Fashion-MNIST
FNN_7	60	$\langle 3072, 40, 20, 10 \rangle$	CIFAR-10
FNN_8	1024	$\langle 3072, 512, 512, 10 \rangle$	CIFAR-10
FNN_9	896	$\langle 3072, 512, 256, 128, 10 \rangle$	CIFAR-10

5.2 Evaluation Metrics

The experiments in this paper consist of two parts. The first part assesses the importance of neurons, aiming to confirm the conjectures in Subsect. 4.1 and demonstrate the feasibility of the AccMILP approach. This experiment is evaluated using two metrics: neural network prediction accuracy, which includes both overall accuracy and accuracy for each label, and the time required for neuron evaluation.

The second part evaluates the AccMILP approach. We measure the size of the NNV model by the number of constraints coded on the hidden layer neurons before and after ablation, and the speed of verification by the NNV model solution time. Since only one neuron importance evaluation is required for each network and verification on that network does not require this step, we treat the neuron importance evaluation process as a pre-processing step. Therefore, the NNV model solution time here does not include the neuron importance evaluation process. In addition, AccMILP is an approximate solution approach, and there may be instances where robustness is judged as non-robust. For further performance evaluation, we documented the quantity of samples slated for verification during the experiments and the count of adversarial samples discovered. This allowed us to gauge the method's effectiveness.

5.3   Evaluation of Neuron Importance

We first conducted neuron ablation in FNN_1 to evaluate the impact of removing neurons on network accuracy. Specifically, we ablated all neurons in the hidden layer of FNN_1. Figure 4 illustrates the impact of neuron ablation on the overall prediction accuracy of network. The results showed that the prediction accuracy of the neural network tended to decrease after ablating most of the neurons. However, in cases where only a few neurons were ablated, the prediction accuracy either improved slightly or remained relatively stable. Additionally, the experiment indicated that neurons in hidden layers closer to the output layer had a more significant impact on the network's prediction accuracy. This observation underscores the importance of neurons in different layers and their contributions to overall network performance.

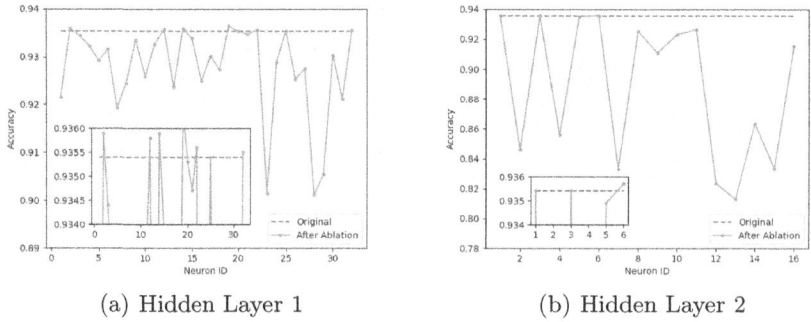

(a) Hidden Layer 1   (b) Hidden Layer 2

Fig. 4. Result of Neuron Ablation on Prediction Accuracy of Neural Network

Figure 5 illustrates the impact of neuron ablation on the prediction accuracy for each label within three hidden layer neurons. Notably, the results vary for different labels when ablating the same neuron, leading to both increases and decreases in prediction accuracy. Most labels experience only marginal changes in prediction accuracy, with a small subset of neurons causing significant reductions, typically within the range of 10%–20%.

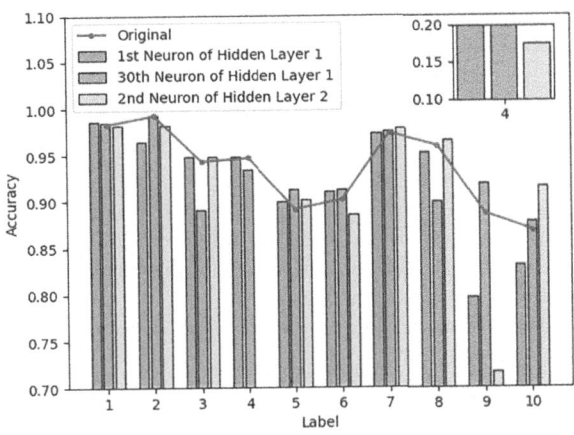

Fig. 5. Result of Neuron Ablation on Prediction Accuracy of Each Label

To validate our conjectures in Subsect. 4.1 and demonstrate the feasibility of the AccMILP framework, we conducted neuron ablation experiments on various FNNs listed in Table 2, and the results are presented in Table 3. The results show that there are instances where neuron ablation increases or decreases the prediction accuracy, both for all data and for data with specific labels. Neurons whose ablation improves prediction accuracy are referred to as unimportant neurons. Notably, across different networks trained on diverse datasets with varying structures, unimportant neurons constitute a significant proportion of all hidden layer neurons. This proportion tends to increase with the number of hidden layers and the number of neurons in the hidden layers. When considering label-specific predictions, the percentage of unimportant neurons is even higher. This finding further underscores the feasibility of our approach, particularly because it incorporates label-specific validation in the assessment of neural network robustness.

As the number of layers and neurons in a neural network increases, importance assessment using the full test set takes longer. A sampling of the testing dataset results in a certain amount of randomness, resulting in different results for neuron importance using different samples of data. Therefore, we further evaluated the evaluation time and stability of neuron importance at different sampling rates in FNN_2. We performed ten neuron importance calculations at each sampling rate and used the standard deviation as the stability indicator,

Table 3. Impact of neuronal ablation on network accuracy.

Model	MIA[a]	MDA[b]	MIS[c]	MDS[d]	UIRA[e]	UIRS_min[f]	UIRS_max[g]
FNN_2	0.0008	0.0013	0.0079	0.0157	54.20	66.6	91.99
FNN_3	0.0012	0.0139	0.0179	0.0785	68.08	70.31	84.60
FNN_4	0.0001	0.1088	0.067	0.948	5	45	70
FNN_5	0.0012	0.005	0.084	0.117	39.55	59.47	91.8
FNN_6	0.0013	0.0124	0.037	0.094	58.04	66.07	86.83
FNN_7	0.0001	0.2166	0.378	0.6	15	41.67	56.67
FNN_8	0.0048	0.0074	0.08	0.102	62.99	61.62	68.36
FNN_9	0.0036	0.0146	0.098	0.158	49.11	57.37	74

[a]MIA: Maximum enhancement value of accuracy.
[b]MDA: Maximum decrease value of accuracy.
[c]MIS: Maximum enhancement of accuracy on individual labels.
[d]MDS: Maximum decrease of accuracy on individual labels.
[e]UIRA: Ratio of unimportant neurons to all hidden layer neurons based on changes in accuracy across all test samples.
[f]UIRS_min: Minimum ratio of unimportant neurons to all hidden layer neurons based on changes in accuracy across individual label test samples.
[g]UIRS_max: Maximum ratio of unimportant neurons to all hidden layer neurons based on changes in accuracy across individual label test samples.

the smaller the standard deviation, the higher the stability, and the results are shown in Fig. 6.

As the sampling rate increases, the time of neuron importance evaluation increases linearly and the standard deviation decreases at a slower rate. At the sampling rate of 10%, the standard deviation is approximately 0.01, which indicates that at this sampling rate, the randomness of the sampling has only a 1% effect on the neuron importance calculation, and the sampling rate is sufficient for the neuron importance calculation at this time, which is sufficient to calculate the neuron importance in 10% of the time it would take to use the entire test set, approximately 60 s. There may be cases where some categories of samples are not sampled when the sampling rate is too small, and the data from the sampled testing dataset cannot be used as a basis for neuron importance assessment.

5.4   Evaluation of AccMILP

While several well-established neural network verification methods exist, including Venus [3], Neurify [30], MIPVerify [28], and Reluplex [16], AccMILP is primarily rooted in MILP and can be considered an enhancement of the MIPVerify approach. Consequently, our comparative evaluation with the MIPVerify method primarily focuses on aspects such as model validation accuracy, model size, and solution time.

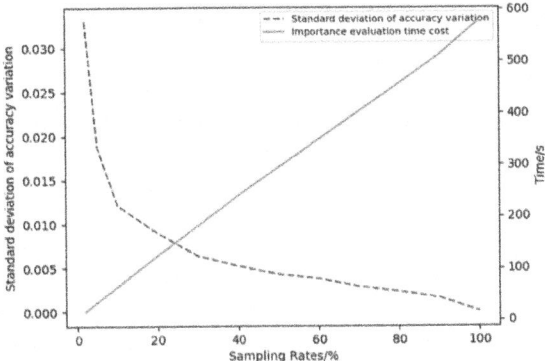

Fig. 6. Evaluation Time and Stability of Neuron Importance at Different Sampling Rates

Similar to the MIPVerify, when we use AccMILP to find an adversarial sample in the perturbation range, the model has the following solutions, *infeasible*, *positive number*, or *zero*.

- The *infeasible* solution denotes that the model can not find an adversarial sample in the perturbation range. If we can not find the adversarial sample which can be predicted as all labels except itself in the perturbation range, the network is robust to the input with the perturbation range.
- The *positive number* solution denotes that the model has found an adversarial sample in this perturbation range. At this point, it is hard to judge whether the network is robust or not, and this kind of result only appears in the approximate solution approach.
- The *zero* solution denotes that the model is too relaxed to solve.

Table 4. Experiment Results on FNNs.

Model	AccMILP				MIPverify			
	A_c[a]	N_s[b]	N_a[c]	A_t[d]	A_c[a]	N_s[b]	N_a[c]	A_t[d]
FNN$_1$	175	50	30	2.42	249	50	50	13.91
FNN$_4$	174	50	35	4.01	249	50	47	18.06
FNN$_7$	189	50	50	116.59	249	50	50	1612.09

[a]A_c: Average number of constraints used to encode the hidden layer neurons.
[b]N_s: Total number of samples to be verified.
[c]N_a: Number of computed adversarial examples or samples judged as robust.
[d]A_t: Average time (in second) on verification.

Table 4 depicts the performance of AccMILP and MIPVerify on three networks mentioned on Table 2 in terms of the number of encoding constraints for the hidden layers, the number of computed adversarial examples or samples judged as robust, and the average model solving time. In the neural networks used in this experiment, the number of neurons in the hidden layers is significantly smaller than the total number of neurons in the entire neural network. Therefore, we used the number of encoding constraints for the hidden layers as an evaluation criterion for comparing model size. Compared to MIPVerify, AccMILP is capable of reducing the average verification model size by approximately 30%. Furthermore, the number of computed adversarial examples or samples judged as robust can also be maintained at over 60% of the MIPVerify method, with specific outcomes depending on the model size and the training dataset used. Additionally, AccMILP exhibits an average solving time reduction of 80% to 93% when compared to MIPVerify across various models.

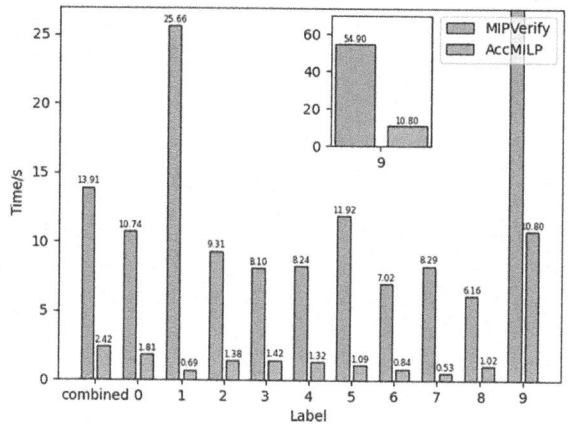

Fig. 7. Solution Time of Different approaches

Figure 7 compares the solution time on FNN_1 between the MIPVerify and AccMILP when the solution is a positive number. Overall, AccMILP reduces the average model solution time by about 83%. For finding adversarial samples in a specific label, AccMILP can reduce the average model solution time in the range of 80% to 98%.

AccMILP is also equally effective in the other networks as shown in Table 2. For example, in FNN_2, the MIPVerify method may take more than 45 h for solving, whereas the AccMILP method can complete the solution in around 5 h. Due to the excessively long solving times and limited experimental samples, the results are not recorded in Table 4. Based on the experimental results mentioned above, it is evident that AccMILP can reduce the average size of the verification model by 30%, significantly reduce the solving time by at least 80%, and maintain performance equal to or greater than 60% of MIPVerify.

Therefore, by evaluating the importance of neural network neurons and encoding neurons with varying degrees of relaxation based on their importance, the AccMILP approach significantly reduces the size of the NNV model without compromising verification accuracy. This capability opens the door to the verification of DNNs. It's worth noting that this approach can be extended to other neural network verification methods that incorporate abstraction. Since both the heuristic methods for neuronal importance assessment and activation function relaxation coding are employed in this study, further improvements to the AccMILP approach can be expected with more accurate methods for assessing neuronal importance and more suitable coding approaches for activation function relaxation.

6   Conclusion

This paper proposes an original approach called AccMILP, aimed at expediting the Neural Network Verification (NNV) process and enhancing its scalability. Firstly, we conduct ablation experiments on neurons and introduce the concept of neuron importance. Secondly, we combine the exact solution approach with the approximate solution approach based on neuron importance, resulting in a new neural network verification method called AccMILP. This method strikes a balance between solution accuracy and verification scale. Finally, we assess the performance of AccMILP on several FNNs using standard datasets like MNIST, FashionMNIST and CIFAR10.

The results show that AccMILP approach can reduce the average size of the verification model by about 30%, decrease the solution time by at least 80%, and maintain performance equal to or greater than 60% of MIPVerify while reducing the verification model size.

The AccMILP approach can also be used in other abstract approaches of the ReLU function. Users can make their own substitutions to the abstraction approach and select the classification approaches based on the importance index of neurons.

References

1. Akintunde, M., Lomuscio, A., Maganti, L., Pirovano, E.: Reachability analysis for neural agent-environment systems. In: Sixteenth International Conference on Principles of Knowledge Representation and Reasoning (2018)
2. Ashok, P., Hashemi, V., Křetínský, J., Mohr, S.: DeepAbstract: neural network abstraction for accelerating verification. In: Hung, D.V., Sokolsky, O. (eds.) ATVA 2020. LNCS, vol. 12302, pp. 92–107. Springer, Cham (2020). https://doi.org/10.1007/978-3-030-59152-6_5
3. Botoeva, E., Kouvaros, P., Kronqvist, J., Lomuscio, A., Misener, R.: Efficient verification of ReLU-based neural networks via dependency analysis. In: Proceedings of the AAAI Conference on Artificial Intelligence, vol. 34, pp. 3291–3299 (2020)

4. Bunel, R., Mudigonda, P., Turkaslan, I., Torr, P., Lu, J., Kohli, P.: Branch and bound for piecewise linear neural network verification. J. Mach. Learn. Res. **21**, 1–39 (2020)

5. Bunel, R.R., Turkaslan, I., Torr, P., Kohli, P., Mudigonda, P.K.: A unified view of piecewise linear neural network verification. In: Advances in Neural Information Processing Systems, vol. 31 (2018)

6. Carlini, N., Wagner, D.: Towards evaluating the robustness of neural networks. In: 2017 IEEE Symposium on Security and Privacy (SP), pp. 39–57. IEEE (2017)

7. Cheng, C.-H., Nührenberg, G., Huang, C.-H., Ruess, H.: Verification of binarized neural networks via inter-neuron factoring. In: Piskac, R., Rümmer, P. (eds.) VSTTE 2018. LNCS, vol. 11294, pp. 279–290. Springer, Cham (2018). https://doi.org/10.1007/978-3-030-03592-1_16

8. Cheng, C.-H., Nührenberg, G., Ruess, H.: Maximum resilience of artificial neural networks. In: D'Souza, D., Narayan Kumar, K. (eds.) ATVA 2017. LNCS, vol. 10482, pp. 251–268. Springer, Cham (2017). https://doi.org/10.1007/978-3-319-68167-2_18

9. Deng, L.: The MNIST database of handwritten digit images for machine learning research [best of the web]. IEEE Sig. Process. Mag. **29**(6), 141–142 (2012). https://doi.org/10.1109/MSP.2012.2211477

10. Dutta, S., Jha, S., Sankaranarayanan, S., Tiwari, A.: Output range analysis for deep feedforward neural networks. In: Dutle, A., Muñoz, C., Narkawicz, A. (eds.) NFM 2018. LNCS, vol. 10811, pp. 121–138. Springer, Cham (2018). https://doi.org/10.1007/978-3-319-77935-5_9

11. Dvijotham, K., Stanforth, R., Gowal, S., Mann, T.A., Kohli, P.: A dual approach to scalable verification of deep networks. In: UAI, vol. 1, p. 3 (2018)

12. Ehlers, R.: Formal verification of piece-wise linear feed-forward neural networks. In: D'Souza, D., Narayan Kumar, K. (eds.) ATVA 2017. LNCS, vol. 10482, pp. 269–286. Springer, Cham (2017). https://doi.org/10.1007/978-3-319-68167-2_19

13. Fazlyab, M., Morari, M., Pappas, G.J.: Safety verification and robustness analysis of neural networks via quadratic constraints and semidefinite programming. IEEE Trans. Autom. Control **67**(1), 1–15 (2020)

14. Gehr, T., Mirman, M., Drachsler-Cohen, D., Tsankov, P., Chaudhuri, S., Vechev, M.: AI2: safety and robustness certification of neural networks with abstract interpretation. In: 2018 IEEE Symposium on Security and Privacy (SP), pp. 3–18. IEEE (2018)

15. Goodfellow, I.J., Bengio, Y., Courville, A.: Deep Learning. MIT Press, Cambridge (2016). http://www.deeplearningbook.org

16. Katz, G., Barrett, C., Dill, D.L., Julian, K., Kochenderfer, M.J.: Reluplex: an efficient SMT solver for verifying deep neural networks. In: Majumdar, R., Kunčak, V. (eds.) CAV 2017, Part I. LNCS, vol. 10426, pp. 97–117. Springer, Cham (2017). https://doi.org/10.1007/978-3-319-63387-9_5

17. Kochenderfer, M.J.: Decision Making Under Uncertainty: Theory and Application. MIT Press, Cambridge (2015)

18. Krizhevsky, A., Hinton, G.: Learning multiple layers of features from tiny images. In: Handbook of Systemic Autoimmune Diseases, vol. 1, no. 4 (2009)

19. Krizhevsky, A., Sutskever, I., Hinton, G.E.: ImageNet classification with deep convolutional neural networks. Commun. ACM **60**(6), 84–90 (2017). https://doi.org/10.1145/3065386

20. Madry, A., Makelov, A., Schmidt, L., Tsipras, D., Vladu, A.: Towards deep learning models resistant to adversarial attacks. stat **1050**, 4 (2019)

21. Meyes, R., Lu, M., de Puiseau, C.W., Meisen, T.: Ablation studies to uncover structure of learned representations in artificial neural networks. In: Proceedings on the International Conference on Artificial Intelligence (ICAI), pp. 185–191. The Steering Committee of The World Congress in Computer Science, Computer Engineering and Applied Computing (2019)
22. Mnih, V., et al.: Playing atari with deep reinforcement learning. ArXiv abs/1312.5602 (2013). https://api.semanticscholar.org/CorpusID:15238391
23. Mnih, V., et al.: Human-level control through deep reinforcement learning. Nature **518**(7540), 529–533 (2015)
24. Narodytska, N.: Formal analysis of deep binarized neural networks. In: IJCAI, pp. 5692–5696 (2018)
25. Narodytska, N., Kasiviswanathan, S., Ryzhyk, L., Sagiv, M., Walsh, T.: Verifying properties of binarized deep neural networks. In: Proceedings of the AAAI Conference on Artificial Intelligence, vol. 32 (2018)
26. Raghunathan, A., Steinhardt, J., Liang, P.S.: Semidefinite relaxations for certifying robustness to adversarial examples. In: Advances in Neural Information Processing Systems, vol. 31 (2018)
27. Singh, G., Gehr, T., Püschel, M., Vechev, M.: An abstract domain for certifying neural networks. Proc. ACM Program. Lang. **3**(POPL), 1–30 (2019)
28. Tjeng, V., Xiao, K.Y., Tedrake, R.: Evaluating robustness of neural networks with mixed integer programming. In: International Conference on Learning Representations (2018)
29. Vaswani, A., et al.: Attention is all you need. In: Proceedings of the 31st International Conference on Neural Information Processing Systems, NIPS 2017, pp. 6000–6010. Curran Associates Inc., Red Hook (2017)
30. Wang, S., Pei, K., Whitehouse, J., Yang, J., Jana, S.: Efficient formal safety analysis of neural networks. In: Advances in Neural Information Processing Systems, vol. 31 (2018)
31. Wong, E., Kolter, Z.: Provable defenses against adversarial examples via the convex outer adversarial polytope. In: International Conference on Machine Learning, pp. 5286–5295. PMLR (2018)
32. Xiao, H., Rasul, K., Vollgraf, R.: Fashion-MNIST: a novel image dataset for benchmarking machine learning algorithms. ArXiv abs/1708.07747 (2017). https://api.semanticscholar.org/CorpusID:702279
33. Yin, B., Chen, L., Liu, J., Wang, J.: Efficient complete verification of neural networks via layerwised splitting and refinement. IEEE Trans. Comput. Aided Des. Integr. Circ. Syst. **41**(11), 3898–3909 (2022). https://doi.org/10.1109/TCAD.2022.3197534
34. Zhang, H., Weng, T.W., Chen, P.Y., Hsieh, C.J., Daniel, L.: Efficient neural network robustness certification with general activation functions. In: Advances in Neural Information Processing Systems, vol. 31 (2018)
35. Zhang, Y., et al.: QVIP: an ILP-based formal verification approach for quantized neural networks. In: Proceedings of the 37th IEEE/ACM International Conference on Automated Software Engineering, pp. 1–13 (2022)
36. Zhu, Y., Wang, F., Wan, W., Zhang, M.: Attack-guided efficient robustness verification of ReLU neural networks. In: 2021 International Joint Conference on Neural Networks (IJCNN), pp. 1–8. IEEE (2021)

A.I. for Software Engineering

Word2Vec-BERT-bmu: Classification of RISC-V Architecture Software Package Build Failures

Shitian Ma[1], Hui Li[2,3(✉)], Jiaxin Zhu[2,3,4,5], Xiaohui He[6], Shuyang Zhang[2],
and Junfeng Zeng[1]

[1] School of Computer, Electronics and Information, Guangxi University, Nanning, Guangxi,
China
{mashitian,zengjunfeng}@st.gxu.edu.cn
[2] Institute of Software, Chinese Academy of Sciences, Beijing, China
{lihui2012,zhujiaxin,zhangshuyang}@otcaix.iscas.ac.cn
[3] Key Laboratory of System Software (Chinese Academy of Sciences) and State Key Laboratory
of Computer Science, Institute of Software, Chinese Academy of Sciences, Beijing, China
[4] University of Chinese Academy of Sciences, Nanjing, China
[5] Nanjing Institute of Software Technology, Nanjing, China
[6] Beijing Sinohowe Technology Co. Ltd., Beijing, China
hexiaohui@sinohowe.com

Abstract. In recent years, the emerging open source RISC-V architecture has
gradually attracted wide attention. In order to support the compilation of multiple
Linux operating system distribution images and packages, developers need to build
and adapt the packages. Due to the complexity of software packages and the diversity of developer experience levels, the success of software package construction
is uncertain. Existing research lacks automatic classification methods for the reasons of RISC-V architecture software package construction failures. Therefore, an
automatic classification model Word2Vec-BERT-bmu is proposed to effectively
and automatically locate software package construction failures. Firstly, two popular Linux distribution building platforms, OpenSuse and OpenEuler, were selected
as the sources of build failure log data, and 10 types of build failures were manually analyzed and summarized. Secondly, the Word2Vec-BERT-bmu model is
proposed to construct the failure classification using an automated software package with multi-feature concatenation. Experimental results show that the Macro
F1 value is improved by 2–4% compared with other models. In addition, for realworld software packages, the effectiveness and accuracy of our model we proposed
are further verified by manual repair of software packages.

Keywords: Build failure logs · Logs classification · RISC-V · BERT · Natural
language processing

1 Introduction

In recent years, although X86 and ARM architectures dominate the instruction set market, innovation is limited due to their closed nature, high supply chain dependence
and expensive licensing fees. With the development of technology and the progress of

G. Bai et al. (Eds.): ICECCS 2024, LNCS 14784, pp. 111–124, 2025.
https://doi.org/10.1007/978-3-031-66456-4_6

instruction set architecture, RISC-V architecture has attracted wide attention because of its openness and simplicity [1]. Globally, many companies are focusing on porting software packages to the RISC-V architecture. However, when adapting the RISC-V architecture, developers face a series of challenges, including software rebuild efficiency [2], compatibility, and build failures. Build failures may cause software to fail to be successfully ported to RISC-V architecture, resulting in abnormal operation, performance degradation, or increased resource consumption.

OBS (Open Build Service) is a major open source software building platform, focusing on compiling RPM (Redhat Package Manager) packages for different operating systems (such as OpenEuler, OpenSuse, etc.) and providing a pre-compiled software environment [3]. As the system updates and the number of users increases, the number of build logs skyrockets. When build failures occur, developers often spend a lot of time manually analyzing the failure logs, trying to fix them, and building again. However, sorting through the large number of build failure logs can take a lot of time and effort even for experienced developers. Therefore, the classification analysis of package build failures is crucial for developers to fix build failures. Construction log is unstructured data, which itself has high complexity and uncertainty. Traditional classification techniques are often difficult to meet the user's requirements for efficiency and accuracy in this scenario. Traditional machine learning methods have some obvious limitations, such as limited understanding of the structure and semantics of text, requiring manual design of feature engineering, and being difficult to deal with large-scale and complex failure log data [4]. Deep learning methods and natural language processing are effective means to deal with unstructured data. Therefore, this paper uses deep learning methods to automatically learn semantic features to improve the accuracy of constructing misclassification.

In order to achieve efficient detection of construction failure causes in the process of software package porting to RISC-V, this paper collects 1057 real software package construction failure logs from OpenSuse and OpenEuler, manually analyzes and summarizes the construction failure causes and repair solutions. Secondly, taking the build failure log of RISC-V software package as the text mining object, an effective deep learning method is proposed to realize the automatic classification of build failure causes. At the same time, the accuracy, advancement and applicability of the model are verified.

The main contributions of this paper are as follows:

- Manually analyze and mine the build failure logs of real software packages ported to RISC-V architecture, and form a high-quality RISC-V build failure log dataset to provide standard analysis and reference examples for developers. The dataset formed by the research in this paper is currently open source. (https://github.com/mstsky115/Build_error_logs_data)
- Based on the failure log dataset built by RISC-V, the deep learning method is used to achieve more accurate automatic failure classification, so as to provide effective automatic classification tools for developers.
- In the scenario of software package construction failure log classification, Word2Vec-BERT-bmu model is proposed, which improves the Macro F1 value by 2–4% compared with other BERT variants distilBERT and RoBERTa.

The structure of this paper is as follows. Section 2 introduces the background and research motivation, Section 3 will introduce the construction of software package construction failure dataset and summarize the reasons for RISC-V software package construction failure classification. In Sect. 4, we will experimentally evaluate and fix build misclassifications. Section 5 describes related work and the final section concludes the paper.

2 Background

Build failures are one of the most common problems with package builds [5] when porting packages to the RISC-V architecture. There are many reasons for failures, such as compiler compatibility problems, dependency mismatch, and so on. These problems often require in-depth analysis and resolution to ensure that the software is successfully built and run on the RISC-V architecture.

Take the OpenSuse build platform as an example [6], OBS is an open source package build platform for packaging software and distributing it across multiple operating systems. The platform enables developers to build packages for different architectures, including X86,ARM, and RISC-V. To summarize the construction process of OBS, it can be divided into nine stages, and the detailed construction process explanation is shown in Fig. 1:

1) Source Code Checking: Check the source code of the package to ensure that they conform to the specification and can be compiled correctly.
2) Establishing the Build Environment: OBS build server sets up the build environment for the package, including installing the required dependencies, development tools, etc.
3) Executing(%prep): Decompress and prepare the source code of the package.
4) Executing(%build): Compile the package.
5) Executing(%install): Install the compiled files to a specific directory.
6) Executing(%doc): Install the documentation and help files of the package.
7) Executing(%license): Install the license file of the package.
8) Executing(%clean): Cleans up temporary files generated during the build process.
9) Executing(%rmbuild): Delete all files generated during the build process.

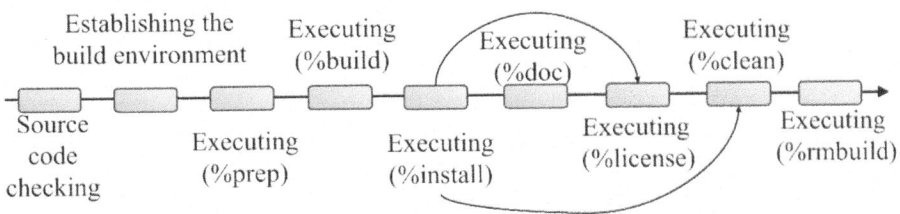

Fig. 1. Building Process of OBS

Build failures may occur in any construction phase of OBS. This paper constructs an failure log through empirical analysis to help developers classify and understand the causes of failures, and classify and understand different types of failures in the log. Specifically, by carefully reading and analyzing typical software package failure logs, key information such as failure types, failure locations, and failure-related descriptions are obtained.

The complexity of the reasons for RISC-V package construction failures and the difference of developer experience levels face great challenges in the stability and efficiency of successful package construction with the increase of package size. It is of great significance to quickly locate the causes of software package build failures and guide developers to fix them efficiently. It is of great significance to study automatic build failure classification methods.

3   Dataset Construction

3.1   Data Sources

This paper focuses on OBS, a software packaging platform that supports various Linux distributions. Because of its openness and data availability, it is widely used by related researchers. To ensure the reliability and practicability of the research, OpenSuse and OpenEuler, two popular Linux distributions, are selected as the sources of construction failure log data. In this paper, a log collection tool is designed to collect build failure logs from the OBS build platform [7, 8]. Finally, a total of 839 build failure logs were collected, including 416 from OpenSuse, 602 from OpenEuler, and 39 from OBS(Tarsier) that maintained OpenEuler-RISC-V in the later period [9].

3.2   Analysis and Classification

It analyzes the characteristics of build logs to find out the characteristics related to build failures, so as to optimize and improve the software package build process. For example, developers can analyze compile failures, test failures, missing dependencies, and other information in the build logs to improve the efficiency of the successful build of the package.

An excerpt of the RISC-V build failure log for aegisub in the dataset is shown in Fig. 2, where the package build failure because package does not support target architecture (RISC-V).

```
1.  [  555s] lj_arch.h:362:2: error: #error "No target architecture defined"
2.  [  555s]   362 | #error "No target architecture defined"
3.  [  555s]       | ^~~~~
4.  [  555s] Makefile:269: *** Unsupported target architecture. Stop.
5.  ......
6.  [  614s] error: Bad exit status from /var/tmp/rpm-tmp.DIBhbc (%build)
7.  [  614s]
8.  [  614s] RPM build errors:
9.  [  614s]     Bad exit status from /var/tmp/rpm-tmp.DIBhbc (%build)
10. [  614s] ### VM INTERACTION START ###
11. [  614s] [  609.344902][   T1] sysrq: Power Off
12. [  614s] [  609.346900][  T18] reboot: Power down
13. [  614s] ### VM INTERACTION END ###
14. [  614s] sheep85 failed "build aegisub.spec" at Sat Feb 11 02:43:15 UTC 2023.
```

Fig. 2. Code Snippet of aegisub with Building Failure

For each log file, investigate information related to the build failure, including a description of the build failure message line and the software package that produced the log. Finally, the classification of RISC-V software package construction failure reasons is shown in Table 1, including software package construction failure categories, main reasons, software package examples, and data quantity. The details are as follows:

a) Build configuration problem (E1): Loss of dependency: log features usually contain information about 'dependencies' no module, failded build 'dependencies', etc. Dependent versions do not match: its error log signature usually contains' dependency '>= information about a version, etc.

b) Test case problem (E2): Test case execution failed: The error log features typically include Test failed, Test error, Test failures, and so on.

c) Incompatibility problem(E3): The compiler is incompatible: its logs usually contain information such as compiler. Architecture incompatibility: its log characteristics generally include Arch, architecture and other information. Function updates: The logs usually contain information about functions that have been updated, deprecated, and so on. File conflicts: Its logs usually contain information such as file conflicts.

d) Source code problem(E4): Variables/functions, etc., undefined: Their logs usually contain information about undefined variables, etc. Function misuse: its error log usually contains information related to function error. Code spelling/grammar failures: The error log usually contains information such as error for a variable.

e) Files/directories are missing problem(E5): The build process could not find the relevant file or directory: The error log feature usually contains information such as No such file or directory.

f) Memory problem(E6): Memory overflow: Its error log characteristics generally contain information related to memory, Kill process, etc.

g) Time-out problem (E7): The code execution time exceeds the threshold. Procedure: The characteristics of the error log generally include information such as after seconds of inactivity and timeout.

h) Plug-in problem(E8): Plugin cannot compile: The error log feature usually contains information about the Plugin error, etc.

i) Network problem(E9): Unable to download dependencies: Its error log characteristics generally contain relevant information such as "Could not transfer artifact from/to".

j) Other(E10): None of the above categories can be classified: It is impossible to manually determine the specific error. The number of samples is too small (less than 5) to form an independent class.

Table 1. Classification of building failure cause

Failure patterns		Software package examples	Data quantity
Build configuration problem (E1)	Loss of dependency	dom4j	485
	Dependent versions do not match	dnf	
Test case problem (E2)	Test case execution failed	apache-commons-pool2	164
Incompatibility problem(E3)	The compiler is incompatible	ghc-pandoc-lua-marshal	113
	Architecture incompatibility	boxfort	
	Function update	monero-gui	
	File conflicts	dconf	
Source code problem(E4)	Variables/functions, etc., undefined	coturn	89
	Function misuse	janus-gateway	
	Code spelling/grammar failures	matrix-quaternion	
Files/directories are missing problem(E5)	The build process could not find the relevant file or directory	kubernetes1.21	72
Memory problem(E6)	Memory overflow	amsynth	35
Time-out problem (E7)	The code execution time exceeds the threshold. Procedure	atlas	27
Plug-in problem(E8)	Plugin cannot compile	hadoop	15
Network problem(E9)	Unable to download dependencies	hive	20
Other(E10)	None of the above categories can be classified	file-roller	37
		total	1057

4 Evaluation and Validation

4.1 Data Pre-processing

Aiming at the construction log failure information, firstly, the key line information of the failure log is extracted, and the preprocessing process is used to transform the log data into clean, structured and consistent data, which helps to reduce the interference of noise and redundant information, extract meaningful features, and enhance the interpretability and comparability of the text. Thus, the quality and effect of the classification model are significantly improved. Taking the Alto xml package [10] as an example, the key line information extracted from the failure log is "ModuleNotFoundError: No module named 'javapackages'", after data cleaning, stop words removal, lemmization, and preserving the characteristics of build failures, the failure line information is processed into "modulenotfounderror module name javapackages".

4.2 Evaluation Indicators

In order to evaluate the performance of building failure log classification models, it is necessary to use some metrics to measure the accuracy and effect of the model. In this experiment, we use three main evaluation metrics, namely Macro Precision, Macro Recall, and Macro F1 value. These metrics are mainly used to measure the average performance of the classification model over the whole dataset.

Macro Precision is the average of the precision for each class, Macro Recall is the average of the recall for each class, and Macro F1 is the weighted harmonic mean of macro precision and macro recall. The formula is as follows:

$$P_{macro} = \frac{1}{n} \sum_{i=1}^{n} \frac{TP_i}{TP_i + FP_i} \tag{1}$$

Here, n denotes the number of classes, TP_i denotes the number of correctly predicted samples in class i, and FP_i denotes the number of incorrectly predicted samples in class i. Here, n denotes the number of classes, TP_i denotes the number of correctly predicted samples in class i, and FP_i denotes the number of incorrectly predicted samples in class i.

$$R_{macro} = \frac{1}{n} \sum_{i=1}^{n} \frac{TP_i}{TP_i + FN_i} \tag{2}$$

Here, FN is the number of samples whose true class is positive but is incorrectly predicted to be negative.

$$F1_{macro} = \frac{2*P_{macro}*R_{macro}}{P_{macro} + R_{macro}} \tag{3}$$

Macro F1 score is an evaluation metric that comprehensively considers macro precision and macro recall, which can evaluate the effect of the algorithm as a whole and prevent the gap between macro precision and macro recall from getting too large.

4.3 Word2Vec-BERT-bmu Model

Through private basic model experiments, TextRNN, TextCNN, Bi-LSTM, BERT and their variants are used, and the BERT-bmu (BERT-base-multilangual-uncased) model with the best failure log classification effect is selected as the baseline model. The BERT-bmu model has some limitations in some cases, such as the lack of domain-specific knowledge, and the challenge of dealing with linguistic ambiguity and polysemy. In order to further improve the performance of the model, this paper conducts in-depth research on feature engineering, aiming to explore other possible feature representation methods and make up for the shortcomings of the BERT-bmu model in constructing failure log classification scenarios. In this process, we try methods such as Word2Vec and N-gram to expand the feature space and enrich the input information of the model, and propose the Word2Vec-BERT-bmu model.

The structure diagram of Word2vec-BERT-bmu is shown in Fig. 3. The input log is processed by Word2Vec and BERT in parallel, and the Word2Vec model generates embedding vectors for each word through the context information of the vocabulary, which retains the semantic information in the text. At the same time, BERT-bmu (12-layer Transformer Encoder) also preprocesses the input log to learn rich language expressions. In the next step, the output of BERT-bmu and the output of Word2Vec are concatenated to create a hybrid embedding vector, so as to capture the characteristics of the input log more comprehensively. This multi-level and multi-source embedding enables the Word2Vec-BERT-bmu model to have stronger semantic understanding and context awareness in text classification. Finally, after feature extraction and regularization, the results are passed to a Softmax layer that maps the features to a probability distribution over 10 classes.

The Word2Vec parameter is defined as follows: vector_size = 128, window = 5, min_count = 1, workers = 4; Learning rate: The model's optimizer uses the Adam optimizer with a learning rate of 5e-5. Loss function: The model uses a cross-entropy loss function to measure the failure between the model's prediction and the actual label; Vector concatenation: The vector of Word2Vec is concatenated with the CLS output of BERT-bmu (dimension 768), and the final input vector dimension is 896. The correspondence of each component in the control vector mainly depends on the order of concatenation. In this paper, the vector of Word2Vec is in the front, and the CLS output of BERT-bmu is in the back. Dropout rate: After the concatenated embedding vectors of the model, a Dropout rate of 0.5 is used to reduce overfitting. Output Layer: The output layer of the model is a fully connected layer with 10 neurons and uses softmax activation function for multi-class classification.

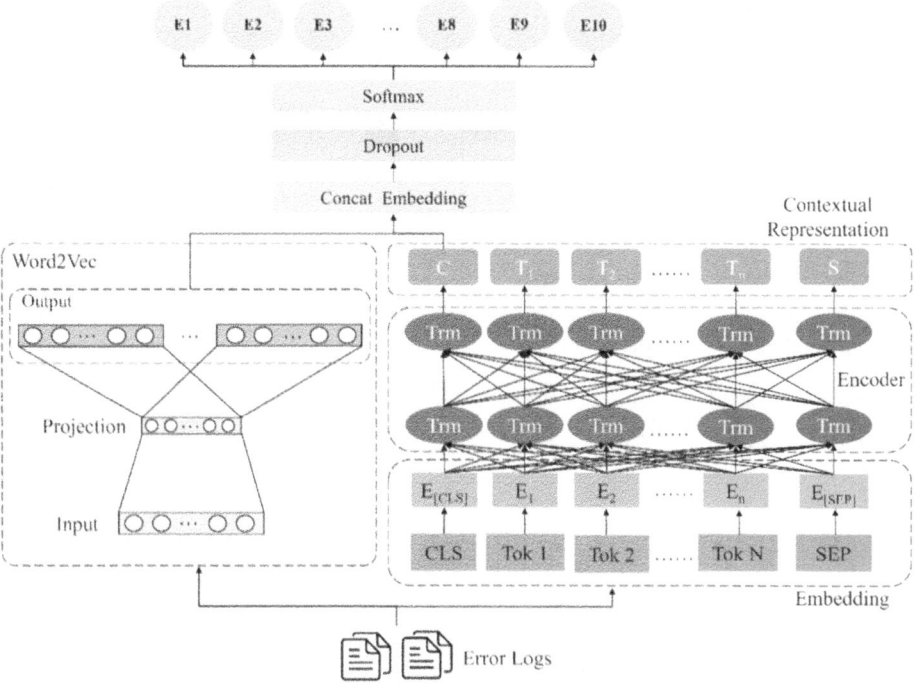

Fig. 3. Word2Vec-BERT-bmu Structure Diagram

4.4 Experimental Environment

The experimental environment information is as follows: Python Tensorflow2.12 was used to complete the construction of the model framework, model training and prediction. The operating system environment is Linux, the hardware environment is 16 core CPU: Intel(R) Xeon(R) Gold 5215 CPU @ 2.50 GHz, 128 GB memory, and the GPU is NVIDIA Quadro P5000 with 16 GB memory.

In order to evaluate the performance of different models on the same dataset, and the training results of different models are comparable, the log dataset is divided into training set and test set according to the ratio of 8:2, and the training parameters are kept consistent, while performing 5-fold cross validation.

4.5 Analysis of Results

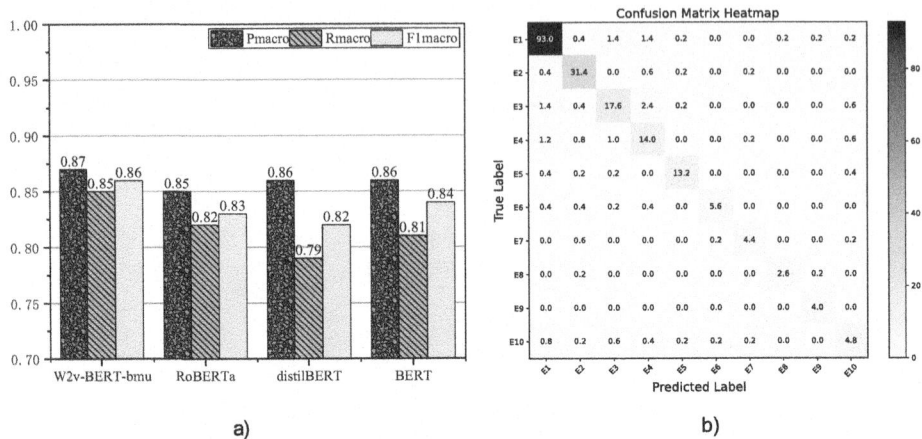

a)   b)

Fig. 4. Experimental Results

As shown in Fig. 4-a, after horizontal comparison of BERT and its variants BERT-bmu, distilBERT and RoBERTa, Word2Vec-BERT-bmu shows better results than using BERT and its variants alone in the failure log classification scenario built based on software packages. The Pmacro scored 87%, Rmacro and F1macro outperformed other models by 1% to 6%. At the same time, The confusion matrix heatmap of the Word2Vec-BERT-bmu model is shown in Fig. 4-b, where the diagonal values are much larger than the off-diagonal values, indicating that the overall performance of the model is good. Due to the uneven distribution of the number of samples, the colors are lighter further back in the category.

The reason why this improvement can improve the classification effect is analyzed as follows: 1) The word vector of Word2Vec provides the model with richer and more diverse word representations, which enhances the understanding ability of the model. Compared with the BERT-bmu model, which is limited by the vocabulary size, Word2Vec can represent rare words or professional terms more accurately, because BERT-bmu is limited by the vocabulary size. Word2Vec can more accurately represent rare words or technical terms; 2) Although the BERT-bmu model adopts a Transformer-based self-attention mechanism, which can capture the context information within and between sentences, however, because the input of BERT-bmu is based on WordPiece token-ization, some words may be segmented into sub-words, which makes it lack of expressive power. By combining with Word2Vec, the representation of the whole sentence can be enhanced by averaging or concatenating the Word2Vec vectors of words, making up for the lack of BERT-bmu in word level expression. We complement Word2Vec's word-level representation with BERT-bmu's sentence-level representation. This study shows that the optimization through feature engineering provides an effective optimization strategy for solving the problem of failure log classification, and has certain guiding significance for similar text classification tasks.

4.6 Fix Validation

In order to further verify the effectiveness of the Word2Vec-BERT-bmu model in the automatic classification of software package construction failures, this paper fixes the Word2Vec-BERT-bmu model according to the model classification results. For real software packages, the collected software packages with construction failures were manually repaired and verified by means of experience summary [11]. In this paper, the software packages that have not been repaired in three communities in the data set are randomly selected for repair, and the specific repair results are shown in Table 2.

Table 2. Repair validation with random sampling

Count	Failure	Fix
1	Incompatibility problem(E3)	Add support for RISC-V architecture
1	Build configuration problem (E1)	spec files add missing dependencies
1	Source code problem(E4)	Import relevant definitions/headers

By randomly extracting faulty software packages for manual repair, the repair scheme corresponds to the failure classification, which successfully proves the accuracy and effectiveness of the proposed method for classifying the causes of RISC-V software package construction failures.

5 Related Work

With the open source expansion of log datasets and the gradual maturity of machine learning and deep learning methods, significant progress has been made in the field of log anomaly detection and log parsing. Zhao [12] and Du [13] improved the LSTM model and used different feature extraction for log detection. He [14] proposed an online log parsing method (Drain), which uses a fixed depth parse tree and specially designed rule encoding for parsing to improve the efficiency of log parsing. Xie [15] combined deep learning model and knowledge graph technology to realize the failure prediction and analysis of Hadoop cluster. Zhu [16] evaluated the performance of 13 previous automatic log parsers on 16 different datasets to understand their properties such as accuracy, robustness, and efficiency. Yu [17] evaluate the basic algorithm and deep learning algorithm on five public log detection datasets, and propose a framework for optimizing training time, inference time and performance analysis, which can be compared between anomaly detection models. However, the log structure of the above data set is obvious and basically contains time series, while the failure information of the build failure log for the RISC-V architecture is diverse and unordered, and does not contain time series, so the previous methods cannot be directly used.

The study of Seo [18] explored compiler failures and solutions in the build phase of the software development process, and the study revealed the frequency of builds, the failure rate, and common failures related to dependencies between components. This study focuses on the problems in the build process of RISC-V packages, and the types of build failures are also very different from the Seo build phase failure messages. At the same time, based on the analysis of failure types, this study further proposes an automatic classification model of failure causes to improve the efficiency of software development. Li [19] proposed a text classification method based on dynamic fusion of multiple features, which focuses on global semantic features, while RISC-V needs more context understanding for the specific features that are constructed incorrectly. Chen [4] proposed a BERT-based local feature convolutional network model to improve the accuracy of Chinese news long text classification, but its design focus does not meet the complexity of construction failures, such as software construction usually designs multiple compilers, programming languages and dependencies. Alagarsamy [20] 's hybrid LSTM and RNN model eliminates the connection classification layer in deep learning models, but it does not consider the specific characteristics of construction failures. Jiang [21] 's hybrid sentiment classification model is mainly used to capture sentiment information, but does not consider the context requirements of RISC-V construction failures. The TextCNN-SVM model by Dong [22] improves the classification ability of TextCNN, but it is not designed for the specific requirements of construction failures. In contrast, the focus of this study is to explore multiple neural network models for solving the classification task of building failure causes in RISC-V software packages. By conducting a comprehensive horizontal comparison of multiple models, considering their performance in building failure classification ability and generalization ability, and optimizing based on the BERT model with the best application effect. In addition, the Word2Vec-BERT-bmu model is proposed to further improve the construction failure classification effect. Alghanmi [23] and Srinarasi [24] used a similar word vector combination model, which provides a reference for further optimization and improvement of subsequent models.

6 Conclusion

This study focuses on solving the problem of software package construction misclassification in RISC-V architecture. In the face of challenges such as low efficiency of manual classification, high dependence on expertise and experience, and lack of literature, we construct a real RISC-V construction failure dataset, and introduce deep learning technology to solve the classification problem. The Word2Vec-BERT-bmu model is proposed for automatic construction of failure classification. Compared with a variety of models, the accuracy is 7% higher than BERT. Finally, the accuracy and effectiveness of the classification are verified by manual repair of actual software packages, which provides support for developers to solve problems. In the future, we will conduct tests on large models, such as GPT, LLAMA2, etc. At the same time, we plan to carry out automated repair recommendation to further improve the construction efficiency, and expand the scope of research to other construction platforms, such as Fedora, to provide comprehensive and practical solutions for RISC-V and other architecture construction problems.

Acknowledgments. This work is supported by the Strategic Priority Research Program of the Chinese Academy of Sciences, Grant No. XDA0320102, Youth Innovation Promotion Association of the Chinese Academy of Sciences (YICAS), Grant No. 2023121.

References

1. Riedel, S., Schuiki, F., Scheffler, P., Zaruba, F., Benini, L. Benin.: Banshee: A Fast LLVM-based RISC-V binary translator. In: 2021 IEEE/ACM International Conference On Computer Aided Design (ICCAD), pp. 1–9. Munich, Germany (2021)
2. Lamb, C., Zacchiroli, S.: Reproducible builds: increasing the integrity of software supply chains. IEEE Softw. **39**(2), 62–70 (2022)
3. Becker, B., Jeannerod, N., Marché, C., Régis-Gianas, Y., Sighireanu, M., Treinen, R.: The CoLiS platform for the analysis of maintainer scripts in Debian software packages. Int. J. Softw. Tools Technol. Transfer **24**(5), 717–733 (2022)
4. Chen, X., Cong, P., Lv, S.: A long-text classification method of Chinese news based on BERT and CNN. IEEE Access **10**, 34046–34057 (2022)
5. Courtès, L.: Building a secure software supply chain with GNU guix. Art Sci. Eng. Program. **7**(1), 1–26 (2022)
6. openSUSE Build Service. https://build.opensuse.org/. Accessed 23 May 2023
7. openSUSE:Factory:RISCV. https://build.opensuse.org/project/show/openSUSE:Factory: RISCV. Accessed 23 May 2023
8. openEuler:Mainline:RISC-V. https://build.openeuler.openatom.cn/project/show/openEuler: Mainline:RISC-V. Accessed 23 May 2023
9. Show openEuler:23.03. https://build.tarsier-infra.com/project/show/openEuler:23.03. Accessed 23 May 2023
10. Build Log for Package aalto-xml. https://build.openeuler.openatom.cn/package/live_b uild_log/openEuler:Mainline:RISC-V/aalto-xml/advanced_riscv64/riscv64. Accessed 09 Aug 2023
11. Gao, Y., et al.: An empirical study on crash recovery bugs in large-scale distributed systems. In: Proceedings of the 2018 26th ACM Joint Meeting on European Software Engineering Conference and Symposium on the Foundations of Software Engineering, pp. 539–550 (2018)
12. Zhao, Z., Xu, C., Li, B.: A LSTM-based anomaly detection model for log analysis. J. Sign. Process Syst. **93**(7), 745–751 (2021)
13. Du, M., Li, F., Zheng, G., Srikumar, V.: Deeplog: Anomaly detection and diagnosis from system logs through deep learning. In: Proceedings of the 2017 ACM SIGSAC Conference on Computer and Communications Security, pp. 1285–1298 (2017)
14. He, P., Zhu, J., Zheng, Z., Lyu, M.R.: Drain: an online log parsing approach with fixed depth tree. In: 2017 IEEE International Conference on Web Services (ICWS), pp. 33–40. Honolulu, HI, USA (2017)
15. Xie, Y., Yang, K., Luo, P.: LogM: log analysis for multiple components of hadoop platform. IEEE Access **9**, 73522–73532 (2021)
16. Zhu, J., et al.: Tools and benchmarks for automated log parsing. In: 2019 IEEE/ACM 41st International Conference on Software Engineering: Software Engineering in Practice (ICSE-SEIP), pp. 121–130. Montreal, QC, Canada (2019)
17. Yu, B., et al.: Deep learning or classical machine learning? An empirical study on log-based anomaly detection. In: Proceedings of the 46th IEEE/ACM International Conference on Software Engineering. pp. 1–13 (2024)

18. Seo, H., Sadowski, C., Elbaum, S., Aftandilian, E., Bowdidge, R.: Programmers' build errors: a case study (at google). In: Proceedings of the 36th International Conference on Software Engineering, pp. 724–734 (2014)
19. Li, Y., Zhang, S., Lai, C.: Agricultural text classification method based on dynamic fusion of multiple features. IEEE Access **11**, 27034–27042 (2023)
20. Alagarsamy, S., James, V.: RNN LSTM-based deep hybrid learning model for text classification using machine learning variant XGBoost. Int. J. Performability Eng. **18**(8), 545–551 (2022)
21. Jiang, X., Song, C., Xu, Y., Li, Y., Peng, Y.: Research on sentiment classification for netizens based on the BERT-BiLSTM-TextCNN model. PeerJ. Comput. Sci. **8**, e1005 (2022)
22. Dong, X., Hu, R., Li, Y., Liu, M., Xiao, Y.: Text sentiment polarity classification based on TextCNN-SVM combination model. In: 2021 IEEE International Conference on Artificial Intelligence and Computer Applications (ICAICA), pp. 325–328. Dalian, China (2021)
23. Alghanmi, I., Anke, L. E., Schockaert, S.: Combining BERT with static word embeddings for categorizing social media. In: Proceedings of the Sixth Workshop on Noisy User-Generated Text (w-nut 2020), pp. 28–33 (2020)
24. Srinarasi, S., et al.: A Combination of enhanced WordNet and BERT for semantic textual similarity. In: Proceedings of the 2021 2nd International Conference on Control, Robotics and Intelligent System, pp.191–198 (2021)

Test Architecture Generation by Leveraging BERT and Control and Data Flows

Guangyu Wang, Ji Wu$^{(\boxtimes)}$, Haiyan Yang, Qing Sun, and Tao Yue ⓘ

School of Computer Science and Engineering, Beihang University, Beijing, China
{wgy,wuji,yhy,sunqing,yuetao}@buaa.edu.cn

Abstract. In practices, test cases are often designed by test engineers based on the functionalities of the System under Test (SUT) in parallel and independently. This can lead to a lack of a comprehensive overview of the test architecture, hampering the reuse of test functions when implementing new test cases. To address this challenge, we propose *ATAG*, an automated test architecture generation approach, which employs an optimization algorithm to retrieve highly cohesive and loosely coupled test functions based on control flows and data flows of test cases. We also equip *ATAG* with a newly proposed BERT-based model, i.e., FunBERT, for generating test function names. We conducted an empirical study with three industrial datasets to evaluate the effectiveness of *ATAG* and FunBERT. Results show that test architectures generated with *ATAG* (benefiting from both control and data flows) improved, on average, $\approx$ 26–35% coupling and $\approx$ 28–50% cohesion of the original test architectures manually constructed by test engineers from our industrial partner. FunBERT achieves 97.9%, 98.3%, and 98.1% in Precision, Recall, and F1-score, and significantly outperforms the best baseline method BERT.

Keywords: Test architecture generation · Cohesion · Coupling

1 Introduction

Test automation is widely acknowledged as a fundamental way of ensuring software quality and enhancing testing efficiency [19]. A typical test process involves collecting and analyzing test requirements, designing test cases, developing test scripts, executing them, and analyzing test results [32]. Notably, a test case is often implemented as a top-level test function that calls lower-level test functions, which are often shared across test cases, to realize the specified test actions. Properly architecting test functions to implement a given set of test cases is practically needed, especially considering that test cases and test functions might need to evolve over time along with the evolution of the system under test (SUT) [7,35]. To this end, it is necessary to holistically analyze the SUT and the given set of test cases to identify their relations and common test actions. However, in industrial practices [22], test cases are usually identified and specified according

© The Author(s), under exclusive license to Springer Nature Switzerland AG 2025
G. Bai et al. (Eds.): ICECCS 2024, LNCS 14784, pp. 125–145, 2025.
https://doi.org/10.1007/978-3-031-66456-4_7

to functionalities of the SUT, such that test engineers can design and implement them in parallel and independently. Consequently, test engineers often lack an overview of the whole test architecture, which hinders effective communication among themselves.

In software architecture development, hierarchical design is a common practice [38], where modules are identified and connected hierarchically by maintaining the proper cohesion and coupling. The hierarchical design has also been applied in test generation [?], e.g., developing test functions with a hierarchical structure to implement test cases. However, developing a hierarchical test architecture has never been trivial in practice, as it requires test engineers to possess comprehensive knowledge of test cases and expertise in architectural design. This is typically not the case in industry, as test engineers usually focus only on developing specific test cases and test functions to cover specific features of the SUT. Moreover, along with the natural evolution of the SUT and constant changes in task assignments, some test functions become bloated with too many test actions included, and relations among test functions gradually become incomprehensible, difficult to maintain, and hard to reuse. After analyzing the test function set (containing, in total, 18015 test functions) from our industrial partner, we observed that 12.28% of test functions exceed the recommended size of 30 lines of code (LoC), among which 12.34% test functions exceed 100 LoC, violating the widely accepted "30 LoC coding principles" [21]. Moreover, around 20% of test functions exhibit high fan-out calling of more than 10 or even hundreds of other test functions, which exceeds 7, a commonly acknowledged lower bound for assessing a test function as a complex one [25].

Lacking a proper hierarchical test architecture causes that test actions shared among test cases are implemented as different test functions with the same or similar functionalities and names. When test engineers select a test function from existing ones to implement a new test case either manually or automatically with, e.g., SRTEF [22], choosing the right one becomes a challenge. Two extreme strategies for designing a test architecture are: 1) implementing each test case as a single test function without considering shared test actions across test cases, and 2) implementing each test action as a test function and then implementing each test case by calling the test functions according to the execution order of the corresponding test actions. The former results in test functions containing similar or duplicated code blocks that could have been reused across test functions, while the latter results in many simple test functions. Both discourage the healthy maintenance of test functions and consequently lead to the overall low efficiency of a test process. An expected strategy should balance these two extreme strategies by keeping test functions with proper sizes, desired cohesion, and appropriate degree of coupling among themselves. A cost-effective solution is therefore needed to implement such a strategy to automatically generate test architecture, which can then be followed to design test functions by conforming to the high-cohesion and low-coupling principles.

In this paper, we propose an Automated Test Architecture Generation method, named *ATAG*, which takes a suite of step-wise textual test cases as the input and generates a test architecture and test function names. A generated test architecture has hierarchically organized test functions that satisfy the principles of high cohesion and low coupling. For function names, $ATAG$ has the FunBERT model, which is built on the Bidirectional Encoder Representations from Transformers (BERT), a well-known deep learning model designed for NLP tasks.

$ATAG$ is equipped with three optimization strategies: control-flow based ($ATAG_{cf}$), data-flow based ($ATAG_{df}$), and both combined ($ATAG_{cf+df}$). To evaluate $ATAG$, we employ three datasets of various sizes from our industrial partner. Results show that test architectures generated with $ATAG_{cf+df}$ improved, on average, $\approx$26–35% coupling and $\approx$28–50% cohesion of the original test architectures manually constructed by test engineers from our industrial partner. Results of our experiment show that data flows contribute more to the overall performance of $ATAG$ than control flows, and $ATAG_{cf+df}$ performed the best across all datasets. Regarding the effectiveness of FunBERT, we compare it with five commonly used techniques in terms of Precision, Recall, and F1-score. Results show that FunBERT achieves the highest scores in all three metrics, specifically 97.9%, 98.3%, and 98.1%.

Structure. $ATAG$ is discussed in Sect. 2. The experiment design is presented in Sect. 3. Section 4 presents results and analyses. The related work is presented in Sect. 5. At last, Sect. 6 concludes the paper.

2   Approach

In this section, we first discuss the overall application context of $ATAG$, define key concepts, and formulate the problem to be solved as an optimization problem (Sect. 2.1). We then present the overview of $ATAG$ in 2.2, followed by its two key phases: Test Architecture Generation (TAG) and Test Function Name Generation (TFNG) in Sects. 2.3 and 2.4, respectively.

2.1   Context, Definitions, and Problem Formulation

Context. In practice, test case design is usually followed by test function implementation to realize its behavior [22]. Black-box testing requires test engineers to design test cases according to the requirement specification of the SUT, which is often specified in restricted natural languages such as Keyword-Driven Testing [1], Restricted Test Case Modeling [36], and Behavior-Driven Development. Subsequently, all the designed test cases would be implemented in a specific programming language, such as Ruby, in the case of our industrial partner. The separation of test case design and test function implementation enables test cases and test functions to be managed as reusable libraries, i.e., test case and test function libraries [22].

Concepts. For a given SUT, its $TestArchitecture$ defines which $TestFunctions$ should be designed and hierarchically structured to implement a given set of

TestCases for testing a specific set of functionalities of the SUT according to test requirements. A *TestCase* describes an independent test behavior and consists of a sequence of *TestActions*. For example, the following test case contains five test action descriptions:

a. Query business
b. Deactivate business
c. Activate business
d. Delete business
e. Clear environment

If two *TestFunctions* implement a common sequence of *TestActions*, duplication would be present in their implementations, which is discouraged when considering healthy maintenance of test functions. For example, test function tf_1 realizes three test actions *abc*: query, deactivate, and activate business. Test function tf_2 realizes test actions *bcd*. To avoid duplication, another test function tf_3 can be identified to realize *bc* so that tf_1 and tf_2 hierarchically call tf_3. A *TestFunction* can invoke another *TestFunction* at a specific offset via *AggregationAction*. Here, the invoking *TestFunction* acts as the 'upper' function while the invoked one serves as the 'lower' function. Thus, the 'upper' *TestFunction* aggregates the behavior of the 'lower' *TestFunction*. A *TestAction* visits one or more data entities typed as *DataObject* of two kinds: *InternalDataObject* and *ExternalDataObject*. An *InternalDataObject* is initialized and accessed exclusively within a *TestFunction*, while an external one is accessible to multiple *TestFunctions* either via parameter passing or as a returned value.

Problem Formulation. Given a set of *TestCases* in natural language, the test architecture generation process suggests a set of *TestFunctions* to implement the *TestCases*. The problem can be further transformed into an optimization problem to extract *TestFunctions*, realizing a sequence of *TestActions* and construct hierarchical relations among *TestFunctions* with the objectives of maximizing the cohesion within each test function $Coh(tf_i)$ and minimizing the coupling among test functions $CBF(tf_i|TF)$, as formulated as in Eq. 1, where TF represents a set of *TestFunctions*, tf_i is one of them, $CBF(tf_i|TF)$ measures coupling among tfs as formulated in Eq. 3, and $Coh(tf_i)$ measures tf_i's cohesion as formulated in Eq. 5. The optimization objectives are subject to the constraint as shown in Eq. 2, where $CALL_{tf_i,tf_j}$ is 1 if there is a calling relationship from tf_i to tf_j, 0 otherwise. The constraint requires that calls between test functions should be unique and acyclic such that there is no death loop of calling or uncontrolled recursions.

$$\left. \begin{array}{l} \text{Minimize} \sum CBF(tf_i|TF) \\ \text{Maximum} \sum Coh(tf_i), tf_i \in TF \end{array} \right\} \qquad (1)$$

$$\left. \begin{array}{l} \text{s.t. } CALL_{tf_i,tf_j} * CALL_{tf_j,tf_i} = 0 \\ tf_i, tf_j \in TF, CALL \in \{0,1\} \\ TF = \{tf_1, tf_2, \ldots, tf_n\} \end{array} \right\} \qquad (2)$$

2.2   Overview of *ATAG*

As shown in Fig. 1, for a given set of test cases, *ATAG* first parses each of them to retrieve a sequence of test actions, which forms the initial test function set. By taking it as the input, *ATAG* starts the iterative test architecture generation process with one of the three chosen optimization strategies defined based on control flows extracted among the test actions and data flows identified between the test actions and data objects. The optimization objective is to maximize the cohesion within each test function and minimize the coupling among all test functions in a test architecture. The iterative process terminates when the given coupling and cohesion thresholds are met. The second stage of *ATAG* is to generate test function names by using the proposed FunBERT model.

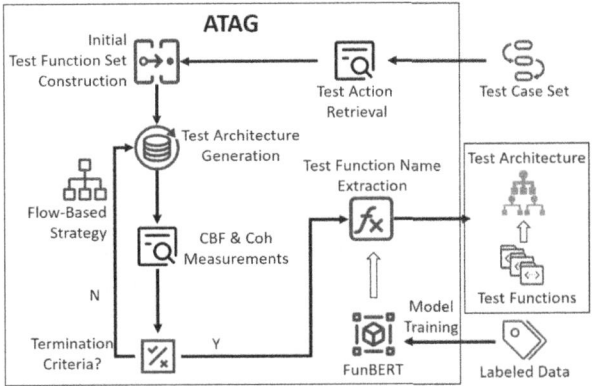

Fig. 1. The Overall Process of *ATAG*

2.3   Test Architecture Generation

Process. *ATAG* takes a set of test cases as input. Each has a sequence of test actions specified in natural language, and produces a set of hierarchically structured test functions as the output, i.e., test architecture. *ATAG* starts by obtaining all test actions specified in all test cases and derives an initial set of disconnected test functions. Then, the sequence of test actions (initially just one action) realized in these test functions is updated based on the control flow and data flow specified in test cases to achieve higher cohesion and lower coupling. The newly formed test functions are evaluated for avoiding calls that lead to death loops or uncontrolled recursions based on the specified constraints (Eq. 2). This process continues as these new test functions are further formed until the termination criterion is met: the fluctuation of coupling and cohesion converges to given thresholds, indicating no need for further optimization.

Three Optimization Strategies. To guide $ATAG$, we formulate three optimization strategies to iteratively increase the cohesion and decrease the coupling.

$ATAG_{cf}$ is a control-flow-based strategy that starts with a single test action, identifies adjacent sequences of shared test actions, and gradually increases the calling of test functions while preventing cyclic call loops. $ATAG_{df}$ is a data-flow-based strategy that tries to confine data objects within a test function as much as possible to reduce potential data flow interactions across test functions, consequently improving the cohesion of each test function. $ATAG_{df}$ identifies objects accessed by adjacent test functions to extend the data flow, and when a test function accesses more than one object, i.e., when multiple data flows intersect, $ATAG_{df}$ might suggest adding a new test function to combine the involved data flows into one.

For instance, there are five test functions $(tf_1\text{-}tf_5)$, each containing a test action $(ta_1\text{-}ta_5)$, where one object is accessed by ta_3 and ta_4, and another is accessed by ta_4 and ta_5. When there is a test requirement to execute test functions tf_1 to tf_5 sequentially, $ATAG_{cf}$ identifies ta_1, ta_2, and ta_3 as a control flow, and generates a high-level test function tf_{123} based on the shared control flow, which further calls tf_1, tf_2, and tf_3; while $ATAG_{df}$ identifies that ta_3 and ta_4, ta_4 and ta_5 are two intersected data flows, and generates high-level test functions tf_{34}, tf_{45} and tf_{345}, where tf_{34}, and tf_{45} are based on independent data streams. and tf_{345} is based on the intersection of data flows.

$ATAG_{cf+df}$ is a hybrid strategy that combines $ATAG_{cf}$ and $ATAG_{df}$ to update test functions. Specifically, $ATAG_{cf+df}$ begins with a single test action and identifies adjacent shared test actions. It combines the objects from these shared test actions into internal objects. It further generates mid-level test functions that invoke the function responsible for executing the shared test action. Subsequently, $ATAG_{cf+df}$ identifies adjacent data flows based on the objects traversed in the mid-level function and generates higher-level functions that call the functions associated with the corresponding data flow. In the previous example, $ATAG_{cf+df}$, based on one control flow (tf_1, tf_2, tf_3) and two data flows $(tf_3$ to tf_4, tf_4 to $tf_5)$, generates high-level test functions tf_{123}, tf_{1234} and tf_{12345}, where tf_{123} is based on shared control flow, tf_{1234} and tf_{12345} are based on the intersection of data flows.

The three strategies all ensure high cohesion and low coupling of test functions while building test architecture, but their focuses differ. $ATAG_{cf}$ encapsulates functions based on shared action sequences, effectively reducing repetitive test functions. However, it may result in tight interdependencies among the test functions, i.e., high coupling, as it interrupts the data flow, leading to an increased reliance on external data objects. Also, the number of test functions generated depends heavily on the number of shared action sequences in the test case dataset. If the dataset covers a diverse set of SUT features, there are possibly fewer shared action sequences, and hence, $ATAG_{cf}$ may generate fewer test functions. $ATAG_{df}$ covers entire data operations, making it less dependent on the diversity of features in the dataset. This also raises a concern: when data flows intersect or run in parallel, generated test functions must cover multiple data flows, which may cause unnecessarily duplicated functional blocks. $ATAG_{cf+df}$ combines these two strategies to ensure complete data operations while reducing duplication of shared test action sequences across test functions.

Coupling and Cohesion Metrics. Coupling between objects [10] is a classic metric for counting how many classes can be reached from a single class. We adapt it for Coupling Between test Functions (CBF) to measure the coupling among test functions, calculated with Eq. 3. Specifically, CBF counts the number of two kinds of unidirectional dependencies: the number of control dependencies among functions (N_{CD}) through function calls and the number of data dependencies between test functions manifested through shared access to external data objects (N_{DD}). For the generated test architecture, its whole CBF will be calculated as the average of all CBFs with Eq. 4.

$$CBF(tf|TF) = \sum_{\substack{tf_j \in TF \\ tf_j \neq tf}} N_{CD}(tf, tf_j) + N_{DD}(tf, tf_j) \tag{3}$$

$$AveCBF(TF) = \frac{\sum_{tf_i \in TF} CBF(tf_i|TF)}{|TF|} \tag{4}$$

To measure the cohesion of a test function, we adapt the classical class cohesion metric Coh [6]. As shown in Eq. 5, nta is the number of test actions in a test function; ndo is the number of data objects visited by the test actions in the test function; and $v(ta, o)$ counts whether a test action ta visits a data object o in the test function with 1 indicating yes, 0 otherwise. To address the problem of the absence of data objects in a test function, we use $ndo + 1$ instead of ndo in the equation.

$$Coh(tf_i) = \frac{\sum_{ta \in tf_i, o \in tf_i} v(ta, o)}{nta(tf_i) * (ndo(tf_i) + 1)} \tag{5}$$

2.4   Test Function Name Generation

Problem. A test function should be specified by its function name, parameter list, return, and functional description. This paper focuses on extracting the test function name from the literal test action descriptions, which can be defined as a keyword extraction and concatenation problem to solve.

We use an example to explain the problem further. A test action description is "Specify the path to issue the service AS @SDH2", containing nine words, where "AS @SDH2", as a keyword, defines a data object with the name of "SDH2" to refer to the "service". Tagging keywords and positional information returns the label sequence $\{A, O, B, C, D, O, E, O, O\}$, where A to E indicates the positions at which each of the five words (e.g., "Specify") should be placed in the function name, and O (e.g., "the") indicates that the word is not present in the function name. Hence, the function name can be defined as "specify_path_to_issue_service," which carries the essential information about what the function does.

Actually, it is quite normal for engineers to use domain terminologies and/or their abbreviations (e.g., SDH for Synchronous Digital Hierarchy) to specify test actions. When they are used, it isn't easy to comprehend the meaning of a test action without knowing the context, i.e., the test case owning the test action. Since all test actions despite *AggregationAction* are from a test case, and eventually, the test function will be called to realize the test case, it is necessary to comprehend the test function in the context of the test case. We chose the test case with the highest test function percentage, which is the ratio of the number of test actions realized in the test function to the number of actions in the test case, as the contextual test case.

FunBERT Model. The FunBERT model extracts keywords and their sequence from a given sequence of the descriptions of the test actions of the test function and the corresponding context, i.e., the test case. To do so, FunBERT employs two parallel BERT models: one for extracting local keywords and their sequence based on the sequential test action descriptions within the test function and the other for extracting global keywords and their sequence in the test case. After fusing the obtained information, the extracted keywords are concatenated as the function name according to their positions. FunBERT takes keyBert as its technical base to ensure the effectiveness of keyword extraction.

In the data pre-processing stage, each data item of the descriptions of the test actions of the test function, the descriptions of the test actions of the test case, and the labeled test function name are segmented and padded into fixed-length word embedding. Subsequently, the word embedding with position embedding is fed into two pre-trained BERT models, $BERT_{tf}$ and $BERT_{tc}$, to generate feature encoding, respectively. To maximize the utilization of semantic features in both the description of test actions of the test function and the description of test actions of the test case, FunBERT employs vector concatenation to generate the integrated semantic features denoted as Q, as defined in Eq. 6.

$$Q = BERT_{tf}\left(TF_{embedding}\right) \oplus BERT_{tc}\left(TC_{embedding}\right) \qquad (6)$$

During the training phase, the number of tags can be dynamically adjusted to fine-tune the weights of two encodings of the test function description and the test case description. Finally, FunBERT utilizes a label prediction model, primarily including a fully connected layer and Softmax. For each word in a given input vector, FunBERT computes the probability of the word belonging to each label specified in the training data set, and the label with the maximal probability is selected as the output. As shown in Eq. 7, p represents the normalized label probability, W represents weights, and b represents bias.

$$p = Softmax(W^T Q + b) \qquad (7)$$

Based on the predicted label with the highest probability for each word, it identifies whether it is a keyword and its position in a function name and ultimately assembles the function name.

2.5 *ATAG*'s Overall Process

As shown in Algorithm 1, steps 1–7 generate test functions through the action and function initialization, update test functions to improve cohesion and coupling by satisfying constraints based on a given optimization strategy, and terminate the optimization with the given criteria, i.e., once changes of CBF and Coh reach the pre-defined thresholds (τ_{cbf} and τ_{coh}). As shown in Fig. 2, when the generated TF does not satisfy the optimization constraints, i.e., it possesses lower Coh and higher CBF (top right), $ATAG$ returns to the previous state to find a better solution (bottom right). To update test functions TF obtained in the last iteration, Algorithm 1 uses the configured strategy $OPT(TF)$ to generate test functions, then either retrieve a common sequence of test actions or aggregate the test actions therein based on data flow.

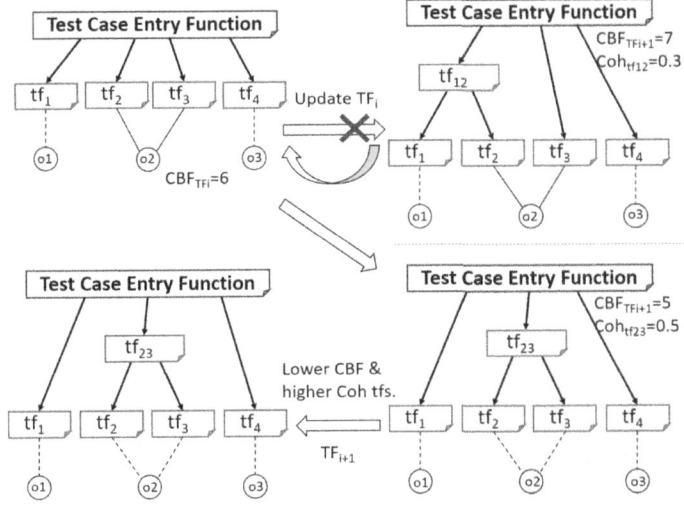

Fig. 2. One Iteration in *ATAG*. Labels *tf* and *o* represent the test function and the visited object. Objects linked by solid (dashed) lines are external (internal) objects.

Steps 8–11 generate test function name with the concatenated descriptions of the contained test actions both in test function (i.e. $tf.desc$) and the chosen contextual test case (i.e. $tc_{tf}.desc$) using FunBERT, as introduced in Sect. 2.4, where $tf.size$ and $tc.size$ are the number of test actions contained in tf and tc_{tf} respectively. Eventually, $ATAG$ returns the set of test functions TF.

Algorithm 1. *ATAG*

Input: *TC*, test case set; *OPT*, optimization strategy; ΔCBF threshold τ_{cbf}; ΔCoh threshold τ_{coh};

Output: *TA*, test architecture structuring a set of test functions *TF*;

1: Extract each test action *ta* from *TC* as the initial *tf* to add into *TF*;
2: Set iteration $i = 0$;
3: **repeat**
4:    Compute cbf_i among *tf*s and coh_i of *tf* in TF_i;
5:    Update TF_i by $TF_{i+1} \leftarrow OPT(TF_i)$;
6:    Compute cbf_{i+1} and coh_{i+1} of *tf*s in TF_{i+1};
7: **until** $|cbf_{i+1}\text{-}cbf_i| \leq \tau_{cbf}$ or $|coh_{i+1}\text{-}coh_i| \leq \tau_{coh}$
8: **for** *tf* in *TF* **do**
9:    $tc_{tf} = \text{argmax}(tf.size/tc.size)$ for any *tc* in *TC*;
10:   $tf.name = \text{FunBERT}(tf.desc, tc_{tf}.desc)$;
11: **end for**
12: **return** *TA* containing the hierarchically structured *TF*

---

3 Experiment Design

In this section, we discuss the baselines, the employed datasets, evaluation metrics, research questions, parameter settings, and experiment execution.

3.1 Baselines

ATAG primarily deals with two tasks: test architecture generation and test function name generation. For the test architecture generation, we take the practice of Huawei as the baseline for evaluating *ATAG*. The practice is that a test engineer is usually assigned a small subset of test cases to design and implement them as test functions in Ruby. Incrementally, more test cases are assigned to test engineers to achieve higher test coverage, fault detection rate, etc. Test engineers often design and implement each test case without knowing or considering the overall picture of the test suite. The practice is time-wise efficient. Therefore, we observe that most manually crafted test functions consist of only a single test action. Moreover, there are a lot of test actions realized by multiple test functions. Therefore, to compare with *ATAG*, we take the manually crafted test function set with this practice as the baseline.

There exists a lot of keyword extraction techniques, including supervised and unsupervised algorithms. To evaluate *ATAG*'s FunBERT, we employ five techniques as the baselines: the statistical-based TF-IDF [30], the graph-based TextRank [26], the RNN with CRF-based BiLSTM-CRF, and the transformer-based models GPT2 [29] and BERT, which are now commonly used mainstream keyword extraction methods applied in numerous studies [8,14].

3.2 Industrial Datasets

We utilize three industrial datasets (named DS-S, DS-M, DS-L) from our industrial partner Huawei to evaluate *ATAG*. The datasets are formed in the context of testing Huawei's Network Cloud Engine-Transport (NCE-T) product, the core enabling component of "Intelligent Optical Network", which is applied in various transmission and networking scenarios, such as backbone networks, metropolitan networks, and enterprise access over the last ten years.

Each dataset consists of a set of test cases, a terminology dictionary, and a set of manually programmed test functions. Test cases in these three datasets are used in different application scenarios, independently testing the Time-Delay Services, Synchronous Digital Hierarchy Service, and Client Business of NCE-T. The descriptive statistics of the datasets are shown in Table 1, including the number of test actions, the number of unique test actions, the number of test cases, the number of control flows, the number of data flows, the average and maximum control flow and data flow length in a test case.

From Table 1, one can observe that the three datasets vary significantly. Specifically, DS-S has the fewest test actions, test cases, control flows, and data flows, while DS-L has the most, and DS-M is in between. We can also see that the number of unique test actions is only a small portion of the total number of test actions specified in all test cases, i.e., 20.1% (=148/735), 23.7% (=473/1990) and 6% (=1220/20241) for DS-S, DS-M, and DS-L, respectively. DS-M and DS-L are comparable in terms of the average control length, maximal control length, and maximal data flow length. DS-M has the highest average data flow length.

In the terminology dictionary, there are 2,194 Chinese-English words and phrases specifically related to NCE-T terms and transport network terms.

Table 1. Descriptive Statistics of the Three Industrial Datasets

	DS-S	DS-M	DS-L
# of test actions	735	1990	20241
# of unique test actions	148	473	1220
# of test cases	60	95	1744
# of control flows	111	216	2066
# of data flows	358	961	12834
Control flow length (average)	6.62	9.21	9.79
Control flow length (maximum)	18	40	45
Data flow length (average)	5.41	9.78	6.34
Data flow length (maximum)	16	37	40

3.3   Evaluation Metrics

Regarding evaluating the overall quality of generated test architectures, we select CBF (Eq. 4) and Coh (Eq. 5) as the evaluation metrics since they are commonly used for evaluating software architectures. We also use the number of test actions, the number of internal and external objects, the depth of calls, and the number of direct calls to characterize generated test architectures. To evaluate the effectiveness of FunBERT, we adopt Precision, Recall, and F1-score.

3.4   Research Questions (RQs)

RQ1: How effective is $ATAG$ in generating test architectures compared to the baseline? *Rationale.* RQ1 aims to ensure that test architectures generated with $ATAG$ have high quality and offer advantages over the baseline. *Task.* To answer this question, we set Huawei's method as a baseline and conducted an experiment involving three different datasets of varying scales, scenarios, and design styles. We use metrics CBF and Coh to evaluate the outputs quantitatively.

RQ2: How effective are the optimization strategies of $ATAG$ for generating test architectures? *Rationale.* RQ2 is designed to investigate which of the three flow-based optimization strategies of $ATAG$ (Sect. 2.3) is the most suitable for which dataset. *Task.* We configure $ATAG$ as control flow-based $ATAG_{cf}$, data flow-based $ATAG_{df}$, and a hybrid one with control flow followed by data flow $ATAG_{cf+df}$. These configurations are applied to the three different datasets of varying characteristics, and the results are compared against the baseline.

RQ3: How do the keyword extraction methods differ in generating test function names? *Rationale.* In this RQ, we compare FunBERT with five keyword extraction techniques regarding their contributions to the overall effectiveness of $ATAG$. *Task.* We compare $ATAG$'s FunBERT with the five baselines in terms of the standard metrics of Precision, Recall, and F1-scores, and also compare the content and comprehensibility of the extracted function names.

3.5   Parameter Settings and Experiment Execution

We set thresholds τ_{cbf} and τ_{coh} being 0.01 and 0.001 based on our pilot study. For training FunBERT, we use the BERT-base-uncased model and set its dropout ratio 0.1, a commonly used configuration. Each Transformer layer has 768 hidden units, and the Feedforward Neural Network has 2048 hidden units. FunBERT is optimized by the Adaptive Moment Estimation with decoupled weight decay (AdamW) optimizer [23]. By following the popular setting of Adam, we keep the learning rate value of 3e–5. We configure the BERT model to use 12 encoder layers, corresponding to the BERT-base architecture and set the batch size to 4 for the training. We divide the data into training, validation, and test sets in a 6:2:2 ratio, commonly applied dataset splits [24].

The experiment was run on a PC with AMD Ryzen 5 5600X 6-core CPU, 16-GB RAM, and NVIDIA GeForce RTX 2060 GPU, running Windows 10 (64-bit). The algorithms, models, and experimental procedures can be found in our replication package: https://github.com/WGYbuaa/ATAG.

4   Results and Analyses

4.1   RQ1: Assessing $ATAG$'s Overall Performance in Test Architecture Generation

The descriptive statistics of test functions generated by $ATAG_{cf+df}$, $ATAG_{cf}$, and $ATAG_{df}$ for five runs on the three datasets are presented in Table 2. As

Table 2. Descriptive statistics of generated test functions and overall time cost (in seconds) by $ATAG$ and the baseline (average of five runs) - RQ1

	DS-S	DS-M	DS-L
# of test functions in the baseline	362	1578	5440
Avg. # of test functions generated by $ATAG_{cf}$	171	584	2132
Avg time cost of $ATAG_{cf}$	10	23	224
Avg. # of test functions generated by $ATAG_{df}$	212	719	5251
Avg time cost of $ATAG_{df}$	10	23	227
Avg. # of test functions generated by $ATAG_{cf+df}$	212	762	5976
Ave time cost of $ATAG_{cf+df}$	10	23	228

the table shows, the baseline contains more test functions than what the three $ATAG$ strategies generated for DS-S and DS-M. The main reason is that, in these two datasets, the baseline contains many duplicates; in other words, the two test architectures have a large space to be optimized. Dataset DS-L, the largest dataset, has many data flows (12834, Table 1), which consequently requires the generation of a large number of middle-layer test functions. We can also notice that the number of generated test functions and the required average time cost for all three strategies and the baseline consistently increase along with the growth of the dataset size (DS-S → DS-L), as expected.

Furthermore, we compare each $ATAG$ optimization strategy with the baseline in terms of the coupling and cohesion of the respective test architecture, and results are presented in Table 3. From the table, we can observe that, in terms of average coupling (Avg CBF defined in Eq. 4) among test functions, both $ATAG_{cf+df}$ and $ATAG_{df}$ outperformed the baseline consistently for all datasets. The average improvement of $ATAG_{cf+df}$ over the baseline regarding the reduction of coupling among test functions (CBF) ranges from 25.75% to 35.29%. $ATAG_{df}$'s CBF improvement over the baseline ranges from 26.51% to 41.17%. $ATAG_{df}$ performed a bit better than $ATAG_{cf+df}$ because, in some instances, test functions generated based on control flows (with $ATAG_{cf+df}$) might increase data flow dependencies.

On the other hand, $ATAG_{cf}$ outperformed the baseline for DS-S and DS-M with very slight improvement (i.e., 6.25% for DS-S and 5.88% for DS-M). One can also notice that for DS-L, $ATAG_{cf}$ negatively contributed to the average CBF (i.e., –1.51%). This is because $ATAG_{cf}$, without considering data flows, generated some test functions that visit shared data objects, etc., consequently increasing coupling. These results clearly indicate that information captured in data flows contributes much more than that captured in control flows to the performance of $ATAG$, consistently for all the three datasets. Regarding the sum of cohesion (Sum Coh) of all test functions, we can observe the same pattern as for CBF: $ATAG_{cf+df}$ performed the best among all strategies, and data flows contributed significantly to the overall cohesion of test functions.

When looking through the three datasets, the baseline performed the worst for DS-M (1.7 in Avg CBF and 243 in Sum Coh). In contrast, $ATAG_{df}$ achieved the best *Avg CBF* improvement (35.29%), and $ATAG_{cf+df}$ achieved the best *Sum Coh* improvement (49.87%) for DS-M. The main reason is that DS-M has the highest average data flow length (9.78, Table 1), which provides the opportunity for $ATAG_{cf+df}$ and $ATAG_{df}$ to explore their potential to maximize cohesion and minimize coupling. These results indicate that the data flow information between test actions and data objects mainly determines the potential of optimizing a test architecture.

Table 3. Results of comparing $ATAG_{cf+df}$ and the baseline regarding CBF and Coh of generated test architectures – RQ1.

DS[1]	Char	Baseline	$ATAG_{cf+df}$	Imp	$ATAG_{cf}$	Imp	$ATAG_{df}$	Imp
DS-S	Ave CBF	1.6	1.1	31.25%	1.5	6.25%	1.1	31.25%
	Sum Coh	395.3	516.8	30.73%	433.5	9.66%	531.4	34.42%
DS-M	Ave CBF	1.7	1.1	35.29%	1.6	5.88%	1.0	41.17%
	Sum Coh	243	364.2	49.87%	281	15.63%	346.2	42.46%
DS-L	Ave CBF	1.32	0.98	25.75%	1.34	-1.51%	0.97	26.51%
	Sum Coh	10129.7	12929.8	27.64%	10587.5	4.51%	12590.4	24.29%

DS - Dataset, Char - Characteristic, Imp - Improvement

4.2   RQ2: Comparing the Three Optimization Strategies of *ATAG*

Table 4. Characteristics (in average values) of generated test architectures with $ATAG_{cf+df}$, $ATAG_{cf}$, and $ATAG_{df}$ - RQ2.

	Characteristic	$ATAG_{cf+df}$	$ATAG_{cf}$	$ATAG_{df}$
DS-S	Avg. # of test actions	1.97	1.3	2.3
	# of external objects	1.4	1.17	1.4
	# of internal objects	0.5	0.1	0.9
	depth of calls	1.39	1.0	1.6
	# of direct calls	2.64	3.08	2.48
DS-M	Avg. # of test actions	3.6	1.3	4.0
	# of external objects	2.0	1.2	1.9
	# of internal objects	1.6	0.06	2.1
	depth of calls	1.8	1.0	1.99
	# of direct calls	2.4	2.58	2.57
DS-L	Avg. # of test actions	2.2	1.07	2.2
	# of external objects	1.5	1.06	1.4
	# of internal objects	0.7	0.01	0.8
	depth of calls	1.4	1.0	1.5
	# of direct calls	2.48	2.14	2.56

As shown in Table 4, in terms of the average number of test actions and number of internal data objects in generated test functions, test functions generated by $ATAG_{cf}$ have smaller sizes than those generated by $ATAG_{df}$ and $ATAG_{cf+df}$. This is because the strategy of $ATAG_{cf}$ intends to capture sequences of test actions shared across test cases. However, in the three datasets, various test cases were designed to cover different functionalities of the SUT and do not exhibit many opportunities for $ATAG_{cf}$ to identify shared test actions. Consequently, $ATAG_{cf}$ generates fewer and shorter test functions than $ATAG_{cf+df}$ and $ATAG_{df}$. We can also observe that shared test action sequences in test functions generated by $ATAG_{cf}$ cover two data flows on average, as they only visit approximately one external data object (e.g., 1.17 for DS-S). Moreover, the test architectures generated by $ATAG_{cf}$ for the three datasets consistently have an average depth of calls between test functions of 1, indicating that there is no shared sequence of test actions among the called test functions that can be identified to increase cohesion further.

As observed in RQ1, data flow information largely contributes to high cohesion and low coupling of generated test architectures. Since both $ATAG_{cf+df}$ and $ATAG_{df}$ benefit from data flow information, as shown in Table 3, the characteristics of the test architectures generated by them are very similar. The main difference lies in using control flow information in $ATAG_{cf+df}$. We recommend using $ATAG_{df}$ if a given set of test cases contains numerous long data flows. $ATAG_{cf}$ is optimized for test cases where data flows are limited or less complex. This is, however, rare as executing or testing a system requires interaction with or processing of data objects.

4.3 RQ3: Comparing the Keyword Extraction Methods in generating Test Function Names

To evaluate FunBERT of $ATAG$, we compare its performance with the five baselines: statistical-based TF-IDF, the graph-based TextRank, the RNN with CRF-based BiLSTM-CRF, and the transformer-based models GPT2 and BERT. Results are reported in Table 5. From the table, we can observe that unsupervised methods (i.e., TF-IDF and TextRank) consistently lag behind supervised ones (i.e., BiLSTM-CRF, GPT2, BERT, FunBERT). Among all methods, FunBERT achieved the best performance in terms of precision, recall, and F1-score.

As unsupervised methods, TF-IDF and TextRank are suitable for longer texts with sufficient semantic information rather than short texts, which is our case for test function descriptions. As a result, their performance is comparatively poorer. Thanks to the Transformer, GPT-2, BERT, and FunBERT exhibited good performance. Additionally, due to the incorporation of contextual information, precisely the descriptions of test cases, FunBERT consistently achieved the highest values among these three well-performed techniques, achieving 97.9% precision, 98.3% recall, and 98.1% F1-score, respectively.

We further selected one well-performing model from each of the two types of supervised methods (RNN with CRF-based and transformer-based) to gain in-depth insights into why FunBERT performed the best in generating test function names. Note that the two unsupervised methods are not considered due to their poor performance. From Table 6, we can observe that, in Case 1, both BiLSTM-CRF and BERT extracted "SDH" as part of the function name because the information provided by the test action itself is limited and does not clearly specify the meaning of "SDH." However, FunBERT can capture contextual information, indicating that "SDH" is one of the abbreviations related to a specific business domain and that "business" is more suitable as part of the function name ("create_business"), which is very close to the manually crafted function name ("create business"). In Case 2, contextual information is more pronounced. FunBERT not only extracted noun phrases like the other two models but also identified the key action "set" in the contextual information. This aligns better with the naming conventions commonly used by humans.

Table 5. Results of comparing FunBERT with the baselines – RQ3.

Technique	Precision (%)	Recall (%)	F1 (%)
TF-IDF	64.5	23.3	34.3
TextRank	60.5	7.0	12.5
BILSTM-CRF	76.0	75.2	75.6
GPT2	92.8	90.6	91.7
BERT	94.1	95.5	94.8
FunBERT	**97.9**	**98.3**	**98.1**

Table 6. Examplifying test function names generated by BiLSTM-CRF, BERT, and FunBERT – RQ3.

	Case 1	Case 2
Test Action	create SDH as@sdh	@bod global convergence point configuration
Contextual Information	...@sdh deletes SDH business...	...@bod sets the convergence point,...
Generated test function name		
Human Benchmark	create business	set convergence point
BiLSTM-CRF	create_SDH	convergence_point_configuration
BERT	create_SDH	global_convergence_point_configuration
FunBERT	create_business	set_ convergence_point_configuration

FunBERT exhibits outstanding performance regarding Precision, Recall, and F1-score, demonstrating its superiority in semantic feature representation and extraction compared to the baseline techniques. We also observed that test function names generated by FunBERT are well aligned with the naming conventions commonly used by test engineers, thus demonstrating the effectiveness of the proposed structural optimization in fusing contextual information.

4.4   Threats to Validity

Internal validity pertains to experimental bias and errors that can arise from the employed NLP tools in parsing natural language statements and the chosen termination criteria of the TAG algorithm. *ATAG* parses test case descriptions and extracts test actions and data objects. The precision of the extraction is paramount, as it directly influences subsequent steps. Hence, we employ the industrial-strength NLP tool spaCy. Selecting the threshold of the termination criterion also affects the quantity and quality of generated test architectures. A larger threshold may lead to premature termination, resulting in fewer test functions generated and, hence, fewer test actions, possibly insufficiently covering control and data flows. Instead, a smaller threshold could result in an overly large number of test functions, potentially impacting the overall cohesion of the test function set. The threshold set in our experiment was based on the results of the pilot study we conducted, which should have been defined by domain experts or via empirical studies. In the future, we will conduct dedicated empirical studies to devise recommendations on its selection. *External validity* refer to the generalizability of *ATAG*. For this study, we only used three datasets from the same domain. However, they are industrial datasets and, hence, representative of real-world practice. Moreover, the largest dataset contains 1744 test cases, which is considered large-scale. In the future, we plan to apply *ATAG* on more datasets for different domains. Regarding *Construct validity*, we chose CBF derived from the classical coupling metric [10] and Coh, a quantitative calculation of cohesion [13,18]. Precision, Recall, and F1-score are also representative metrics, commonly employed as a performance measure for models or algorithms [34,37]. Also, the same metrics were applied to *ATAG* and the baselines.

5   Related Work

Test architecture commonly refers to the overall structure and design of tests. In this paper, we define test architecture as a hierarchically structured list of test functions implementing a given set of test cases to achieve a specific test objective. To the best of our knowledge, we are not aware of any approach that directly generates such test architecture. There are, however, works on software architecture generation and optimization. In these works, evolutionary algorithms (e.g., Genetic Algorithms) and machine learning are often applied s [5,31]. For instance, Barnes et al. [5] proposed a data migration scenario-based approach for planning software architecture evolution – recommending architecture evolution paths based on evolution constraints and evaluations.

In terms of architecture and code refactoring, Lin et al. [20] presented an interactive, and guided recommendation tool, based on the metaphor of the GPS route navigation, which considers the current implementation as the starting point to recommend the desired high-level structure and the corresponding refactoring path. Ivkovic et al. [17] introduced a framework to suggest the most suitable refactoring action using UML models and annotating refactoring context. Ouni et al. [27] proposed a multi-objective search-based method to recommend code refactoring actions, which had been validated with industry cases. Al Dallal et al. [2] conducted a systematic literature review on code refactoring and concluded that most refactoring approaches focus on enhancing and optimizing a program in terms of its cohesion, coupling, complexity, size, and inheritance. Instead, *ATAG* focuses on generating optimized test architectures by utilizing cohesion and coupling metrics. We also observe that Large Language Models (LLMs) demonstrate their capability in handling small-scale code generation tasks [9,12,33], but the infeasibility of dealing with architectural designs and generated code often struggle to achieve high accuracy [11,16].

Currently, function name generation and extraction have two types based on the information used: source code of functions [3,4,28], and natural language descriptions [15]. Source code provides all necessary information for understanding a function's behavior; hence, it would be a sound source from which keywords can be retrieved to formulate function names. Some studies retrieve keywords by transferring source code into sequences based on the corresponding syntax trees [4,28], while some treat source code as natural text by using convolution attention neural network to extract keywords [3]. Source code-based function name generation approach works well when the names of variables and data structures carry the necessary semantic information. Therefore, code comments have become the major source of information recently to generate function names since they are written for human understanding. Gao et al. [15] defined the function name generation as a sequence-to-sequence problem and employed an RNN encoder-decoder model with attention and copying mechanism to extract representational words from function description in natural language. FunBERT considers the function name generation as a sequence-to-sequence task as well. It captures and fuses the local keywords from the literal description of a function and the global keywords from the test case by using two parallel BERT models.

6   Conclusions

This study introduces *ATAG* for generating test architectures and test function names. The test architecture generation component utilizes three optimization strategies to parse test cases, identify test actions and data objects, and construct test functions optimally in terms of the coupling among them and cohesion in each of them. For generating test function names, *ATAG* trained its own BERT model, named FunBERT, which leverages contextual information to achieve better performance as compared to state-of-the-art keyword extraction techniques, including BERT itself. The effectiveness of *ATAG* is evaluated with three industrial datasets. Results show that *ATAG* performed better than the baseline in test architecture generation, and the FunBERT model achieves much better performance than the five baselines in test function name extraction.

References

1. ISO/IEC/IEEE international standard - software and systems engineering – software testing – part 5: Keyword-driven testing. ISO/IEC/IEEE 29119-5 First edition 2016-11-15, pp. 1–69 (2016). https://doi.org/10.1109/IEEESTD.2016.7750539
2. Al Dallal, J., Abdin, A.: Empirical evaluation of the impact of object-oriented code refactoring on quality attributes: a systematic literature review. IEEE Trans. Software Eng. **44**(1), 44–69 (2017)
3. Allamanis, M., Peng, H., Sutton, C.: A convolutional attention network for extreme summarization of source code. In: International Conference on Machine Learning, pp. 2091–2100. PMLR (2016)
4. Alon, U., Zilberstein, M., Levy, O., Yahav, E.: code2vec: learning distributed representations of code. Proc. ACM Program. Lang. **3**(POPL), 1–29 (2019)
5. Barnes, J.M., Pandey, A., Garlan, D.: Automated planning for software architecture evolution. In: 2013 28th IEEE/ACM International Conference on Automated Software Engineering (ASE), pp. 213–223. IEEE (2013)
6. Briand, L.C., Daly, J.W., Wüst, J.: A unified framework for cohesion measurement in object-oriented systems. Empir. Softw. Eng. **3**, 65–117 (1998)
7. Brooks, P.A., Memon, A.M.: Automated GUI testing guided by usage profiles. In: Proceedings of the 22nd IEEE/ACM International Conference on Automated Software Engineering, pp. 333–342 (2007)
8. Campos, R., Mangaravite, V., Pasquali, A., Jorge, A., Nunes, C., Jatowt, A.: Yake! keyword extraction from single documents using multiple local features. Inf. Sci. **509**, 257–289 (2020)
9. Chen, K., Yang, Y., Chen, B., López, J.A.H., Mussbacher, G., Varró, D.: Automated domain modeling with large language models: a comparative study. In: 2023 ACM/IEEE 26th International Conference on Model Driven Engineering Languages and Systems (MODELS), pp. 162–172. IEEE (2023)
10. Chidamber, S., Kemerer, C.: A metrics suite for object oriented design. IEEE Trans. Software Eng. **20**(6), 476–493 (1994). https://doi.org/10.1109/32.295895
11. Dhar, R., Vaidhyanathan, K., Varma, V.: Can llms generate architectural design decisions?-an exploratory empirical study. arXiv preprint arXiv:2403.01709 (2024)
12. Eisenreich, T., Speth, S., Wagner, S.: From requirements to architecture: an AI-based journey to semi-automatically generate software architectures. arXiv preprint arXiv:2401.14079 (2024)
13. Figueiredo, E., Sant'Anna, C., Garcia, A., Bartolomei, T.T., Cazzola, W., Marchetto, A.: On the maintainability of aspect-oriented software: a concern-oriented measurement framework. In: 2008 12th European Conference on Software Maintenance and Reengineering, pp. 183–192 (2008). https://doi.org/10.1109/CSMR.2008.4493313
14. Firoozeh, N., Nazarenko, A., Alizon, F., Daille, B.: Keyword extraction: issues and methods. Nat. Lang. Eng. **26**(3), 259–291 (2020)
15. Gao, S., Chen, C., Xing, Z., Ma, Y., Song, W., Lin, S.W.: A neural model for method name generation from functional description. In: 2019 IEEE 26th International Conference on Software Analysis, Evolution and Reengineering (SANER), pp. 414–421. IEEE (2019)
16. Hou, X., et al.: Large language models for software engineering: a systematic literature review (2024)
17. Ivkovic, I., Kontogiannis, K.: A framework for software architecture refactoring using model transformations and semantic annotations. In: Conference on Software Maintenance and Reengineering (CSMR 2006), p. 10. IEEE (2006)

18. Jin, W., Zhong, D., Cai, Y., Kazman, R., Liu, T.: Evaluating the impact of possible dependencies on architecture-level maintainability. IEEE Trans. Software Eng. **49**(3), 1064–1085 (2023). https://doi.org/10.1109/TSE.2022.3171288
19. Karhu, K., Repo, T., Taipale, O., Smolander, K.: Empirical observations on software testing automation. In: 2009 International Conference on Software Testing Verification and Validation, pp. 201–209 (2009). https://doi.org/10.1109/ICST.2009.16
20. Lin, Y., Peng, X., Cai, Y., Dig, D., Zheng, D., Zhao, W.: Interactive and guided architectural refactoring with search-based recommendation. In: Proceedings of the 2016 24th ACM SIGSOFT International Symposium on Foundations of Software Engineering, pp. 535–546 (2016)
21. Lippert, M., Roock, S.: Refactoring in Large Software Projects: Performing Complex Restructurings Successfully. Wiley, Hoboken (2006)
22. Liu, K., Wu, J., Yang, H., Sun, Q., Wan, R.: Srtef: test function recommendation with scenarios and latent semantic for implementing stepwise test case. IEEE Trans. Reliab. **71**(2), 1127–1140 (2022). https://doi.org/10.1109/TR.2022.3164645
23. Loshchilov, I., Hutter, F.: Decoupled weight decay regularization. In: 7th International Conference on Learning Representations, ICLR 2019, New Orleans, LA, USA, 6–9 May 2019. OpenReview.net (2019), https://openreview.net/forum?id=Bkg6RiCqY7
24. Ma, S., Xing, Z., Chen, C., Chen, C., Qu, L., Li, G.: Easy-to-deploy API extraction by multi-level feature embedding and transfer learning. IEEE Trans. Software Eng. **47**(10), 2296–2311 (2019)
25. McConnell, S.: Code Complete. Pearson Education, London (2004)
26. Mihalcea, R., Tarau, P.: Textrank: Bringing order into text. In: Proceedings of the 2004 Conference on Empirical Methods in Natural Language Processing, pp. 404–411 (2004)
27. Ouni, A., Kessentini, M., Sahraoui, H., Inoue, K., Deb, K.: Multi-criteria code refactoring using search-based software engineering: an industrial case study. ACM Trans. Softw. Eng. Methodol. (TOSEM) **25**(3), 1–53 (2016)
28. Pour, M.V., Li, Z., Ma, L., Hemmati, H.: A search-based testing framework for deep neural networks of source code embedding. In: 2021 14th IEEE Conference on Software Testing, Verification and Validation (ICST), pp. 36–46 (2021). https://doi.org/10.1109/ICST49551.2021.00016
29. Radford, A., Wu, J., Child, R., Luan, D., Amodei, D., Sutskever, I., et al.: Language models are unsupervised multitask learners. OpenAI Blog **1**(8), 9 (2019)
30. Sparck Jones, K.: A statistical interpretation of term specificity and its application in retrieval. J. Doc. **28**(1), 11–21 (1972)
31. Tian, Z., Levitin, G., Zuo, M.J.: A joint reliability-redundancy optimization approach for multi-state series-parallel systems. Reliab. Eng. Syst. Saf. **94**(10), 1568–1576 (2009)
32. Williams, L., Maximilien, E., Vouk, M.: Test-driven development as a defect-reduction practice. In: 14th International Symposium on Software Reliability Engineering, 2003. ISSRE 2003, pp. 34–45 (2003). https://doi.org/10.1109/ISSRE.2003.1251029
33. Xu, X., et al.: LMPA: improving decompilation by synergy of large language model and program analysis. arXiv preprint arXiv:2306.02546 (2023)
34. Yan, A., Kim, J., Raich, R.: Forensics for adversarial machine learning through attack mapping identification. In: ICASSP 2023 - 2023 IEEE International Conference on Acoustics, Speech and Signal Processing (ICASSP), pp. 1–5 (2023). https://doi.org/10.1109/ICASSP49357.2023.10095092

35. Yuan, X., Memon, A.M.: Iterative execution-feedback model-directed GUI testing. Inf. Softw. Technol. **52**(5), 559–575 (2010)
36. Yue, T., Ali, S., Zhang, M.: RTCM: a natural language based, automated, and practical test case generation framework. In: Proceedings of the 2015 International Symposium on Software Testing and Analysis, pp. 397–408 (2015)
37. Zhao, Y., et al.: Adaptive multi-view graph convolutional network for gene ontology annotations of proteins. In: 2022 IEEE International Conference on Bioinformatics and Biomedicine (BIBM), pp. 90–93 (2022). https://doi.org/10.1109/BIBM55620.2022.9995517
38. Zhu, L., Aurum, A., Gorton, I., Jeffery, R.: Tradeoff and sensitivity analysis in software architecture evaluation using analytic hierarchy process. Software Qual. J. **13**, 357–375 (2005)

Less is More: An Empirical Study of Undersampling Techniques for Technical Debt Prediction

Gichan Lee[ID] and Scott Uk-Jin Lee[⊠][ID]

Department of Computer Science and Engineering, Hanyang University,
Ansan 15588, Republic of Korea
{fantasyopy,scottlee}@hanyang.ac.kr

Abstract. Technical Debt (TD) prediction is crucial to preventing software quality degradation and maintenance cost increase. Recent Machine Learning (ML) approaches have shown promising results in TD prediction, but the imbalanced TD datasets can have a negative impact on ML model performance. Although previous TD studies have investigated various oversampling techniques that generates minority class instances to mitigate the imbalance, potentials of undersampling techniques have not yet been thoroughly explored due to the concerns about information loss. To address this gap, we investigate the impact of undersampling on TD model performance by utilizing 17,797 classes from 25 Java open-source projects. We compare the performance of the models with different undersampling techniques and evaluate the impact of combining them with widely-used oversampling techniques. Our findings reveal that (i) undersampling can significantly improve TD model performance compared to oversampling and no resampling; (ii) the combined application of undersampling and oversampling techniques leads to a synergy of further performance improvement compared to applying each technique exclusively. Based on these results, we recommend practitioners to explore various undersampling techniques and their combinations with oversampling techniques for more effective TD prediction.

Keywords: Technical Debt · Undersampling · Class Imbalance

1 Introduction

Technical Debt (TD) is a metaphor representing sub-optimal design or implementation choices that may offer short-term benefits but can incur significant costs when making changes to the software over time [1]. If TD is not repaid and remains in various artifacts of the software, it not only reduces development efficiency but also leads to a gradual increase in maintenance costs, significantly degrading the quality of the software to be delivered in the future [2]. As a result, many tools have been developed to predict TD from the early development phases [3]. In particular, Machine Learning (ML) approaches to predict TD by utilizing the existing software repositories have been actively researched and shown high performance [4–6]. To train the models, they have constructed TD datasets by extracting various metrics or mining the developers' comments related to the existence of TD (i.e., Self-Admitted Technical Debt (SATD) [7]).

© The Author(s), under exclusive license to Springer Nature Switzerland AG 2025
G. Bai et al. (Eds.): ICECCS 2024, LNCS 14784, pp. 146–156, 2025.
https://doi.org/10.1007/978-3-031-66456-4_8

However, TD datasets can often be imbalanced due to the inherent characteristics of TD [8]. ML models fed with such datasets may have a bias towards increasing the overall accuracy by predicting most samples as the majority class (i.e., non-TD instances), rather than utilizing the features of the instances. This results in models that excel in overall accuracy but fail to predict minority class instances that practitioners are actually interested in (i.e., TD instances). Consequently, the usability and generalizability of the models are significantly reduced in real-world applications, even if it has high accuracy in experimental environments [9]. Therefore, it is essential to balance the distributions of instances in the TD dataset to mitigate the impact of biased learning due to the imbalance.

One of the widely-used techniques for imbalanced dataset is data resampling which refers to the process of balancing the distribution of instances in the dataset. It includes two main techniques that are oversampling which increases the number of minority class instances and undersampling which reduces the number of majority class instances [10]. In particular, ML-based Software Engineering (SE) research has frequently utilized SMOTE-based oversampling techniques and analyzed the effects of applying these techniques in SATD detection, code smell detection, and defect prediction [11–13]. They have demonstrated different effects of applying SMOTE-based oversampling techniques for TD-related tasks in terms of model performance. But, despite the various investigations and applications of these oversampling techniques, interest in undersampling methods has been very limited in TD-related fields. Such preference for oversampling is mainly due to the concern that undersampling techniques may result in information loss when removing samples [4, 6].

However, various undersampling techniques can offer different benefits in mitigating class imbalance depending on the specific characteristics of the dataset [9]. For example, instances of a majority class in a TD dataset may contain redundant information to their neighboring instances. Hence, removing these instances through undersampling may not result in significant information loss. Additionally, there may be instances of a majority class in close proximity to the borderline with the minority class. If these instances can be deleted selectively, the ML models can be trained to distinguish TD and non-TD instances more clearly. But, despite such potential of undersampling which may improve the performance of ML models in various ways, there is a lack of research that investigates the effects of applying undersampling techniques for ML-based TD prediction.

Furthermore, there may also be potential synergies when combining SMOTE-based oversampling techniques with undersampling techniques. For example, it is possible to perform SMOTE-based oversampling and undersampling simultaneously to selectively generate or remove instances from both majority and minority class. This approach has a possibility to yield higher performance improvements than exclusive resampling that inadvertently generate or remove the instances in bulk. Since research on ML-based TD prediction is still in its early stages [4], exploring the effectiveness of applying undersampling techniques in this area can be both timely and valuable for practitioners.

In that vein, our goal is to analyze the effects of applying different undersampling techniques to the imbalanced TD dataset consisting of 17,797 classes from 25 Java open-source projects. We evaluate and compare the performance changes in four ML models (i.e., Logistic Regression (LR), Support Vector Machine (SVM), Random Forest

(RF), XGBoost (XGB)) when applying undersampling, SMOTE-based oversampling, and both techniques to the imbalanced training data. Our results revealed that: (i) undersampling techniques significantly improved the performance of ML models in TD prediction with varying effects depending on the model type compared to SMOTE-based oversampling and no resampling; (ii) applying undersampling techniques in conjunction with oversampling techniques generates synergy that further improves the performance of the ML models compared to applying either undersampling or oversampling techniques. Our research contributions can be summarized as follows:

- The first in-depth exploration comparing the impact of various undersampling techniques on the performance of ML models for TD prediction with those of SMOTE-based oversampling techniques.
- Demonstrating the synergistic effects of combining undersampling and SMOTE-based oversampling techniques in improving the performance of ML models for TD prediction.
- Providing practitioners with the insights into the time efficient way of applying undersampling techniques and the parameter tuning for the techniques to achieve optimal performance improvements.
- An online replication package that includes the source code, dataset, and detailed implementation guidelines to enhance the reproducibility of this study [14].

2 Methodology

2.1 Research Questions

- **RQ1:** How does undersampling impact ML performance in TD prediction compared to no resampling and SMOTE-based oversampling?
- **RQ2:** What are the performance improvement effects of combining undersampling and oversampling compared to using them exclusively?

RQ1 aims to minimize the gap in the current literature by offering insights into the impact of each undersampling technique but also statistically comparing their performance with no resampling and the SMOTE-based oversampling techniques in TD prediction. The goal of RQ1 is to assist researchers and practitioners in determining whether undersampling techniques can be competitive alternatives for addressing the imbalance in TD dataset.

RQ2 explores the synergy of applying both undersampling and oversampling techniques simultaneously for TD prediction. This question assesses the joint impact of these techniques on the ML model performance and statistically compares their effectiveness with the exclusive application of oversampling or undersampling technique. The goal of RQ2 is to determine if combining undersampling techniques can further improve the performance of TD prediction models with widely-used oversampling techniques in literature.

Figure 1 illustrates an overview of the experimental design in this study. First, we selected a TD dataset to apply various data resampling techniques and train ML models. Second, we chose different resampling techniques based on the distribution of instances within the TD dataset. Third, we trained representative linear, non-linear, and ensemble

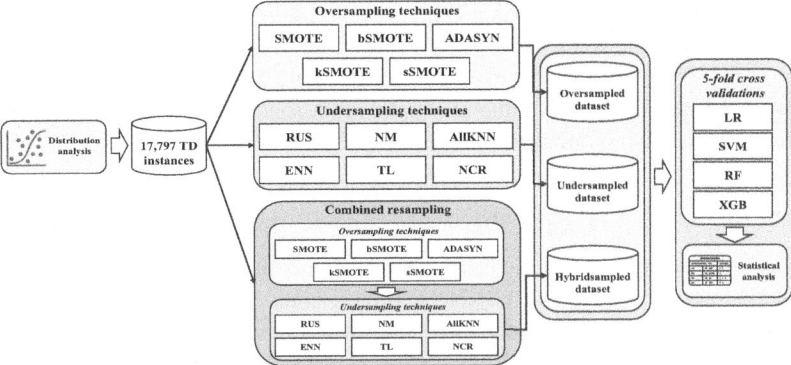

Fig. 1. Overview of the Experiment in this Study

ML algorithms that demonstrated high performance in relevant research and evaluated the selected performance metrics after training and validation of ML models. Finally, we answered the RQs by statistically comparing the performance metrics of ML models across each resampled dataset.

2.2 Context Selection

In this study, we utilized the TD dataset created by Tsoukalas et al. [4] for ML-based TD prediction. They constructed the dataset using an empirical benchmark designed to assess the level of agreement among TD prediction tools for 25 Java open-source projects [15]. The instances were labeled as 1 for high-TD modules and 0 for the rest of the modules, and various metrics were employed to derive the features representing these instances in the dataset. They included 18 features consisting of class-level source code metrics, refactoring-related metrics, densities of duplicates and comment lines, and module-level Git-related metrics in the dataset. After the preprocessing to remove features with missing values and outliers, the TD dataset comprises 16,604 non-TD instances and 1,193 TD instances. Moreover, they validated the effectiveness of the dataset by revealing the discriminative power of each feature. To the best of our knowledge, no dataset for ML-based TD prediction other than the ones created by Tsoukalas et al. has been assembled with a diverse range of features, undergone preprocessing to reduce the possible bias during the training process, and analyzed the relationships between the features and target instances. Consequently, we selected their dataset to investigate the effects of undersampling techniques.

Next, we applied Principal Component Analysis (PCA) to reduce the 18 dimensions in our dataset, retaining three principal components that accounted for 96% of the original variance of dataset. Upon examining the distribution of the dataset, we observed that a large number of non-TD instances are densely and overlappingly distributed within a narrow space, and they also overlap with TD instances near the borderline. Additionally, most of the TD instances form clusters near the majority class, while the remaining TD instances are sparsely distributed and significantly separated from other TD instances. Therefore, we hypothesized that the performance of ML models may decrease even if they form decision boundaries around the observed borderline due to the overlapping instances in the vicinity. We inferred that it is essential to consider not only areas near the borderline but also regions where instances are sparsely distributed. Taking the insights gained from the data distribution into account, we thoroughly selected the oversampling and undersampling techniques.

2.3 Data Resampling Techniques

In this study, we exclusively applied six undersampling techniques (i.e., RUS, NearMiss (NM), All k-Nearest Neighbors (AllKNN), Tomek Links (TL), Neighborhood Cleaning Rule (NCR)) and five SMOTE-based oversampling techniques (i.e., SMOTE, bSMOTE, ADASYN, KMeansSMOTE (kSMOTE), sSMOTE) to our TD dataset. Furthermore, we applied both resampling techniques simultaneously with a total of 30 combinations. We applied the resampling techniques to the TD dataset and to obtain each resampled dataset. When comparing the performance improvement effect of applying undersampling and oversampling techniques exclusively, we did not perform parameter tuning for them to ensure a fair comparison. In particular, since undersampling techniques have unique parameters based on their instance selection algorithms, tuning these parameters of each undersampling technique could potentially distort the performance comparison results. Therefore, we controlled the possible biases may be caused by the parameter tuning.

On the other hand, SMOTE and its variants share the common parameter which is sampling rate. Based on the previous research that shows performance improvements of ML models for defect prediction vary depending on the sampling rate [16], we tuned the sampling rate for SMOTE and its variants when applying them together with under-sampling techniques. Through this, we investigated whether applying various under-sampling techniques to cases that showed the best performance improvements with each SMOTE-based oversampling method can result in further performance improvements. If undersampling brings additional improvements to the best performance obtained by tuned SMOTE-based oversampling, it indicates that there exists synergy. Lastly, since the ratio of majority class to minority class in our dataset was approximately 15:1, we adjusted the sampling rate during oversampling for the ratio to change from 10:1 to 1:1 (i.e., sampling rates ranging 0.1 to 1.0). To implement the resampling techniques, we utilized the imbalanced-learn package [21] and included the relevant scripts in our online replication package [14]. Through this process, we collected a total of 356 differently resampled TD datasets.

2.4 Training and Validation of ML Algorithms

We selected LR, SVM, RF, and XGB as representative linear, non-linear, and ensemble-based ML algorithms to train on the resampled TD datasets. These algorithms have demonstrated high performance in studies and for the dataset we used [4, 11, 13, 17, 18]. Our goal for selecting ML algorithms was to determine whether the performance improvement effect of undersampling could be significant even with the ML algorithms that have already demonstrated excellent performance in the related studies. Further-more, LR, SVM, RF, and XGB have different sensitivities to imbalanced datasets due to their unique configurations. We aimed to enhance the generalizability of this study, which investigates the effects of applying undersampling techniques, by utilizing these algorithms. Similar to the reason for not tuning undersampling techniques, we compared the performance of the models without tuning hyper-parameters because we believe that the improvement in model performance produced by tuning varies among different models and needs to be controlled. We assessed the performance of ML models using 5-fold cross-validation and repeating the process 100 times [22]. When partitioning the 5 folds for evaluating the ML algorithms, we employed stratified sampling to preserve an equal ratio of majority and minority classes. We applied resampling techniques solely to the four training folds and assessed the performance of the ML models using the remaining fold.

When evaluating the performance of ML models for TD prediction, we selected widely-used performance metric, F1-score. After obtaining average F1-scores for models, we also conducted the Wilcoxon signed-rank test [19] and calculated Cliff's delta effect sizes [20] to assess the statistical difference of the F1-scores before and after applying resampling. We then denoted the difference between the F1-score sets with a 'D' when the p-value is less than 0.05 to indicate a significant difference between F1-scores and with an 'S' otherwise to represent no significant difference. We also followed the common approach based on the Cliff's delta effect size to categorize the magnitude of the difference. We labeled the absolute value of delta as negligible (N) when it is less than or equal to 0.147, small (S) when it is greater than 0.147 and less than or equal to 0.33, medium (M) when it is greater than 0.33 and less than or equal to 0.474, and large (L) when it is greater than 0.474.

3 Results

3.1 RQ1: How Does Undersampling Impact ML Performance in TD Prediction Compared to no Resampling and SMOTE-Based Oversampling?

Table 1 presents the average F1-scores of ML models for the original dataset without any resampling and for resampled datasets obtained by applying various undersampling techniques. The F1-scores are presented with different formatting where we marked improved average F1-scores in blue and decreased ones in red, with the highest and lowest scores emphasized in bold. These average F1-scores are calculated from a total of 500 folds for each dataset. The results in Table 1 show that all undersampling techniques except RUS contributed to statistically significant performance improvements of ML models. The performance improvement effect of each undersampling technique varied between the algorithms. NM resulted in the most significant improvement for the LR model, ENN had the same effect on the SVM and XGB models, and both TL and NCR showed similar result for the RF model in equal magnitude. Furthermore, all undersampling techniques except RUS substantially improved the performance of LR and SVM but only ENN, TL, and NCR could slightly improve the performance of RF and XGB. We infer that this phenomenon is due to RF and XGB having inherently robust characteristics against imbalanced datasets, resulting in the relatively weaker effects from undersampling techniques. On the other hand, we also observed application of undersampling techniques causing performance degradation which have been a concern in related studies. RUS significantly reduced the performance of all models, and NM also decreased the performance of XGB even though it significantly improved the performance of LR. This along with the fact that all undersampling techniques demonstrated different degrees of performance improvement depending on the algorithm.

Table 1. ML Models with Undersampling vs. Those with No Resampling.

Under-sampling Techniques	LR	SVM	RF	XGB	(Wilcoxon signed-rank, Cliff's delta)			
Original	0.594	0.62	0.705	0.702	-	-	-	-
RUS	**0.571**	**0.583**	**0.57**	**0.575**	**(D[a], L[b])**	**(D, L)**	**(D, L)**	**(D, L)**
NM	0.67[c]	0.667	0.7	0.628	**(D, L)**	(D, L)	(D, N)	(D, L)
AllKNN	0.647	**0.69**	0.71	0.708	(D, L)	**(D, L)**	(D, N)	(D, N)
ENN	0.645	0.688	0.716	**0.714**	(D, L)	(D, L)	(D, S)	**(D, S)**
TL	0.601	0.635	**0.717**	0.711	(D, S)	(D, S)	**(D, S)**	(D, S)
NCR	0.644	0.686	**0.717**	0.71	(D, L)	(D, L)	**(D, S)**	(D, S)

[a.]D: different distribution, S: same distribution, [b.]L: large, M: medium, S: small, N: negligible,

[c.]Blue: increase, Red: decrease, Bold: best & worst

Based on these observations, we concluded that practitioners should investigate not only SMOTE-based oversampling techniques, but also various undersampling techniques to improve ML models for TD prediction. Furthermore, we recommend that practitioners should compare different undersampling techniques to determine which would be appropriate depending on the type of ML model they use. Considering all the results and their implications, we inferred that finding the most effective undersampling technique suitable for all possible circumstances in TD prediction is unnecessary. Instead, a comparative investigation process is required to determine the appropriate undersampling technique for given circumstances. Also, we compared the best results from applying SMOTE-based oversampling techniques without any tuning with the best results from applying undersampling techniques, in terms of their performance improvements as shown in Table 2. When the sampling rate is not tuned, it is evident that there are more cases where SMOTE-based oversampling techniques significantly deteriorate the performance of ML models when compared with applying undersampling techniques. Furthermore, the best F1-scores obtained through SMOTE-based oversampling techniques are notably lower than those obtained through undersampling techniques. Even after considering that the sampling rate was not tuned for fair comparison, SMOTE-based oversampling techniques showed significantly weaker performance improvements compared to untuned undersampling techniques. This clearly highlights the competitiveness of undersampling techniques for imbalanced TD datasets. Based on these results, we answered to RQ1:

Table 2. ML Models with Oversampling vs. Those with NO Resampling and Best Undersampling.

Oversampling Techniques	LR	SVM	RF	XGB	(Wilcoxon signed-rank, Cliff's delta)			
Original	0.594	0.62	0.705	0.702	-	-	-	-
SMOTE	0.607	0.603	0.706	0.71	(D, S)	(D, M)	(S, N)	(D, S)
bSMOTE	0.578	0.579	0.695	**0.706**	(D, M)	(D, L)	(D, S)	**(D, N)**
ADASYN	**0.563**	**0.556**	**0.691**	0.707	**(D, L)**	**(D, L)**	**(D, M)**	(D, N)
kSMOTE	**0.633**	**0.643**	**0.706**	0.707	**(D, L)**	**(D, M)**	(S, N)	(D, N)
sSMOTE	0.604	0.607	0.703	0.709	(D, S)	(D, M)	(S, N)	(D, S)
Best over.	0.633	0.643	0.706	0.71	**(D, L)**	**(D, L)**	**(D, S)**	**(D, N)**
Best under.	**0.67**	**0.688**	**0.717**	**0.714**				

3.2   RQ2: What are the Performance Improvement Effects of Combining Undersampling and Oversampling Compared to Using Them Exclusively?

Table 3 presents the average F1-scores and corresponding sampling rates when the best performance is achieved for each tuned SMOTE-based oversampling technique. Furthermore, it also shows the average F1-scores, sampling rates, and the corresponding undersampling techniques when the best performance is achieved by combining each SMOTE-based oversampling technique with each undersampling technique.

As a result, SMOTE-based oversampling techniques with the sampling rate tuned resulted in a significantly higher performance improvement of all ML models, while those without the tuning often decreased performance of ML models as illustrated in Table 2. Meanwhile, from the results of applying undersampling techniques combined together with oversampling techniques, we made two key observations. Firstly, there was no performance degradation for the ML models. Secondly, the combined applications of both resampling techniques often resulted in higher performance improvements with statistical significance when compared to the best performance attained by each SMOTE-based oversampling technique with the tuning. This clearly demonstrates the synergy produced from combining undersampling and SMOTE-based oversampling techniques. Moreover, we calculated the statistical difference between the best performance achieved for each model when undersampling or oversampling is applied exclusively and when they are applied together.

Answer to RQ1:
Undersampling techniques significantly improve the performance of ML models for TD prediction when compared to no resampling and SMOTE-based oversampling. However, since the effectiveness of each undersampling technique varies by each model, it requires a tailored approach to determine the best undersampling technique for various circumstances.

Table 3. ML Models with Tuned Oversampling vs. Those with Combined Resampling

Over.+ under.	LR	SVM	RF	XGB	(Wilcoxon signed-rank, Cliff's delta)			
Original	0.594	0.62	0.705	0.702	-	-	-	-
[a]t-SMOTE	0.681 (0.2)[b]	0.701 (0.2)	0.729 (0.2)	0.711 (0.3)	(D, L)	(D, L)	(D, L)	(D, S)
t-SMOTE +best under.	0.690 (0.1, NM)[c]	0.706 (0.1, NM)	0.729 (0.1, TL)	0.712 (0.1, TL)	(D, S)	(D, N)	(D, N)	(S, N)
t-bSMOTE	0.680 (0.2)	0.685 (0.2)	0.724 (0.2)	0.710 (0.2)	(D, L)	(D, L)	(D, M)	(D, S)
t-bSMOTE +best under.	0.690[d] (0.1, NM)	0.699 (0.1, NM)	0.727 (0.1, TL)	0.711 (0.1, NCR)	(D, S)	(D, M)	(D, N)	**(S, N)**
t-ADASYN	0.682 (0.2)	0.691 (0.2)	0.725 (0.1)	0.710 (0.9)	(D, L)	(D, L)	(D, M)	(D, S)
t-ADASYN +best under.	0.689 (0.1, NM)	0.699 (0.1, NM)	0.727 (0.1, TL)	0.710 (0.1, NCR)	(D, L)	(D, S)	(D, N)	**(S, N)**
t-kSMOTE	0.633 (1.0)	0.643 (1.0)	0.707 (0.1)	0.707 (0.2)	(D, L)	(D, M)	(D, N)	(D, N)
t-kSMOTE +best under.	0.659 (0.1, NM)	0.691 (0.4, RUS)	0.722 (0.6, RUS)	0.714 (0.1, ENN)	(D, L)	(D, L)	(D, M)	**(D, S)**
t-sSMOTE	0.682 (0.1)	0.696 (0.2)	0.729 (0.2)	0.712 (0.3)	(D, L)	(D, L)	(D, L)	(D, S)
t-sSMOTE +best under.	0.691 (0.1, NM)	0.703 (0.1, NM)	**0.729** (0.2, TL)	0.713 (0.2, TL)	(D, L)	(D, S)	**(S, N)**	(S, N)
Best exclusive.	0.682 (t-ADASYN)	0.701 (t-SMOTE)	0.729 (t-SMOTE)	0.714 (ENN)	**(D, L)**	**(D, N)**	(S, N)	(S, N)
Best combined.	**0.691**	**0.706**	0.729	0.714				

[a]·t: tuned, [b]·(best sampling rate), [c]·(best sampling rate, best undersampling technique),
[d]·Green: statistically significant increase by over+under.

On the one hand, we found that combined application of both undersampling and oversampling techniques further improved the best performance of the LR model where we also observed a slight further improvement for SVM although the effect size was negligible. On the other hand, there was no significant difference between the performance improvement effects for RF and XGB models. Despite the higher performance improvement effects shown by undersampling techniques compared to SMOTE-based oversampling techniques without tuning, we observed no further performance improvement for these two models. This led us to infer that the effect of tuning the sampling rates of SMOTE-based oversampling techniques had reached the maximum performance improvement that RF and XGB could achieve through resampling. We conjectured that since RF and XGB are inherently robust against imbalanced datasets, there might be a limit to the performance improvement that can be achieved through resampling. Based on these results and its overall implications, we conclude that practitioners can pursue further performance improvement for ML models in TD prediction by not only using SMOTE-based oversampling but also experimenting with various undersampling techniques to be combined.

> **Answer to RQ2:**
> Combining undersampling and oversampling techniques creates synergy, which improves the performance of ML models in TD prediction beyond the use of each technique alone. Practitioners should explore not only SMOTE-based oversampling but also undersampling to be combined with for the possible further performance improvements of ML models.

4 Conclusion

In this study, we conducted a comprehensive investigation into the impact of various undersampling techniques on the performance of ML models for TD prediction. We compared the effects of undersampling techniques with SMOTE-based oversampling techniques on the performance improvement of TD prediction models. Furthermore, we investigated whether synergetic effects can be achieved when applying undersampling and oversampling techniques simultaneously.

As a result, our study demonstrated that undersampling techniques can significantly improve the performance of ML models for TD prediction when compared to applying no resampling or SMOTE-based oversampling techniques. Additionally, the combination of undersampling and SMOTE-based oversampling techniques exhibits synergistic effects with no performance degradation and often better results than applying each technique exclusively. Such research results suggests that practitioners should consider exploring various undersampling techniques to achieve better performance of ML models in their TD prediction tasks. Furthermore, we recommend that practitioners should also investigate and compare the different combinations of undersampling and oversampling techniques to obtain the best performance of ML models for identifying TD. In our future research, we aim to explore how unique parameters of various undersampling techniques impact the performance improvement effects.

Acknowledgments. This work was partly supported by Institute of Information & communications Technology Planning & Evaluation (IITP) grant funded by the Korea government(MSIT) (No.RS-2022-00155885, Artificial Intelligence Convergence Innovation Human Resources Development (Hanyang University ERICA)) and the National Research Foundation of Korea(NRF) grant funded by the Korea government(MSIT) (NRF-2023R1A2C1006390).

Disclosure of Interests. The authors have no competing interests to declare that are relevant to the content of this article.

References

1. Cunningham, W.: The WyCash portfolio management system. In: ACM SIGPLAN OOPS Messenger pp. 29–30 (1992)
2. Avgeriou, P., Kruchten, P., Ozkaya, I., Seaman, C.: Managing technical debt in software engineering (dagstuhl seminar 16162). In: Dagstuhl Reports, Wadern, Germany: Schloss Dagstuhl-Leibniz- Zentrum fuer Informatik (2016)
3. Alves, N.S.R., et al.: Prediction and management of technical debt: a systematic mapping study. Inf. Softw. Technol. 100–121 (2016)
4. Tsoukalas, D., et al.: Machine learning for technical debt prediction. In: IEEE Transactions on Software Engineering, pp. 4892–4906 (2021)
5. Albuquerque, D., et al.: Managing Technical Debt Using Intelligent Techniques - A Systematic Mapping Study. IEEE Trans. Softw. Eng. (2022)
6. Sala, I., Tommasel, A., Arcelli Fontana, F.: DebtHunter: a machine learning-based approach for detecting self-admitted technical debt. In: Evaluation and Assessment in Software Engineering, pp. 278–283 (2021)
7. Potdar, A., Shihab, E.: An exploratory study on self-admitted technical debt. In: 2014 IEEE International Conference on Software Maintenance and Evolution, pp. 91–100. IEEE (2014)
8. Thabtah, F., et al.: Data imbalance in classification: experimental evaluation. Inf. Sci. 429–441 (2020)
9. He, H., Garcia, E.A.: Learning from imbalanced data. IEEE Trans. Knowl. Data Eng. 21(9), 1263–1284 (2009)
10. Estabrooks, A., Jo, T., Japkowicz, N.: A multiple resampling method for learning from imbalanced data sets. In: Comput. Intell. 18–36 (2004)
11. Sridharan, M., Mantyla, M, Rantala, L., Claes, M.: Data balancing improves self-admitted technical debt detection. In: 2021 IEEE/ACM 18th International Conference on Mining Software Repositories (MSR). IEEE (2021)
12. Pecorelli, F., Di Nucci, D., De Roover, C., De Lucia, A.: A large empirical assessment of the role of data balancing in machine-learning-based code smell detection. J. Syst. Softw. (2020)
13. Bennin, K.E., Keung, J.W., Monden, A.: On the relative value of data resampling approaches for software defect prediction. Empir. Softw. Eng. 602–636 (2019)
14. Online replication package of "Less is More: An Empirical Study of Undersampling Techniques for Technical Debt Prediction" (2024). https://figshare.com/s/8e8cee8eb5e96c7 2a22c
15. Amanatidis, T., Mittas, N., Moschou, A., Chatzigeorgiou, A., Ampatzoglou, A., Angelis, L.: Evaluating the agreement among technical debt measurement tools: building an empirical benchmark of technical debt liabilities. Empir. Softw. Eng. 25(5), 4161–4204 (2020)
16. Agrawal, A., Menzies, T.: Is "better data" better than "better data miners"? on the benefits of tuning SMOTE for defect prediction. In: Proceedings of the 40th International Conference on Software engineering, pp. 1050–1061 (2018)
17. Li, Y., Soliman, M., Avgeriou, P.: Identifying self-admitted technical debt in issue tracking systems using machine learning. Empir. Softw.Eng. (2022)
18. Tomek, I.: Two modifications of CNN. IEEE Trans. Syst. Man Cybern. SMC-6(11), 769–772 (1976)
19. Woolson, R.F.: Wilcoxon signed-rank test. In: Wiley Encyclopedia of Clinical Trials, pp. 1–3 (2007)
20. Cliff, N.: Dominance statistics: ordinal analyses to answer ordinal questions. Psychol. Bull. (1993)
21. Lemaître, G., Nogueira, F., Aridas, C.K.: Imbalanced-learn: a python toolbox to tackle the curse of imbalanced datasets in machine learning. J. Mach. Learn. Res. 559–563 (2017)
22. Davis, J., Goodrich, M.: The relationship between Precision-Recall and ROC curves. In: Proceedings of the 23rd International Conference on Machine Learning, pp. 233–240 (2006)

Smart Contract

Modeling and Verification of Solidity Smart Contracts with the B Method

Fayçal Baba[1,2]($\boxtimes$)(ID), Amel Mammar[3](ID), Marc Frappier[2](ID),
and Régine Laleau[1](ID)

[1] Université Paris-Est Créteil, 94000 Créteil, France
faycal.baba.kouba@gmail.com, marc.frappier@usherbrooke.ca
[2] Université de Sherbrooke, Sherbrooke, QC J1K2R1, Canada
laleau@u-pec.fr
[3] Telecom SudParis, 91000 Courcouronnes, France
amel.mammar@telecom-sudparis.eu

Abstract. Smart contracts written using the SOLIDITY programming language of the ETHEREUM platform are well-known to be subject to bugs and vulnerabilities, which already have led to the loss of millions of dollars worth of assets. Since smart contract code cannot be updated to patch security flaws, reasoning about smart contract correctness to ensure the absence of vulnerabilities before their deployment is of the utmost importance. In this paper, we present a formal approach for generating correct smart contracts from B specification that verify safety properties. Our approach consists of two phases: first a smart contract and its properties are specified and verified in B, then a set of rules we defined are applied to generate the correct smart contract code in SOLIDITY. The approach is implemented in a tool that can generate SOLIDITY contract from a proven B project. The whole approach is demonstrated by a case study on the ERC-20 (ETHEREUM Request for Comments 20) Wrapped Ether (WETH) contract, which is abstractly specified in B, with invariants stating correctness properties, modeled checked with PROB for temporal properties, implemented in B0, proven correct, and automatically translated into a Solidity contract.

Keywords: Smart contract · Solidity · Blockchain · Formal modeling and verification · B Method · Refinement

1 Introduction

In recent years, blockchain technology has garnered significant attention from both industry and academia due to its potential for decentralization, security, and transparency. Initially introduced through the creation of Bitcoin, blockchain serves as a distributed and immutable ledger that securely records

This work was supported by the ANR projet DISCONT, Public Safety Canada and NSERC.

transactions in a transparent manner [13]. Unlike conventional systems reliant on trusted intermediaries such as banks, blockchain employs consensus among nodes to validate and record cryptocurrency transactions. Ethereum, often regarded as blockchain 2.0, stands out as a pioneering platform that not only features a cryptocurrency called ether, but also enables the execution of self-executing programs known as smart contracts [6].

Smart contracts are self-executing agreements that can be executed on a blockchain when predetermined conditions are met. For example, they can automate agreements among multiple parties, enabling one party to trigger a state change that, in turn, allows another party to access ether stored within the smart contract. To create smart contracts, developers primarily utilize SOLIDITY [9], a Turing-complete high-level language similar to JavaScript. This SOLIDITY code is then compiled into Bytecode, a lower-level code, and deployed on the Ethereum blockchain.

Like any software program, smart contracts are susceptible to bugs and vulnerabilities, with potentially grave financial implications. Compounded by the immutability of blockchain, rectifying post-deployment issues becomes impossible. Ethereum has encountered various attacks and vulnerabilities resulting in significant losses, such as the infamous DAO smart contract attack [5], where an attacker exploited a vulnerability (later known as the reentrancy vulnerability) within the contract code, stealing over $50 million worth of ether.

Given these challenges, ensuring the safety and security of smart contracts before deployment to the blockchain is of paramount importance. This paper presents an approach to the formal modeling and verification of SOLIDITY smart contracts using the formal B method [1]. We adopt a classical approach, commencing with the definition of an abstract B model. This abstraction allows for a simpler expression and verification of smart contract properties and functionalities. The verification process leverages a range of techniques, including specification animation, theorem proving and model checking within the existing B Method toolset. Following successful verification, the B refinement process is employed to transform the abstract specification into a more concrete representation, aligning closely with SOLIDITY data and control structures. Then, through established translation rules, the entire model is subsequently translated into SOLIDITY code. The translation rules are implemented in a tool we developed using *ANTLR*[1] and *Jetbrains MPS*[2].

The remainder of this paper is organized as follows: In Sect. 2, we begin with an overview of the literature related to our contribution in the verification of SOLIDITY smart contracts. Then, in Sect. 3, we present the basic concepts of the B method and the SOLIDITY language. Section 4 describes our approach to modeling, verifying and implementing SOLIDITY smart contracts, up to the automatic translation of B0 code (the implementation language of B) into SOLIDITY. The entire approach is illustrated by a case study concerning the WETH token smart contract. Section 5 concludes and presents some future work.

---

[1] https://www.ANTLR.org/.
[2] https://www.jetbrains.com/mps/concepts/.

2   Related Work

Several works precede ours in the analysis, verification, and modeling of SOLID-
ITY smart contracts, reflecting diverse approaches, techniques, and tools. In this
section, we provide an overview of relevant works categorized by the formal
modeling and verification techniques they employ: i) SOLIDITY code/bytecode
analysis, ii) SOLIDITY code verification, and iii) correct-by-construction SOLID-
ITY code generation.

In the first category, researchers focus on detecting common code patterns
leading to known vulnerabilities (e.g., reentrancy, transaction order dependence,
timestamp dependence) using static analysis techniques that examine program
artifacts (e.g., Control Flow Graph, Abstract Syntax Tree). Prominent examples include the works [8,12,16], which rely on Symbolic Execution. The tool
smartcheck in [19] identifies problematic patterns using XPath queries, and fuzz
testing in [11,15,21] detects various SOLIDITY vulnerabilities. These approaches
may not explore all program paths, potentially leading to false negatives. Purely
syntactic techniques might not fully account for operational semantics or execution environment intricacies [20], compromising the analysis soundness and
completeness.

In the second category, researchers translate SOLIDITY code into formal models and verify properties using various tools and techniques. Model checking was
used in [14], where the authors use *NUSMV* to model blockchain applications
and to verify temporal properties in CTL. In [10], Coloured Petri Nets are used
to model and verify smart contract specifications using LTL properties. Although
model checking is employed in various approaches, its limitation are well known,
and approaches that rely on it, such as [10,14], often face the problem of state
explosion when dealing with complex or infinite models [20]. F* is used in [4] to
formally verifies models of smart contracts, analyzing correctness and gas consumption with tools like Solidity* and EVM*. For now, this work only considers
a restricted subset of SOLIDITY. It is also worth noting that F* program verification faces challenges including complexity, and steep learning curve. In [22], an
EVENT-B model generated from SOLIDITY code is used to formalize smart contracts, the model is then simulated and verified using the tool PROB. Notably,
the translation process only considers subset of the SOLIDITY language.

In the third category, the focus is on creating correct-by-construct smart contracts. For instance, in [18], smart contracts are modeled as FSMs and verified
with LTL properties, later generating SOLIDITY code. The authors of [3] employ
EVENT-B and present a top-down verification and refinement approach that theoretically generates SOLIDITY programs from EVENT-B models, although they
did not provide a concrete implementation or practical demonstration of their
solution. The authors of [17] further explores this approach, introducing EB2Sol,
a tool for modeling, verifying, and generating SOLIDITY smart contracts. It is
worth noting that this work considers a restricted subset of SOLIDITY and can
only generate sequential programs consisting solely of assignments statements,
without support for conditional (if) or iterative (while) statements.

Our work falls within the third category. We use the B method [1] to model and verify SOLIDITY smart contracts. The B method supports animation and model checking (LTL and CTL) using PROB[3], and theorem proving using Atelier B[4]. These tools have an industrial strength and have been routinely used in safety critical industrial projects for over 25 years [7]. These tools allow one to express and verify complex models and properties. Developers can perform comprehensive checks to ensure contract correctness, logical consistency, and avoid potential deadlocks. Our approach encompasses a broader subset of the SOLIDITY language, facilitating the modeling and verification process compared to [17], which does not cover conditional (if) or iterative (while) statements.

3   Background

In this section, we present the main concepts of the B method and the SOLIDITY language that are relevant for this paper.

3.1   The B Method

The B method [1] is a formal method for modeling and verifying software systems. It is based on set theory and first-order logic, allowing for the creation of precise and unambiguous system descriptions. At the most abstract level, the system is represented by an abstract machine that includes both the structural aspect, describing system states, and the behavioral aspect, which describes state evolution.

The structural aspect may define enumerated or abstract user types in the 'SETS' clause. It can also declare variables (resp. constants) in the 'VARI-ABLES' (resp. 'CONSTANTS') clause, with type specifications in the 'PROP-ERTIES' clause for constants and the 'INVARIANT' clause for variables, using first order predicates. Additional properties, like safety or integrity of the machine state, can be specified in the 'INVARIANT' clause.

The behavioral aspect comprises the 'INITIALISATION' clause, which assigns initial values to variables, and the 'OPERATIONS' clause, used to describe operations that can modify variable values through simultaneous and possibly nondeterministic substitutions. An abstract machine can reference other machines through various clauses, with different access rights. The relevant clauses for our purposes are the 'SEES' and 'INCLUDES' clauses. The 'SEES' clause provides read-only access, whereas 'INCLUDES' grants additional access, including the ability to invoke operations. Proof obligations must be discharged to verify invariant preservation through the execution of operations in the abstract model.

One of the strengths of the B method is its refinement process that consists in defining successive refinements in a B model development, each of them adding

---

[3] https://prob.hhu.de.

[4] https://www.atelierb.eu/.

more detail and reducing abstraction of the initial abstract machine. The final stage, B0 implementation, represents a concrete model that can be automatically translated into C or Ada, and is subject to constraints (e.g., it can only use concrete data types and deterministic substitutions), ensuring compatibility with programming languages like C or ADA (and in our case, SOLIDITY). An implementation component largely shares the same clauses as an abstract machine, with some differences. 'VARIABLES' (resp. 'CONSTANTS') are replaced by 'CONCRETE_VARIABLES' (resp. 'CONCRETE_CONSTANTS'), and an additional 'VALUES' clause is used to assign values to implementation constants. The implementation component can use 'SEES' to reference an abstract machine, and the 'INCLUDES' clause is replaced by the 'IMPORTS' clause, which disallows read access on variables. The refinement process generates proof obligations that must be fulfilled to ensure that the implementation component preserves the properties of the abstract model.

Table 1 presents the B elements used in our approach to specify and generate SOLIDITY programs. x and E denote a variable and an expression respectively, T and T' denote *substitutions*, P, G, I are predicates, S and S' are sets. In the B method, a substitution is a construct that allows for the specification of operations.

3.2 Smart Contract and Solidity

The ETHEREUM blockchain defines two primary types of accounts: External Owned Accounts (EOAs) and Smart Contract Accounts (SCAs). EOAs are under the control of blockchain users, each associated with a unique address and a private key for initiating transactions. In contrast, SCAs are governed by the immutable code of deployed smart contracts. Both account types can receive, hold, and send ether. However, only EOAs can initiate transactions, which can be simple ether transfers or a call to a smart contract function which can transfer ether among accounts and also invoke functions of other smart contracts.

SOLIDITY is the most popular programming language for ETHEREUM smart contracts. SOLIDITY is a high-level object-oriented language, based on C++ and JavaScript and is statically typed. The language supports inheritance, libraries and complex user-defined types among other features [9]. To be deployed on the ETHEREUM computing platform EVM (ETHEREUM Virtual Machine), SOLIDITY code needs to be compiled into Bytecode. Once the Bytecode is deployed on the platform, a new smart contract is created and its state (i.e., its variables and balance of ether) is recorded on the blockchain. In SOLIDITY, aside from the constructor which is executed once at the contract deployment, all other functions can serve as entry points. A function can be called by users or other smart contracts, and can read and update the smart contract state. To ensure that the execution of a function terminates, the initiator of a call pays a fee for its execution, measured in units of *gas*.

In our approach, we consider a subset of SOLIDITY which contains the main components of the language and is general enough to express useful contracts:

Table 1. Subset of relevant B elements

B element	B syntax
Enumerated set	$<enum\_name> = \{Id_1, Id_2,..,Id_n\}$
Total function type	$S \rightarrow S'$
Total injective function	$S \rightarrowtail S'$
Array type	$(n..n') \rightarrow S$
Inclusion	$S' \subseteq S$
Belongs	$x \in S$
Lambda Expression	$\lambda\, x.\ (P \mid E)$
Variable declaration	**VARIABLES** v **INVARIANT** $v \in S$
Constant declaration	**CONSTANTS** c **PROPERTIES** $c \in S$ **VALUES** $c = E$
B structure type variable declaration	$x \in \mathbf{struct}(m_1 \in S, \ldots, m_n \in S')$
Assignment substitution	$x := E$
Simultaneous substitution	$T\|T'$
Sequential substitution	$T; T'$
Conditional statement	**IF** P **THEN** T **ELSE** T' **END**
Precondition substitution	**PRE** G **THEN** T **END**
While substitution	**WHILE** P **DO** T **INVARIANT** I **VARIANT** E **END**
Var statement : Declares local variable(s) in an operation body	**VAR** $Identifier^+$ **IN** T **END**

- State variables and constants definitions
- Enumeration type declaration
- Structure type declaration
- Constructor declaration
- Function declaration

We consider the following SOLIDITY types: *int* (integer), *uint* (unsigned integer), *bool* (boolean), *address*, *bytes*, *structtype*, *mapping* (A mapping is a set of pairs (*key, value*)), enumeration type, and *arrays* type. There are types not supported yet, including the *contract* type, the *function* type, and the *fixedpointnumbers* type.

The type *address payable*, an extension of the type *address*, offers additional methods to a smart contract to transfer ether to a variable of *address payable* type. The keyword *payable* can also be added to a function signature and denotes that the function can receive ether from the account of the caller of the function.

A SOLIDITY function is always invoked and executed within a context that contains the invocation message (i.e., the function called with its parameters), denoted by *msg*, and the last block of the blockchain, denoted by *block*. These elements are stored in the global name space of SOLIDITY. In our subset, we consider the following elements of this global name space:

- *msg.sender*: denotes the caller of a function, which can be a user or another smart contract.
- *msg.value*: denotes the value of ether sent with the message.
- *block.timestamp*: denotes the current timestamp in the blockchain.
- *address(this)*: refers to the address of the current smart contract denoted by *this*.
- *x.balance*: returns the balance of ether of address x.

In a function body, we can find standard programming statements (Assignment, If and While statements). Additionally, to initiate ether transaction, we consider the SOLIDITY function *a.transfer(m)*, which sends an amount m of ether from the contract balance to address a. SOLIDITY also provides mechanisms for handling conditions and errors inside a smart contract function. Function *revert()* stops contract execution and reverts state changes to its value before the call to the function. Function *require(c)* is a conditional revert; it evaluates a condition c; if c is true, then the function proceeds normally, otherwise, a revert is executed.

Listing 1.1 shows a simple code of a SOLIDITY contract that represents an electronic wallet. The constructor function sets the owner of the wallet as the sender of the message *msg.sender* when the contract is created. The deposit function allows the owner of the wallet to deposit funds into the wallet, thanks to its keyword *payable* which suffices to denote contract balance update, and thus its body is empty. The withdraw function allows the owner of the wallet to withdraw an amount of money, provided that the owner is the sender of the message and that the amount requested does not exceed the balance of the wallet.

```
1    contract SimpleWallet {
2        address payable owner;
3        constructor() payable{owner = msg.sender;}
4        function deposit() public payable {}
5        function withdraw(uint _amount) public {
6            require(msg.sender == owner);
7            require(address(this).balance >= _amount);
8            owner.transfer(_amount);}
9    }
```

Listing 1.1. Example of a SOLIDITY program

4   Modeling and Verification of Solidity Smart Contracts

In our approach, illustrated in Fig. 1, a SOLIDITY smart contract is modeled as a B project, where the components of the smart contract are modeled first as a

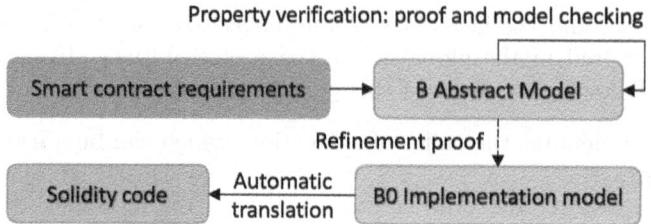

Fig. 1. Process for modeling and verifying a SOLIDITY smart contract

B abstract machine, and contract specific properties are expressed and verified using the B method toolset. The abstract model is then refined to obtain a B0 implementation. This refined model bridges the gap between the abstract B representation and the specific data and control structures of the SOLIDITY programming language. The final step of our approach is the generation of a SOLIDITY implementation using a set of transformation rules that accurately transforms the refined B0 model. The implementation of these rules is achieved through a dedicated tool that uses *ANTLR* for parsing B specifications and *JetBrains MPS* for model-driven development and code generation from B0 to SOLIDITY. The generated SOLIDITY code inherits the correctness and reliability established through the B method formal verification process, providing a solid base ensuring the correctness of a contract.

The following sections present a more detailed explanation of our approach, illustrated by a case study concerning the WETH Token smart contract.

4.1   Case Study Presentation

Wrapped ether (WETH) is a standardized digital representation of ETHEREUM native cryptocurrency, ether, implemented as an ERC-20 token standard[5]. This tokenization process enhances compatibility and ease of use within various ETHEREUM smart contracts. A WETH contract allows users to deposit ether in a contract. The users can then exchange ether (represented as WETH tokens) with each other, without the need of actually transferring ether between their ETHEREUM accounts. The contract keeps track of the balance of each user, called the WETH balance, in a similar way as a bank allows its customers to exchange money between their bank accounts, without actually moving money around. A user can withdraw ether from his WETH account. Thus, transfers between WETH accounts incur lower ETHEREUM transfer **gas** fees and offer faster processing times.

The main components of the WETH smart contract are:

- The *account* variable: variable of type *mapping*; keeps track of the WETH balance of all addresses.

---

[5] https://ethereum.org/en/developers/docs/standards/tokens/erc-20/.

- The *allowance* variable: variable of type *mapping*; keeps track of allowances; an allowance is when an address a allows an address b to transfer an amount of WETH from a account.
- The *deposit* function: deposit ether in a user account.
- The *withdraw* function: withdraw ether from a user account.
- The *transfer* function: transfer WETH to another account.
- The *approve* function: a allows b to transfer WETH up to a certain limit c (i.e., his allowance) from a account.
- The *transferFrom* function: b transfers from a account to an account c, up to an amount previously allowed to him by a.

To further demonstrate the adaptability and practicality of our formal verification approach, we propose an extension to this contract, where the owner has implemented a new behavior: a reward system for early adopters. Specifically, the contract owner offers one WETH token as a reward to each of the first *100* users who successfully accumulate a total balance of *100* tokens in their accounts. The purpose of this new behavior is to demonstrate the ease and flexibility of implementing functionalities that rely on data control structures such as *IF-THEN-ELSE* and *WHILE LOOP* within our formal verification framework using the B method. This demonstration serves as a direct comparison to alternative approaches, notably those employing EVENT-B, where the implementation of such behaviors have not been addressed. To implement this behavior, a new function *rewardTopDepositors* is added. This function can only be executed by the manager of the smart contract, and only once the hundred depositors are selected. The manager must also deposit an amount of ether equal to the amount that is distributed to each of the hundred rewarded depositor.

4.2   The B Model Architecture of SOLIDITY Smart Contracts

Types. Firstly, we define the mapping between SOLIDITY types and B sets. Among the SOLIDITY types defined in Sect. 3.2, some are directly supported in B: $uint \equiv NAT$ or $NAT1$, $int \equiv INT$, SOLIDITY *arrays* $\equiv$ B *arrays* (defined as a function), *bool* (boolean) $\equiv BOOL$, and SOLIDITY structure type as B structure type. A SOLIDITY enumeration type is modeled as a B enumeration set.

For the types not supported in B, we define the following rules: The SOLIDITY types *address*, *bytes*, and *string* are modeled as B sets, and defined in a machine called *Solidity_ Types*. While *bytes* and *string* are modeled as abstract sets, *address* is modeled as an enumerated set containing the values *THIS*, representing the modeled smart contract address value, and *addr_ 0*, representing the *null* address value. The *Solidity_ Types* also defines the constant *USERS*, which represents a subset of the set *address* excluding the values *THIS* and *addr_ 0*. The machine *Solidity_ Types* will be referenced by other components using *SEES* links. Lastly, for each variable of type *mapping*, a B abstract machine is defined and included/imported in the B abstract/implementation model of a smart contract. Such a machine represents the mapping variable, and it defines a variable of type total function, whose domain represents the keys, and range represents the values. The machine also contains the standard operations of the *mapping* type for inserting and reading pairs.

Contract Function Specification. The constructor of a contract is modeled as the initialisation substitution of a B machine, and functions are modeled as operations. Implicit function parameters *msg.sender, msg. value,* and *block.timestamp* are modeled as input parameters of a B operation. These parameters are appropriately typed and included among other potential input parameters in the B precondition substitution. To specify the behavior of a contract function, we use a B substitution **IF** C **THEN** T **END**. Condition C is the condition under which the contract function successfully terminates. Substitution T defines the state changes. In the translation process to *Solidity*, we add an **ELSE** *revert()* **END** statement within this top-level **IF** statement to ensure that the state is reverted to its value before the call when C is not satisfied. Thus, a contract function $f(T_1\ p_1, \ldots, T_n\ p_n)$ is specified in the following form:

$f(msg\_sender, msg\_value, p_1, \ldots, p_n) =$
PRE $msg\_sender \in USERS \land msg\_value \in NAT1 \land$
$\quad p_1 \in T_1 \land \ldots \land p_n \in T_n$ **THEN**
$\quad$ **IF** C **THEN** T **END**
END

Note that in the B world, this is known as a *defensive* operation specification style, which differs from the traditional *offensive* style of the form $f(\boldsymbol{p}) = $ **PRE** *Typing Condition of* $\boldsymbol{p} \land C$ **THEN** S **END**. In B, when an operation f is called, the calling machine must prove that C is satisfied, thus preventing from having to check C in the implementation of f. In our case, there is no other machine that calls our B specification of a contract. Thus, we must establish with the specification of f what happens when C is not satisfied. At the abstract level, when C is not satisfied on a call, an implicit **SKIP** substitution is executed, which leaves the state unchanged. We model this in the implementation by using a call to the *revert* function.

ETHEREUM **Platform Modeling.** To model the *transfer* and *x.balance* SOLID- ITY platform functions, which can be used in a number of machines (or implementations) in a B project, we have introduced an abstract machine named *Platform*. This machine can be included (resp. imported) in abstract machines (resp. implementations) of the B project, leveraging B specification modularization. Within *Platform*, the solidity expression *x.balance* is defined by the variable *balanceOf*, a total function associating each address with its ether balance. *Platform* also defines the *transfer* operation with three input parameters: *sender* address, *receiver* address, and transfer *amount*, along with two essential preconditions. The preconditions validate (1) the existence and distinction of sender and receiver addresses and (2) that the sender balance is greater or equal to the transfer amount. Additionally, it also defines the *get_balanceOf(x)* operation, which retrieves an address ether balance.

The B project modeling the WETH smart contract contains initially 3 machines: (1) The *Solidity_ Types* machine, (2) the *Platform* machine and (3) the abstract machine representing the WETH smart contract. To model the two variables *account* and *allowance* of type `mapping`, two additional machines *account* and *allowance* are created and included in the abstract model of the WETH smart contract. Each of these two machines contains a variable of type *total function*, named (resp.) *accountOf* and *allowanceOf*, where *accountOf* is typed as $ADDRESS \rightarrow NAT$, and *allowanceOf* is typed as $ADDRESS \rightarrow (ADDRESS \rightarrow NAT)$. Each machine also contains two operations; one to insert new values: *set_ accountOf* and *set_ allowanceOf*, and another to read values: *get_ accountOf* and *get_ allowanceOf*. We use naming rules *get_* ⟨variable_name⟩ and *set_* ⟨variable_name⟩ to define mapping machines operations that read (resp. insert) a pair (k, v) in the mapping. Similarly, the operation *set_ accountOf_ abstract* inserts a set of pairs (k, v) instead of a single pair.

B Abstract Specification of the WETH Contract. We specify the operations *deposit, withdraw, transfer, approve,* and *transferFrom*. Each operation changes the state of the machine by modifying the values of the variables *account* and *allowance*. They also define conditions for their execution expressed as a B *IF* substitution that encompasses the body of the operation. For example, one of the conditions to trigger the *withdraw* operation is that the sender (input parameter *msg_ sender*) account must own an amount of tokens greater than or equal to the amount being withdrawn. Additionally, the B method imposes the definition of well-definedness conditions. These conditions ensure, for example, that when an arithmetic expression is used, it must be proven that the operands and the result belong to the defined domain of the operator in B.

To implement the rewarding system, three additional variables have been introduced. (1) *depositors*: This is a subset of the address set, containing users who qualify for the reward system. Users are added to this subset if their total token balance exceeds one hundred. (2) *manager*: This variable represents the contract manager address. (3) *donated*: A Boolean indicating whether the reward has been distributed or not. The *rewardTopDepositors* operation can be invoked by the manager once the specified conditions are met. This operation distributes rewards among the address values stored in set *depositors*.

Listing 1.2 shows an excerpt of the abstract machine WETH

```
1 MACHINE Weth
2 SEES Solidity_Types
3 INCLUDES Platform, account, allowance
4 CONSTANTS threshhold
5 PROPERTIES threshold ∈ NAT
6 VARIABLES manager, depositors, donated
7 INVARIANT
8    depositors ⊆ ADDRESS ∧
9    manager ∈ USERS ∧
10   donated ∈ BOOL ∧
```

```
11   balanceOf(THIS) ≥
12     (Σ(ct).(ct ∈ dom(accountOf)| accountOf(ct))) ∧
13   card(depositors) ≤ threshold
14 INITIALISATION
15     ...
16 OPERATIONS
17 deposit(msg_sender, msg_value) =
18 PRE
19   msg_sender ∈ USERS ∧ msg_value ∈ NAT1
20 THEN
21   IF balanceOf(msg_sender) - msg_value ∈ NAT ∧
22     accountOf(msg_sender) + msg_value ∈ NAT ∧
23     balanceOf(THIS) + msg_value ∈ NAT
24   THEN
25     transfer(msg_sender, THIS, msg_value) ||
26     set_accountOf(msg_sender, accountOf(msg_sender)+
           msg_value) ||
27     IF accountOf(msg_sender) + msg_value
28       ≥ threshold ∧
29       msg_sender ∉ depositors ∧
30       card(depositors) < threshold
31     THEN
32       depositors := depositors ∪ {msg_sender}
33 END  END END
34 ;
35 rewardTopDepositors(msg_sender, msg_value) =
36 PRE
37   msg_sender ∈ USERS ∧ msg_value ∈ NAT
38 THEN
39   IF msg_value = threshold ∧
40     msg_sender = manager ∧
41     card(depositors) = threshold ∧
42     donated = FALSE ∧
43     balanceOf(THIS) + msg_value ∈ NAT ∧
44     balanceOf(manager) - msg_value ∈ NAT ∧
45     ∀xx.(xx ∈ depositors => accountOf(xx) + 1 ∈ NAT)
46   THEN
47     transfer(manager, THIS, msg_value) ||
48     set_accountOf_abstract(λ xx. (xx ∈ depositors |
           accountOf(xx) + 1)) ||
49     donated := TRUE
50 END  END;
51   ...
52 END
```

Listing 1.2. Excerpt of abstract machine WETH_AM

4.3   Verification of B Models of Smart Contracts

Formal verification is an important step of our approach. It ensures that the B model captures all the desired behavioral aspects and satisfies its safety properties ensuring its correctness, and this using the formal verification techniques of the B method. For our case study, it means to make sure that the abstract modeling of the WETH token captures all the desired behaviors such as token issuance, transfers, and ownership management. For this purpose, the ATELIER B theorem prover can be used to discharge proof obligations associated with invariant preservation. The PROB model checker can be used to animate/model check B model. It can exhibit some missing preconditions related to operation calls or discover problems, such as invariant violations or deadlocks. PROB also allows for the verification of temporal properties expressed as LTL or CTL formulae.

In our model, a property we want to verify is that the ether balance of the model (represented by *balanceOf(THIS)*) remains greater or equal to the total supply of tokens (sum of all values in the mapping variable *accountOf*) during conversions between ether cryptocurrencies and WETH Tokens:

$$\text{balanceOf(THIS)} \geq \sum_{ct \in \text{dom(accountOf)}} \text{accountOf(ct)}$$

Proving the abstract B machines with ATELIER B generates 121 proof obligations, of which 111 are automatically proved by the ATELIER B provers, and 10 had to be proved interactively.

Another property that we checked is that the contract doesn't reach a deadlock state, in which case users would be prevented from depositing and withdrawing their ether. This property cannot be verified using ATELIER B; we use instead the PROB model checker. We can also use PROB to simulate the model. The simulation starts by defining an initial state. Then, we can execute the operations if their preconditions are satisfied, and validate whether the model behavior is consistent with the requirements.

Temporal properties can also be expressed and verified using PROB. For instance, we can verify that at all times, if an address has a token balance greater than 0, then the operations *withdraw* and *transferTo* are enabled. This property is formally expressed as the following PROB LTL formula:

$$G(\{\exists(x).(x \in dom(accountOf) \wedge accountOf(x) > 0))\}$$
$$\Rightarrow e(withdraw) \wedge e(transferTo))$$

where the curly brackets {...} are used to express a B predicate, and $e(Op)$ denotes that the operation Op is enabled.

Furthermore, beyond the scope of contract-specific properties verification, our approach also implements precautionary measure to address the reentrancy attack. This type of attack occurs when a *victim contract* initiates a call or ether transfer to a *malicious contract*, which subsequently triggers repeated and recursive calls back to a function within the *victim contract* before the initial call completes, which may cause the different invocations to interact in destructive ways with the *victim contract*. To preempt such detrimental behavior, our strategy involves the modeling and translation of the ether exchange function, specifically by opting to employ the SOLIDITY *transfer* function. Notably, the *transfer* function is adept at mitigating the risk of reentrancy due to its inherently limited allocation of *gas*. This allocation, sufficient for only a single call, effectively blocks potential recursive invocations and protects the contract against reentrancy-based exploits.

4.4   Refinement of the B Models

The B refinement process is used to gradually transform an abstract specification into a more concrete one, closer to SOLIDITY data and control structures. Each step of the process generates proof obligations, that need to be discharged to ensure the preservation of properties.

For our case study, we have developed an implementation machine, shown in Listing 1.3, that refines the B abstract model of the WETH token. This implementation machine, named WETH_i, imports the machines: *Platform, account*, and *allowance*. Within WETH_i, we redefine the operations established in the abstract model. These re-definitions must adhere to our current B translatable subset, which includes the set of expressions and conditions of the B implementation language subsets, along with the assignment, the conditional, the loop and the *VAR-IN* instructions.

At the implementation level, variables and constants of the abstract model are replaced with new concrete ones, and a gluing invariant is added to define a relationship between the abstract variables/constants and their concrete counterparts. For example, the set *depositors* of the abstract machine is replaced by two variables: (1) array *depositors_i* which contains the depositors; (2) variable *index* of type NAT, which denotes the number of depositors currently stored in array *depositors_i*. Consequently, the invariant '*index* = card(*depositors*)' (line 16) equates the value of the implementation variable *index* to the size of the set *depositors* in the abstract machine, and the invariant '*depositors_i*[0..*index* − 1] = *depositors*' (line 17) defines the relation between the content of this set *depositors* and the range of the concrete array *depositors_i*. Variable *depositedOver100*, declared in machine *depositedOver100*, is a mapping from *address* to *boolean*; it serves as an indicator to track whether a qualifying user *address* has already been added to the *depositors_i* array, preventing duplicate entries. It avoids looping over array *depositors_i* in operation *deposit* to find out if a depositor has already been added. It implements the condition of line 29 in Listing 1.2. Since it is not possible to loop over a SOLIDITY mapping, the array *depositors_i* is still needed in function *rewardTopDepositors* to loop over depositors to transfer funds to them. This example shows how consistency between dependent variables can be proved in the implementation. Here, variables *depositors_i* and *depositedOver100* contain the same information (lines 17 and 19), and the implementation invariant ensures that it is consistent with the abstract variable *depositors* (line 19).

Some expressions of the B abstract machine language cannot be used in B0, like the condition $\forall x.x \in$ depositors $\Rightarrow$ accountOf$(x) + 1 \in NAT$ in the abstract operation *rewardTopDepositors*, because it uses a universal quantifier $\forall$. They must be implemented using concrete constructs like the loop from lines 43 to 54 of Listing 1.3.

```
1 IMPLEMENTATION B_weth_i
2 REFINES B_weth
3 SEES Solidity_Types
4 IMPORTS Platform, account, allowance, depositedOver100
5 CONCRETE_CONSTANTS threshold_i
6 PROPERTIES threshold_i ∈ NAT ∧ threshold_i = threshold
7 VALUES threshold_i = 100
8 CONCRETE_VARIABLES manager_i, depositors_i, index, donated_i
9 INVARIANT
10 // TYPING INV
11 index ∈ NAT ∧ index>=0 ∧
12 donated_i ∈ BOOL ∧
13 depositors_i ∈ 0..threshold_i → ADDRESS ∧
14 manager_i ∈ USERS ∧ manager_i = manager ∧ donated_i = donated ∧
15 // GLUING INV
16 index = card(depositors) ∧
17 depositors_i[0..index-1] = depositors ∧
18 (0..index-1) <| depositors_i : 0..index-1 ↣ depositors ∧
19 depositedOver_100~[{TRUE}] = depositors
20 INITIALISATION
21     index := 0;
22     depositors_i := (0..threshold_i) * {addr_0};
23     ...
24 OPERATIONS
25 ...
26 rewardTopDepositors(msg_sender, msg_value) =
27 BEGIN
28  VAR thisBalance, managerBalance IN
29    thisBalance <-- get_balanceOf(THIS);
30    managerBalance <-- get_balanceOf(manager_i);
31    IF msg_value = threshold_i ∧
32    msg_sender = manager_i ∧
33    index = threshold_i ∧
34    donated_i = FALSE ∧
35    thisBalance + msg_value ≤ MAXINT ∧
36    managerBalance - msg_value ≥ 0
37    THEN
38    VAR jj, safe IN
39    //* jj ∈ NAT
40    //* safe ∈ BOOL
41    jj := 0;
42    safe := TRUE;
43    WHILE jj < index ∧ safe = TRUE DO
44      VAR depositorBalance IN
45        depositorBalance <-- get_accountOf(depositors_i(jj));
46        safe := bool(depositorBalance + 1 ≤ MAXINT);
47        jj := jj+ 1
48      END
```

```
49      INVARIANT 0≤index ∧ jj≤index ∧ jj≥0 ∧
50       safe = bool(∀xx.(xx ∈ ran((0..jj-1) <| depositors_i) =>
            accountOf(xx) + 1 ∈ NAT)) ∧
51       donated_i= FALSE  ∧
52       ∀xx.(xx ∈ ran((0..jj-2) <| depositors_i) => accountOf(xx) +
            1 ∈ NAT)
53      VARIANT index - jj
54     END;
55     IF (safe=TRUE) THEN
56      transfer(msg_sender, THIS, msg_value);
57      donated_i := TRUE;
58      VAR ii, depositorBalance IN
59      //* ii ∈ NAT
60      //* depositorBalance ∈ NAT
61      ii := 0;
62      WHILE ii < index DO
63       depositorBalance <-- get_accountOf(depositors_i(ii));
64       set_accountOf(depositors_i (ii), depositorBalance + 1);
65       ii := ii+ 1
66      INVARIANT ii=threshold_i or ii∈ dom(depositors_i) ∧
67       accountOf =
68       accountOf$0<+(λxx. (xx ∈ depositors_i[0..(ii-1)] |
            accountOf$0(xx) + 1)) ∧
69       threshold_i = threshold ∧
70       donated_i = TRUE ∧ safe= TRUE ∧
71       depositors_i[0..(ii-1)]<∈ depositors ∧
72       jj=index ∧
73       ∀xx.(xx ∈ ran((ii+1..index-1) <| depositors_i) => accountOf
            (xx) + 1 ∈ NAT)
74      VARIANT index - ii
75 END END END END END END END
76 ...
77 END
```

Listing 1.3. Excerpt of the Implementation component WETH_IM

4.5 B to Solidity Translation

The final part of our approach is the transformation of the B project into SOLID-
ITY code. For that purpose, we defined a set of translation rules, which have
been implemented in a dedicated tool created with *JetBrains MPS*. The translation rules take into account both the implementation component and the initial
abstract model, since some necessary information are inherited by refinement, and
they must be retrieved from the abstract machine, or some of its refinements. For
instance, in the abstract model, the precondition substitution '*PRE G THEN T*'
of an operation serves to define typing predicates of operation parameters. This
substitution is discarded in the implementation component because the caller of
an operation (i.e., another B machine) must prove that the precondition of the
called operation is satisfied when it is called. Thus, the translation rules retrieve
the typing predicates from the abstract model to generate input parameters of
functions. Other substitutions of operations (i.e. assignment, conditional and loop

substitutions) are directly translated from the implementation component into their equivalent SOLIDITY statement. Table 2 shows an excerpt of the translation rules. The other B0 constructs are easily translated into their equivalent SOLIDITY constructs (assignment, if-then-else, loop, variable declaration).

Table 2. Excerpt of the Translation Rules B to SOLIDITY.

B construct	Solidity Language
Implementation M	**contract** $M\{\ \}$
CONCRETE_CONSTANTS c **PROPERTIES** $c \in S$ **VALUES** $c := E$	$[\![S]\!]$ **constant** $c = E$
CONCRETE_VARIABLES v **INVARIANT** $v \in S$	$[\![S]\!]$ v;
INITIALISATION T	**constructor**() $\{\ [\![T]\!]\ \}$
op_name (input_params) $=$ **BEGIN** T **END**	**function** fun_name (input_params) **public** $\{\ [\![T]\!]\ \}$

The SOLIDITY *payable* keyword is not explicitly represented in the B architecture; it is generated during the translation. If an operation receives ether by calling the operation *transfer* of machine *Platform*, the equivalent function generated in the translation will have the keyword *payable* added in its signature. Also, variables that are used in the *to* parameter of a transfer operation call are casted as *address payable*. For example, if the B model initiates a transfer of an amount m to an address variable *dst*, the variable will be translated with a quick cast to the type *address payable*: *payable(dst).transfer(m)* in order to use the *transfer* function of SOLIDITY.

Any imported machine within the implementation component, with the exception of *Platform*, corresponds to a *mapping* variable and is translated as such in SOLIDITY. The types of both the domain and range of the mapping contained within the machine are translated to their respective SOLIDITY types, as described in Sect. 4.2. Listing 1.4 shows an excerpt of the SOLIDITY code generated.

```
1 contract Weth
2 {
3 mapping (address => uint) private accountOf;
4 mapping (address => mapping (address => uint)) private
     allowanceOf;
5 mapping (address => bool) private depositedOver_100;
6 address[] private depositors;
7 ...
```

```
 8 constructor (){ manager = msg.sender; donated = false;
     index = 0;}
 9 function deposit ( ) payable public {
10   uint senderAccount = accountOf[msg.sender];
11   uint senderBalance = msg.sender.balance;
12   uint thisBalance = address(this).balance;
13   if ( thisBalance+msg.value<=type(uint).max &&
14       senderAccount+msg.value<=type(uint).max &&
15       senderBalance >= msg.value)
16   {
17     accountOf[msg.sender] = senderAccount+msg.value;
18     bool distinct = depositedOver_100[msg.sender];
19     if ( senderAccount+msg.value>=threshold&&distinct==
         false&&index<threshold ){
20       depositors.push(msg.sender);
21       depositedOver_100[msg.sender] = true;
22       index = index+1;
23     }
24   }
25   else { revert();}
26 }
27 ...
28 }
```

Listing 1.4. Excerpt of SOLIDITY code generated from the B WETH token model
project

The Solidity code generated may include conditions within the main *IF* clause
and other code snippets that are implicitly checked by the ETHEREUM virtual
machine. In the B abstract machine, these conditions are needed to prove the
preservation of the machine invariant, to ensure safety properties. They must
be preserved by the B0 implementation and thus appear in the B0 code. For
instance, verifying the sender cryptocurrency balance for an Ether transfer and
preventing overflow or underflow in integer calculations are implicitly checked
in ETHEREUM. As an example, consider line 15 in Listing 1.4, which checks
that the sender balance is greater than or equal to the amount sent. This line is
superfluous because the ETHEREUM platform disallows transfers with insufficient
funds. These conditions and statements could be eliminated using optimization
rules, which we have not done yet.

Currently, our tool lacks a built-in type checker to evaluate the types of
certain components in a B implementation when their types are not explicitly
defined. To address this limitation, we have introduced a temporary rule in which
we explicitly define the types of these components by adding a comment with the
following syntax: //* variable_name : type. This approach is a temporary
solution. We are actively working on the development of a comprehensive type
checker for future tool enhancements.

We can as a final step manually check that the same scenarios executed
on the B model, using PROB, and the generated SOLIDITY program give the

same traces. We use Remix IDE[6], an open source web application that offers an environment for simulating the ETHEREUM blockchain, and used for the development and testing of SOLIDITY smart contracts. Through Remix, we can compile, deploy the generated SOLIDITY code, and simulate execution scenarios of smart contracts. The translation tool and the B proven project associated to this case study can be found in [2]. The implementation generated 131 proof obligations, of which 46 had to be proved interactively.

5   Conclusion and Future Work

This paper proposes a comprehensive approach to develop formally verified SOLIDITY smart contracts using the B formal method. We propose a specific B modeling for a SOLIDITY smart contract taking into account a subset of SOLIDITY and EVM domain variables and functions. We adopt a classical approach, starting from an abstract model to specify the behavior of a contract and its safety properties that needs to be preserved during its execution. The model can be verified, animated and validated using the whole range of tools that support the B method. Finally, after refinement of the abstract machine, the model can be translated into SOLIDITY using the rules we defined and implemented in a tool using *Jetbrains MPS*. The correctness of our translation rules from B0 to SOLIDITY is straightforward to establish, since we are using simple constructs and simple types whose semantics are the same in B0 and in SOLIDITY. In future works, we plan to extend our framework in multiple directions: first by expanding the SOLIDITY subset considered, for instance, function calls between contracts, and use our approach in the other direction, that is, generate a B abstract machine from a SOLIDITY program to prove properties about it.

References

1. Abrial, J.: The B-Book - Assigning Programs to Meanings. Cambridge University Press, Cambridge (1996)
2. B2sol translation tool. https://github.com/ICECCS2024/B2SolPrototype
3. Banach, R.: Verification-led smart contracts. In: Bracciali, A., Clark, J., Pintore, F., Rønne, P.B., Sala, M. (eds.) FC 2019. LNCS, vol. 11599, pp. 106–121. Springer, Cham (2020). https://doi.org/10.1007/978-3-030-43725-1_9
4. Bhargavan, K., et al.: Formal verification of smart contracts: short paper (2016). https://doi.org/10.1145/2993600.2993611
5. Buterin, V.: Critical update re: Dao vulnerability. https://blog.ethereum.org/2016/06/17/critical-update-re-dao-vulnerability/
6. Buterin, V.: Ethereum: a next-generation smart contract and decentralized application platform (2014). https://github.com/ethereum/wiki/wiki/White-Paper
7. Butler, M., et al.: The first twenty-five years of industrial use of the B-method. In: ter Beek, M.H., Ničković, D. (eds.) FMICS 2020. LNCS, vol. 12327, pp. 189–209. Springer, Cham (2020). https://doi.org/10.1007/978-3-030-58298-2_8

---

[6] https://remix.ethereum.org/.

8. Chen, T., et al.: Under-optimized smart contracts devour your money. In: IEEE 24th International Conference on Software Analysis, Evolution and Reengineering, SANER 2017 (2017). https://doi.org/10.1109/SANER.2017.7884650

9. Ethereum: Solidity documentation. release 0.7.0. https://docs.soliditylang.org/en/v0.7.0/

10. Garfatta, I., et al.: Model checking of vulnerabilities in smart contracts: a solidity-to-CPN approach. https://doi.org/10.1145/3477314.3507309

11. Grieco, G., et al.: Echidna: effective, usable, and fast fuzzing for smart contracts. https://doi.org/10.1145/3395363.3404366

12. Luu, L., et al.: Making smart contracts smarter. In: Proceedings of the 2016 ACM SIGSAC Conference on Computer and Communications Security (2016). https://doi.org/10.1145/2976749.2978309

13. Nakamoto, S.: Bitcoin: a peer-to-peer electronic cash system (2008). https://bitcoin.org/bitcoin.pdf

14. Nehai, Z., et al.: Model-checking of smart contracts. In: IEEE International Conference on Internet of Things 2018 (2018). https://doi.org/10.1109/Cybermatics_2018.2018.00185

15. Nguyen, T.D., et al.: sFuzz: an efficient adaptive fuzzer for solidity smart contracts. https://doi.org/10.1145/3377811.3380334

16. Nikolic, I., et al.: Finding the greedy, prodigal, and suicidal contracts at scale. In: 34th Annual Computer Security Applications Conference, ACSAC. https://doi.org/10.1145/3274694.3274743

17. Singh, N.K., et al.: Formal verification and code generation for solidity smart contracts. In: Distributed Computing to Blockchain. https://doi.org/10.1109/COMPSAC51774.2021.00183

18. Suvorov, D., Ulyantsev, V.: Smart contract design meets state machine synthesis: case studies

19. Tikhomirov, S., et al.: Smartcheck: static analysis of ethereum smart contracts. https://doi.org/10.1145/3194113.3194115

20. Tolmach, P., et al.: A survey of smart contract formal specification and verification (2022). https://doi.org/10.1145/3464421

21. Torres, C.F., et al.: Confuzzius: a data dependency-aware hybrid fuzzer for smart contract. https://doi.org/10.1109/EUROSP51992.2021.00018

22. Zhu, J., et al.: Formal simulation and verification of solidity contracts in Event-B. In: IEEE 45th Annual COMPSAC 2021 (2021). https://doi.org/10.1109/COMPSAC51774.2021.00183

Template-Based Smart Contract Verification: A Case Study on Maritime Transportation Domain

Xufeng Zhao[1,3], Qiuyang Wei[1,2], Xue-Yang Zhu[1,3(✉)]📖, and Wenhui Zhang[1,3]

[1] Key Laboratory of System Software (Chinese Academy of Sciences) and State Key Laboratory of Computer Science, Institute of Software, Chinese Academy of Sciences, Beijing, China
{zhaoxf,weiqy,zxy,zwh}@ios.ac.cn
[2] Hangzhou Institute for Advanced Study, University of Chinese Academy of Sciences, Hangzhou, China
[3] University of the Chinese Academy of Sciences, Beijing, China

Abstract. Maritime transportation business suffers from trust issues and burdensome paperwork. Blockchain-based smart contracts are a promising solution. Due to the nature of the blockchain, it is important to verify smart contracts before deployment, especially for its functionality and legality. In this paper, we propose a verification framework that automatically verifies the functionality and legality requirements of maritime transportation smart contracts. Smart contracts of an application, based on a set of templates, are modeled in a network of timed automata; domain-specific requirements are collected and formulated as temporal logic formulas; real-time model checking tool UPPAAL is then used to check whether these requirements are satisfied. We carry out experiments on nine real-world smart contracts to show the effectiveness and feasibility of our framework. We also compare our work with existing tools to show its effectiveness and efficiency.

Keywords: Model checking · UPPAAL · Smart contract · Solidity · Maritime transportation

1 Introduction

Formal verification techniques are widely acknowledged for improving system reliability, but are also notorious for being difficult to use. In this paper, we propose a domain-specific verification method to ease the pain of using them.

Maritime transportation is the backbone of international trade, which accounts for over 90% of the global trade volume [37]. The maritime transportation industry is facing challenges from trustless participants, and burdensome paperwork. For example, a flower export consignment from Kenya to Netherlands requires more than 200 bilateral communications among 20 organizations, and it usually costs about 10 days to deal with the paperwork [9].

© The Author(s), under exclusive license to Springer Nature Switzerland AG 2025
G. Bai et al. (Eds.): ICECCS 2024, LNCS 14784, pp. 179–198, 2025.
https://doi.org/10.1007/978-3-031-66456-4_10

A smart contract is 'a computerized transaction protocol that executes the terms of a contract', defined by Nick Szabo [38] in 1994. Smart contracts rise with emergence of Blockchain platforms, such as Ethereum [14]. Due to the decentralized, tamper-proof and indisputable nature of the blockchain, smart contracts can perform reliable transactions among trustless participants, and thus gained attention from maritime transportation. Smart contracts are introduced to maritime transportation to improve the efficiency, traceability, and transparency [3,18,19,23,34].

Due to the tamper-proof nature of blockchain, a smart contract is hard to modify after deployment. Therefore, verification before deployment is essential. Symbolic execution tools [28,32] are developed to detect specific vulnerabilities patterns. Development frameworks also play a part in avoiding vulnerabilities. For example, smart contracts developed from MariSmart templates [42] are free from re-entrance and transaction order dependency attacks. However, such approaches cannot be used to check various customer requirements.

Formal methods based tools [12,13,24,30] are promising to verify a variety of properties. However, formal methods, i.e. model checking [17], usually require the skills of writing formal models and logic formulas, which makes it challenging to apply formal methods for engineers who lack solid mathematical background.

To ease the difficulty of using model checking techniques, we focus on the automated verification problem of smart contracts specific to maritime transportation domain. Specifically, two challenges are faced when dealing with this problem.

The first challenge is to find the common domain-specific requirements. We collect and extract functionality and legality requirements from maritime transportation related researches [9,19,23,33], reports [21,22], and laws [1,40]. The **functionality requirements** specify the expected behaviors and results of maritime transportation, especially the payment and delivery of the shipment. Besides, maritime transportation usually take place across countries, and must obey multiple international and domestic laws. It is important to ensure that the smart contracts comply with these laws. The **legality requirements** are the legal constraints from maritime transportation laws, such as the time constraints and duties of participants. To the best of our knowledge, we are the first to apply verification on legality issues for maritime transportation smart contracts.

The second challenge is to automatically generate formal models and properties. We propose **MariSmart**, a development and verification framework for smart contracts that monitor and manage maritime transportation business. The goal of the framework is shown in Fig. 1. Users can customize the MariSmart templates for a specific MariSmart application, which can then be verified with MariSmart verification framework. The verification results can be used for revising the code. The design and use of MariSmart templates are introduced in [42]. In this paper, we introduce MariSmart verification framework, a template-based verification technique.

Our main contributions in this paper are as follows.

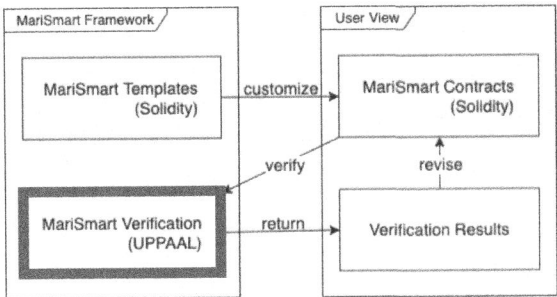

Fig. 1. The goal of the MariSmart framework. The scope of this paper is in the red bold box. (Color figure online)

- We collect and extract common functionality and legality requirements specific to maritime transportation domain.
- We propose an automatic verification framework that transforms the MariSmart application to a network of timed automata, formalizes pre-defined requirements to timed computation tree logic formulas [5], and then verifies them with UPPAAL [12].
- We implement the verification framework, perform case studies on nine real-world smart contract systems, and compare our framework with existing works. The code of the framework and case studies are available on [2].

The reminder of this paper is organized as follows. Section 2 introduces the preliminaries. Section 3 illustrates the formal modeling method. Section 4 introduces the domain-specific requirements and formulates them. Experiments and case studies are shown in Sect. 5. Related work is reviewed in Sect. 6. Section 7 concludes and discusses the future work.

2 Preliminaries

In this section, we introduce the smart contract language Solidity, MariSmart templates and applications, and the back-end verification tool UPPAAL.

2.1 Solidity

Solidity is one of the most popular smart contract languages in Ethereum [14], which can be compiled and executed on the Ethereum Virtual Machine (EVM). A smart contract in Solidity contains declarations of State Variables, Functions, Function Modifiers, Events, Errors, Struct Types, and Enum Types.

Functions are the execution units of the smart contract. There can be a call chain of multiple functions in a transaction. For example in Fig. 2, the function *depart* in (a) calls function *depart* in (b), which calls another internal function *internalTransfer*.

```
1  contract Carrier is Stakeholder {
2     ...
3     function depart(uint _UID)
          public virtual onlyOwner {
4        shipments[_UID].depart();
5     }
6  }
```

(a)

```
1  contract NFTCarrier is Carrier {
2     ...
3     function depart(uint _UID)
          public override onlyOwner {
4        shipments[_UID].depart();
5        /* custom logic begins here */
6        ContainerNFT(ContainerNFT_addr
            ).transferFrom(
7           shipments[_UID].getShipper()
            ,
8        address(this),
9        NFTShipment(address(
            shipments[_UID])).
            getNFTID()
10       );
11    }
12 }
```

```
1  contract Shipment is IShipment {
2     ...
3     modifier pre_depart() virtual
          override {
4        require(msg.sender == carrier)
          ;
5        require(state == State.
            exported);
6        _;
7     }
8     function depart() external
          virtual override pre_depart
          {
9        internalTransfer(consignee,
            shipper, down_payment);
10       depart_time = block.timestamp;
11       state = State.departed;
12       emit ShipmentDeparted(msg.
            sender, block.timestamp);
13    }
14 }
```

(b)

(c)

Fig. 2. Code fragments of MariSmart application. (a) *Depart* function of Carrier template; (b) corresponding function of Shipment template; (c) customized *depart* function of NFTCarrier contract.

Modifiers are used to specify the pre-conditions of the functions. Note that the whole transaction will revert if any of the *require* statement throws an exception.

Inheritance is important feature of Solidity. A Solidity contract can inherit from multiple contracts with the keyword *is*. Users can either add new functions, modifiers and variables to the contract, or override the existing ones with keyword *override*. For example, contract *NFTCarrier* in Fig. 2(c) inherits contract *Carrier* in Fig. 2(a), and overrides function *depart*.

Time is captured with an unsigned integer variable *block.timestamp*. Variables assigned with *block.timestamp* are called *time-related variables*. For example, *depart_time* shown in Fig. 2(b) is a time-related variable.

Other features required in this paper are introduced when they are mentioned.

2.2 MariSmart Application

An application of the maritime transportation developed with MariSmart templates [42] is called a *MariSmart application*. A MariSmart application usually includes several Stakeholder contracts and one Shipment contract.

The Stakeholder contracts implement the activities of participants of the transportation. Six stakeholders are formulated in the templates: *Shipper*, *Carrier*, *Consignee*, *Pre_Shipment_Inspector*, *Export_Port_Operator* and *Import_Port_Operator*. For example, the activity of *Carrier* departing the export part is implemented as function *depart* in Fig. 2(a).

The Shipment contract indicates how the states of the shipment change. There are 12 states: *created*, *signed*, *inspected*, *exported*, *departed*, *lost*, *arrived*, *imported*, *rearranged*, *received*, *claimed*, and *closed*. A change occurs when its corresponding function is called and the function's pre-condition regulated by a modifier is satisfied.

The MariSmart templates can be customized for specific workflow and activities. For example, the customized contract in Fig. 2(c) overrides the function *depart* with additional logic. Please refer to [42] for more details about MariSmart templates and applications.

2.3 UPPAAL

UPPAAL [12] is a real-time model checking tool. The model language of UPPAAL is based on the theory of *timed automaton* (TA). Its specification language is a variation of *timed computation tree logic* (TCTL).

Definition 1. (Syntax of UPPAAL TA). *A TA is a tuple* (L, l_0, A, C, V, E, I), *where*

- L *is a set of locations,*
- $l_0 \in L$ *is the initial location,*
- $A \subseteq Action$ *is a set of actions.* $Action := \{a?|a \in Chan\} \cup \{a!|a \in Chan\}$ *includes input actions (a?) and output actions (a!). Chan is a set of channels.*
- C *is a set of clocks,*
- V *is a set of bounded integer variables,*
- $E \subseteq L \times G(C, V) \times A \times U(C, V) \times L$ *is a set of edges, where* $G(C, V)$ *is a set of linear constraints over* C *and* V *and* $U(C, V)$ *is a set of updates over* C *and* V. *Clocks can only be reset to zero.*
- $I : L \to G(C, V)$ *is a mapping from locations to constraints.*

A *network of timed automata* (NTA) includes a set of TA, sharing a set of global channels, variables and clocks.

The semantics of NTA in UPPAAL is defined as a transition system over the states of TA, where two kinds of transitions are included, namely *delay transitions* and *discrete transitions*. In delay transitions, no edge of TA is fired and the clocks increase; in discrete transitions the edge is fired and clocks are frozen. Specially, edges in different TA may fire synchronously, when there is a pair of actions $c?$ and $c!$ in the edges.

Compared to the theory of TA [6], additional features of NTA in UPPAAL mainly includes the following.

- Bounded integer variables are declared and used in guards, invariants and assignments.

- Committed locations are introduced in UPPAAL. When a NTA is at a *committed* location, it cannot delay and the next transition must involve an outgoing edge of the location.
- User functions are supported in UPPAAL with a syntax similar to C [27].

The properties involved in this work can be divided into two types, namely liveness properties and safety properties. Let β be any propositional formula. Liveness properties are usually in the form of $A\Diamond\ \beta$, meaning that on any execution path β will eventually be true, and $\beta_1 \longrightarrow \beta_2$, meaning that once β_1 true β_2 will eventually be true. Safety properties are usually in the form of $A\Box\ \beta$, meaning that on any execution path β is always true.

3   Contract Modeling

In this section, we illustrate how to model a MariSmart application as a NTA. We first define the structures to represent the MariSmart application, then illustrate how to formally model it. Finally we discuss how to deal with the time-related statements.

3.1   System Description

As is introduced in Sect. 2.2, a MariSmart application includes a Shipment contract and several Stakeholder contracts. Before modeling them as TA, we first abstract the Solidity function, Shipment contract, Stakeholder contracts and the MariSmart application as tuples for further interpretation.

Let *Var* be a set of variables and Var_t be a set of time-related variables.

Definition 2. (Solidity Function). *A public function is defined as a tuple* $FUNC := (Name, Para, Stmt, Stmt_r, Stmt_t, Var, Var_t)$, *where Name is the function name, Para is a set of parameters, Stmt is a set of statements,* $Stmt_r \subseteq Stmt$ *is a set of* require *statements,* $Stmt_t \subseteq Stmt$ *is a set of time-related statements, and* $Var_t \subseteq Var$.

Take function *depart* in Fig. 2(b) for example. Its name is *depart*, *Var* and *Stmt* are taken from both *depart* and its modifier *pre_ depart*, $Stmt_r$ contains the statements in Line 4–5, $Stmt_t$ contains the statement in Line 10, and $Var_t = \{depart\_time\}$.

Definition 3. (Shipment Contract). *A Shipment contract is defined as a tuple* $SHC := (State, Var, Var_t, Func)$, *where State is a set of shipment states,* $Var_t \subseteq Var$, *and Func is a set of functions in the contract.*

Definition 4. (Stakeholder Contract). *A stakeholder contract is defined as a tuple* $STC := (Type, Var, Var_t, Func)$, *where Type is the type of the stakeholder,* $Var_t \subseteq Var$, *and Func is a set of functions in the contract.*

In SHC and STCs, we especially define Var_t and $Stmt_t$, because time-related variables and statements require special modeling in UPPAAL, which is proposed in Sect. 3.3. As a MariSmart application contains one Shipment contract and several Stakeholder contracts, it can be represented by a set of SHC and STCs.

Definition 5. (MariSmart Application). *A MariSmart Application with n stakeholders is defined as a tuple* $MA := (SHC, STC_1, STC_2, ..., STC_n)$.

3.2   Formal Modeling

We model a MariSmart application in a template-based way. That is, the behaviour of each element in the application, Shipment contract or Stakeholder contract, is modeled with a TA. The overview of the NTA of a MariSmart application is shown in Fig. 3. The Shipment contract is modeled as *Shipment TA*, where the states of shipment are modeled as locations, and transitions of the states are modeled as edges. Each Stakeholder contract is modeled as a *Stakeholder TA*. For example, Carrier contract is modeled as Carrier TA, where the function *depart* is modeled as two edges. One from *idle* to *depart_called* models the modifier and statements of the function. The other one from *depart_called* back to *idle* models calling the Shipment function. It fires synchronously with the edge of Shipment TA through channel *chan_depart*.

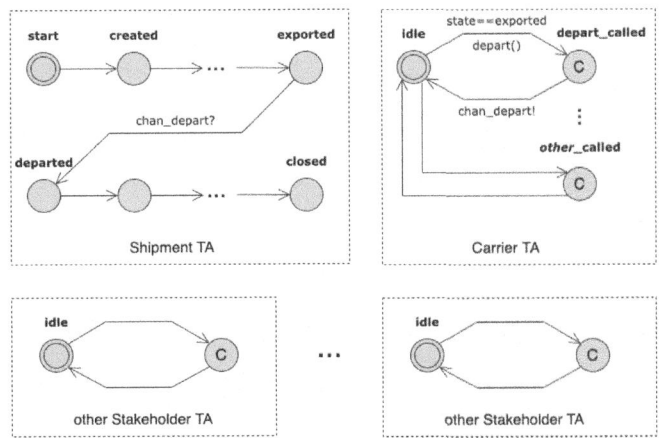

Fig. 3. Overview of NTA of a MariSmart application.

Definition 6. (Shipment TA). *The behavior of Shipment contract $SHC = (State, Var, Var_t, Func)$ is Shipment TA* $TA_{SHC} := (L, l_0, A, C, V, E, I)$, *where*

- $L = State \cup \{start\}$, *where start denotes the state before the shipment is created.*
- $l_0 = start$.

- $A = \{chan\_n?|n = f.Name, f \in Func\}$.
- $C = Var_t$.
- $V = Var \setminus Var_t$.
- *for each function $f \in Func$, there is one or more edges in E, s.t. the source is specified in $f.Stmt_r$, the destination is specified in $f.Stmt$, the guard condition is $f.Stmt_r$, and the action is chan_name?, where name = $f.Name$.*
- *I is empty.*

A Shipment TA is demonstrated in Fig. 4, where states of the shipment are modeled as locations and functions of Shipment contract is modeled as edges. The source and destination of edges are extracted from the modifier and function body. For example, function *depart* in Fig. 2(b) is modeled as the edge from location *exported* to location *departed* in Fig. 4.

Generally speaking, there is one edge in Shipment TA for each function of Shipment contract. However, there are exceptions. For example, function *inspect* can update the shipment state from *signed* to *inspected* when the inspection is passed, or from *signed* to *closed* alternatively. We introduce an auxiliary variable *inspect_flag* to determine which edge to fire. During the transformation, we first scan the Shipment contract to find if there are multiple assignments for the shipment state in one function. If so, an assignment for the auxiliary variable is inserted and transformed in UPPAAL User Function later.

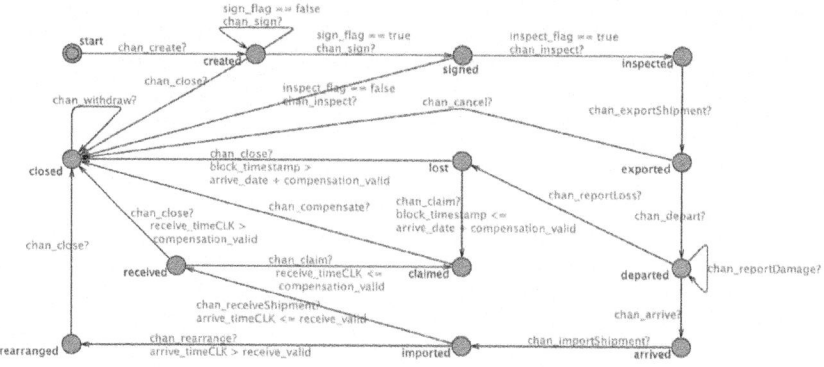

Fig. 4. Overview of TA_{SHC}.

A Stakeholder contract is modeled as a Stakeholder TA, where the function f is modeled as edges between *idle* and *f_ called*. The statements in functions are modeled as UPPAAL User Functions, which supports most statements, such as assignment, **for** statement, **if** statement. Time-related statements need to be modeled in a special way, which is illustrated in Sect. 3.3. We also introduce an auxiliary clock *waitCLK*. It is reset when Stakeholder TA enter location *idle*, and should not exceed a constant limit at *idle*. This way, Stakeholder TA cannot permanently stay at location *idle*, so that the liveness properties can be verified properly.

Definition 7. (Stakeholder TA). *The behavior of Stakeholder contract* $STC = (Type, Var, Var_t, Func)$ *is Stakeholder TA* $TA_{STC} := (L, l_0, A, C, V, E, I)$, *where*

- $L = \{n\_called | n = f.Name, f \in Func\} \cup \{idle\}$.
- $l_0 = idle$.
- $A = \{chan\_n! | n = f.Name, f \in Func\}$.
- $C = \{tCLK | t \in Var_t\} \cup \{waitCLK\}$.
- $V = Var \setminus Var_t$.
- *for each function* $f \in Func$, *there are edges from idle to* n_called, $n = f.name$, *s.t. the value of* $f.Para$ *is selected, the guard is* $f.Stmt_r$, *the update resets clocks and trigger the User Function modeled from* f.
- *for each function* $f \in Func$, *and each Shipment function* g *called by* f, *there are edges from* n_called *to idle, s.t. the update resets* $waitCLK$, *the action is* $chan\_m!$, $m = g.Name$.
- $I(idle) = \{waitCLK \leq Constant\}$.

Take the contract shown in Fig. 5(a) for example. Consignee contract is modeled as Consignee TA shown in Fig. 5(b), where function *receiveShipment* is modeled as edges in Consignee TA in Fig. 5(c). In the edge from *idle* to *receiveShipment_ called*, the value of parameter *received* is selected (in the yellow line), the *require* statement is examined in guard (in the green line) and the User Function (see in Fig. 5(c)) is triggered in update (in the blue lines). Note that *state* is a reserved word in UPPAAL, so we replace it with *status* in Stakeholder TA.

For each function of the Shipment contract, there is usually one corresponding function in the Stakeholder contracts. Consignee contract shown in Fig. 5(a) has an exception, where function *receiveShipment* can either call functions *receiveShipment* or *close*. When the framework transforms this function to a UPPAAL User Function, two auxiliary variable *call_ close* and *call_receiveShipment* are introduced and assigned before the corresponding call statements. In this way, once the marked statements in Fig. 5(c) are executed, only the corresponding edge in Fig. 5(b) can be fired.

Definition 8. (MariSmart Application NTA). *The behavior of a MariSmart application* $MA = (SHC, STC_1, STC_2, ..., STC_n)$ *is modeled with an NTA* $NTA_{MA} := TA_{SHC} \parallel TA_{STC_1} \parallel TA_{STC_2} \parallel ... \parallel TA_{STC_n}$ *with global clocks.*

3.3 Dealing with Time

Modeling of time-related statements in smart contract is not as straightforward as other statements. In Solidity, current time is captured with an unsigned integer variable *block.timestamp*. We model *block.timestamp* and other variables assigned with it as clocks in UPPAAL. However, clocks cannot be used in UPPAAL User Functions. Therefore, we need to model the statements that contain time-related variables in specific ways. Three kinds of such statements need to be modeled specifically. They are demostrated in Table 1.

```
1   contract Consignee is Stakeholder{
2     ...
3     function receiveShipment(
4       uint _UID,
5       bool received)
6       public override onlyOwner {
7       if (received) {
8         shipments[_UID].
                receiveShipment(false);
9       } else {
10        shipments[_UID].close();
11    }}
12    function sign(...){...}
13    function withdraw(...){...}
14    function claim(...){...}
15  }
```

(a)

(b)

(c)

Fig. 5. An example of modeling Stakeholder contract as TA. (a) Consignee contract; (b) Consignee TA, some edges are simplified for brevity; (c) User Function modeled from *receiveShipment*, where the marked statements determine which edge to fire.

The first kind is assignment statements. When a variable v is assigned with *block.timestamp*, the corresponding clock v_CLK is reset to zero in the edge from location *idle* to location *activity_ called*. An example is shown in the first row of Table 1, in which the time-related variable is *depart_ time*.

The second kind is **if** statements with time-related condition expressions. We replace such conditions with a set of auxiliary Boolean variables named by *TIMED_ CONDITION_ #*. The rest of the statements are modeled as the UPPAAL User Function. The value of each condition expression is arbitrary, and all possible values are modeled as guard conditions of separate edges from location *idle* to location *activity_ called*. Take function *arrive* in the second row of Table 1 for example, we split the edge from location *idle* to location *arrive_ called* into two, denoting the two possible value of *TIMED_ CONDITION_ 0* accordingly. For functions that include multiple time-related condition expressions, all combinations of their possible values are modeled similarly.

The third kind is **require** statements. Time-related require statements are modeled as guards of the edge from location *idle* to location *activity_ called*. For example in the third row of Table 1, the require statements in modifier *pre_ receive* is modeled as the guard in TA. Notably, the clock *arrive_ timeCLK* is equivalent to *block.timestamp-arrive_ time*, so we transform the third require statement to *arrive_ timeCLK* <=receive_valid.

4   Requirement Extraction and Formalization

In this section, we introduce how we collect and extract functionality and legality requirements and how to formalize them. The collected requirements can be checked with one-button verification in our framework.

Table 1. Modeling time-related statements.

Time-Related Statements	Transitions in TA

``
function depart() external virtual
    override pre_depart{
  internalTransfer(consignee, shipper,
    down_payment);
  depart_time = block.timestamp;
  state = State.departed;
  emit ShipmentDeparted(msg.sender,
    block.timestamp);}
``
 	
The *block.timestamp* is assigned to *depart_ time*
	status == status_exported
	depart_timeCLK = 0,
	msg_sender = carrier, depart()
	idle waitCLK<WAITMAX depart_called
	
	Reset the corresponding clock named as *function_ nameCLK* in the Stakeholder TA (as marked above).

``
function arrive() external virtual
    override pre_arrive {
internalTransfer(shipper, carrier,
    transportation_fee);
  arrive_time = block.timestamp;
  state = State.arrived;
  if (block.timestamp > arrive_date)
    {
      is_delayed = true;
      emit ShipmentArrivedDelayed(
        msg.sender, block.
        timestamp);
    } else emit ShipmentArrivedInTime(
        msg.sender, block.timestamp);}
``
 	
The *block.timestamp* is used in a condition expression *block.timestamp* > *arrive_ date*.
	``
	// UPPAAL User Function:
	void arrive(){
	  balances[shipper] -=
	      transportation_fee;
	  balances[carrier] +=
	      transportation_fee;
	  status = status_arrived;
	  if(TIMED_CONDITION_0){
	    is_delayed = true;}}
	``
	status==status_departed&& block_timestamp<=arrive_date
	TIMED_CONDITION_0 = false, arrive_timeCLK = 0,
	msg_sender = carrier, arrive()
	idle
	waitCLK<WAITMAX arrive_called
	status==status_departed&& block_timestamp>arrive_date
	TIMED_CONDITION_0 = true, arrive_timeCLK = 0,
	msg_sender = carrier, arrive()
	
	Replace the expression with a Boolean variable, and assign it with all possible values in separate edges (as marked above).

``
modifier pre_receiveShipment() virtual
    override {
  require(msg.sender == consignee);
  require(state == State.imported);
  require(block.timestamp <=
      arrive_time + receive_valid);
  _;}
``
 	
The third *require* statement contains both *block.timestamp* and a time-related variable *arrive_ time*.
	_is_damaged:int[0,1] , _is_passed:int[0,1]
	status==status_imported && arrive_timeCLK<=receive_valid
	receive_timeCLK = 0,
	msg_sender = consignee,
	receiveShipment(_is_damaged,_is_passed)
	idle waitCLK<WAITMAX receiveShipment_called
	
	Note that *arrive_ timeCLK* is equivalent to *block.timestamp-arrive_ time*. Reshape the expression and insert it into the guard of the edge (as marked above).

4.1 Functionality Requirement

Functionality requirements are mainly collected from research papers [9,19,23, 33] and reports [21,22]. They are divided into three kinds.

The first kind of functionality requirements is about workflow, listed as Requirement 1 to 6 in Table 2. They depict the order and dependence of transportation activities. For example, Requirement 1 requires that Shipment contract to be created before any other activity takes place.

Basically, properties of *something always happens* are modeled as safety properties such as $A\square \ \beta$, while properties of *something will eventually happen* are modeled as liveness properties such as $A\diamond \ \beta$ or $\beta_1 \longrightarrow \beta_2$. For example, Requirement 6 *'Shipment will eventually close'* are modeled as $A\diamond \ Shipment.closed$, where *Shipment.closed* denotes that Shipment TA is at Location *closed*.

For Requirements 1 to 3, we introduce a set of auxiliary Boolean variables like *already_ create* to trace previous activities. They are initiated as false, and are set to be true when the corresponding activity takes place. Moreover, conjunctions of locations are abstracted in Requirement 1 to 3 for conciseness. Please refer to the repository in [42] for full formulas.

The second kind of functionality requirements is about the payment system. In the MariSmart application, stakeholders first deposit Ethers (the currency of Ethereum) into the Shipment contract and withdraw them after the order is closed. During the transportation, the transferring between stakeholders is implemented as editing their balances instead of calling *send* or *transfer*. We collect the safety and correctness requirements of such balances as Requirement 7 to 10.

Formalizing these requirements is similar to the workflow requirements, except that a new auxiliary array *net* is introduced to trace the net income for the shipment contract (denoted by net[0]) and stakeholders (denoted by net[1–6]). The values of *net* are set to zero at the beginning, and are updated when the corresponding stakeholders pay or receive Ethers. With array *net*, we can check whether the balances correctly record the transferring results.

The third kind of functionality requirements is about handling accidents and compensation, including Requirement 11 to 16. Accidents include delayed arrival, shipment loss and damage. A MariSmart application should correctly record them and process the compensation.

We introduce clocks *block_timestamp* and *arrive_ timeCLK* while formalizing the third kind of requirements. Clock *block_ timestamp* is set to zero at the beginning. Clock *arrive_ timeCLK* is set to zero when the carrier arrives.

4.2   Legality Requirement

We focus on two typical issues in maritime laws, namely compensation and rearrangement. We collect the legality requirements from Hamburg Rules [40], which was signed in 1978 and has been adopted by 35 countries, and Maritime Codes of PRC. [1]. They are listed as Requirement 17 to 26 in Table 2.

When accidents occur, compensation is usually claimed by the Shipper or Consignee, and paid by the Carrier. The amount of the compensation and time period to claim for it are specified differently in maritime laws.

Take Hamburg Rules [40] as example, we extract Requirement 17 and 19 according to Paragraph 1(a) of Article 6 in [40], which writes 'The liability of the carrier for loss resulting from loss of or damage to goods according to the provisions of article 5 is limited to an amount equivalent to 835 units of account per package or other shipping unit or 2.5 units of account per kilogram of gross weight of the goods lost or damaged, whichever is the higher.' Note that the units in this paragraph is defined as special drawing right (SDR) in Paragraph 1 of Article 26.

Similarly, we extract Requirement 18 on compensation for loss of delay, according to Paragraph 1(b) of Article 6 in [40]. For period in which consignee should claim for compensation, we extract Requirement 20 from Paragraph 2 of Article 19.

When the consignee fails to receive the shipment within a specific period after arrival, the carrier is permitted to resell or auction the shipment, which we denote as rearrangement issues. In Maritime Code of the People's Republic of China [1], for example, it rephrases in Article 88, 'If the goods under lien in accordance with the provisions of Article 87 of this Code have not been taken delivery of within 60 days from the next day of the ship's arrival at the port of discharge, the carrier may apply to the court for an order on selling the goods by auction; where the goods are perishable or the expenses for keeping such goods would exceed their value, the carrier may apply for an earlier sale by auction.' We extract Requirement 26 according to it.

Since all of legality requirements should always be guaranteed, we model them as safety properties, as is shown in Table 2.

Notably, articles in these two maritime laws share a similar structure, which makes it convenient to extract and model them. There are only differences on specific figures between Requirement 17 to 21 from Hamburg Rules, and Requirement 22 to 26 from Maritime Codes of PRC. For example, the period for claiming for compensation is 15 days in Requirement 20, and it comes to 7 days in Requirement 25.

5   Experiments

To show the effectiveness and feasibility of our framework, we rewrite 9 real-world maritime transportation smart contracts to MariSmart application and verify them. We also compare our framework with two existing tools [30,43].

We perform the experiments on a Linux server with 32 Cores, 2.90GHz CPU and 384G RAM. Our tool and all studied MariSmart cases can be accessed from [2]. We also present a Web application for our framework at [41].

5.1   Case Studies on Real-World Smart Contracts

We apply our verification framework on 9 real-world maritime transportation smart contracts, which are denoted as Medical [3], NFT1 [19], NFT2 [18], Recall [35], PPE [34], ShipChain [36], eth-shipment [11], LNG [26] and IoT [23]. Original contracts range from 100 to 500 lines. After modeled as NTA, the scale of the Shipment TA varies with the complexity of the workflow, and the edges of Shipment TA range from 6 to 21. The detailed information is included in first two parts of Table 3.

Table 2. Requirements and their TCTL formulas.

No.	Requirements	TCTL Formulas	Timed
	Functionality Requirement		
1	Shipment is always created before any other activities.	$A\square$ (*Any Stakeholder not at idle*) imply already_create	
2	Stakeholders always sign the shipment before taking any other activities.	$A\square$!Shipment.closed and (*Any Stakeholder not at idle or sign_called*) imply already_sign	
3	Activities on departure side take place before departing.	$A\square$ (*Any Stakeholder acts on departure side*) imply !already_depart	
4	After departure, the shipment either arrives or is lost.	Shipment.departed $\longrightarrow$ Shipment.lost or Shipment.arrived	
5	Activities on destination side take place after arriving.	$A\square$ (Carrier.rearrange_called or Consignee.receiveShipment_called or ImportPortOperator.importShipment_called) imply already_arrive	
6	Shipment will eventually close.	$A\lozenge$ Shipment.closed	
7	Stakeholders' balances are always non-negative.	$A\square$ (sum(i:int[1,6])balances[i]) ≥ 0	
8	Sum of shipment contract and stakeholders' net proceeds is always zero.	$A\square$ (sum(i:int[0,6])net[i]) ==0	
9	Sum of stakeholders' balances is always equal to Ethers in shipment contract.	$A\square$ (sum(i:int[1,6])balances[i]) == net[0]	
10	After closing the shipment, balances will eventually be withdrawn.	Shipment.closed $\longrightarrow$ (forall(i:int[1,6])balances[i]==0)	
11	Delayed arrival is recorded correctly.	$A\square$ (is_delayed imply block_timestamp > arrive_date) and (block_timestamp > arrive_date and arrive_timeCLK==0 imply is_delayed)	✓
12	Delayed arrival leads to either claiming for compensation or closing shipment after compensate period	is_delayed $\longrightarrow$ Shipment.claimed or Shipment.closed and arrive_timeCLK>compensation_valid	✓
13	Loss of shipment is recorded correctly.	$A\square$ (is_lost imply already_reportLoss) and (already_reportLoss imply is_lost)	
14	Shipment loss leads to either claiming for compensation or closing shipment after compensate period	is_lost $\longrightarrow$ Shipment.claimed or Shipment.closed and block_timestamp>arrive_date+compensation_valid	✓
15	Damage of shipment is recorded correctly.	$A\square$ already_reportDamage imply is_damaged	
16	Shipment damage leads to either claiming for compensation or closing shipment after compensate period	is_damaged $\longrightarrow$ Shipment.claimed or Shipment.closed and arrive_timeCLK>compensation_valid	✓
	Legality Requirement		
17	Compensation limit is always less than 835 SDR or 2.5 SDR/kg, whichever is the higher. [40]	$A\square$ compensation_limit <835 or compensation_limit<2.5*weight	
18	Compensation actually paid for delay should be less than 2.5 times of transportation fee. [40]	$A\square$ is_delayed and !is_damaged and !is_lost imply net[carrier]+balances[carrier]$\geq$transportation_fee*(1-2.5)	
19	Compensation actually paid for loss or damage should be less than 835 SDR or 2.5 SDR/kg, whichever is the higher. [40]	$A\square$ (is_damaged or is_lost) imply (net[carrier]+balances[carrier]$\geq$transportation_fee-835 or net[carrier]+balances[carrier] $\geq$transportation_fee-weight*2.5)	
20	Consignee should claim for compensation within 15 days after arrival. [40]	$A\square$ Consignee.claim_called imply receive_timeCLK$\leq$15	✓
21	Consignee should receive shipment within 90 days after arrival, otherwise carrier can resell or auction it. [40]	$A\square$ Carrier.rearrange_called imply arrive_timeCLK>90	✓
22	Compensation limit is always less than 666.67 SDR or 2 SDR/kg, whichever is the higher. [1]	$A\square$ compensation_limit<666.67 or compensation_limit<2*weight	
23	Compensation actually paid for delay should be less than transportation fee. [1]	$A\square$ is_delayed and !is_damaged and !is_lost imply net[carrier]+balances[carrier]$\geq$transportation_fee*(1-1)	
24	Compensation actually paid for loss or damage should be less than 666.67 SDR or 2 SDR/kg, whichever is the higher. [1]	$A\square$ (is_damaged or is_lost) imply (net[carrier]+balances[carrier]$\geq$transportation_fee-666.67 or net[carrier]+balances[carrier]$\geq$transportation_fee-weight*2)	
25	Consignee should claim for compensation within 7 days after arrival. [1]	$A\square$ Consignee.claim_called imply receive_timeCLK$\leq$7	✓
26	Consignee should receive shipment within 60 days after arrival, otherwise carrier can resell or auction it. [1]	$A\square$ Carrier.rearrange_called imply arrive_timeCLK>60	✓

*SDR denotes Special Drawing Right.

Table 3. Experimental results.

Case	Medical		NFT1		NFT2		Recall		PPE		ShipChain		eth-shipment		LNG		IoT	
	\multicolumn{18}{c}{Code Lines of Original Smart Contract}																	
#	210		521		325		252		125		130		193		111		229	
	\multicolumn{18}{c}{Edges in Shipment TA}																	
#	13		13		11		9		6		7		9		21		18	
	\multicolumn{18}{c}{Verification Result}																	
Requirement	result	time	result	time	result	time	result	time	result	time	result	time	result	time	result	time	result	time
1	✗	1	✓	8470	✓	9	✓	255	✓	600	✗	1	✓	28	✓	542	✓	779
2	✗	0	✗	1	✓	8	✗	1	✗	4	✗	0	✓	28	✓	546	✓	776
3	✓	528	✓	9243	✓	7	✓	256	✓	675	✗	1	✓	28	✓	586	✓	769
4	✗	2	✗	8	✗	3	-	-	-	-	✗	0	✓	45	✓	590	✓	871
5	✓	470	✗	0	✗	3	-	-	-	-	✓	0	✓	29	✓	554	✓	774
6	✗	1	✗	1	✓	2	✗	5	✗	3	✓	1	✓	48	✓	13	✗	203
7	✓	656	✓	7465	✓	8	✓	208	✓	542	✓	0	✓	29	✓	514	✓	785
8	✓	578	✓	7724	✓	9	✓	218	✓	628	✓	0	✓	29	✓	527	✓	787
9	✓	628	✓	6790	✓	7	✓	184	✓	532	✓	0	✓	29	✓	576	✓	782
10	✗	1	✗	2	✓	23	✗	6	✗	29	✗	1	✗	2	✓	2079	✓	1266
11	✗	1	✓	9083	✓	8	✓	234	✓	642	✗	1	✓	29	✓	580	✓	776
12	-	-	-	-	-	-	-	-	-	-	-	-	-	-	✓	623	✓	878
13	-	-	✓	7727	✓	8	✓	202	✓	594	✓	1	✓	30	✓	732	✓	798
14	-	-	-	-	-	-	-	-	-	-	-	-	-	-	✓	564	✓	878
15	-	-	✓	7803	✓	8	✓	222	✓	544	✓	1	✓	28	✓	530	✓	798
16	-	-	-	-	-	-	-	-	-	-	-	-	-	-	✓	639	✓	904
17	-	-	-	-	-	-	-	-	-	-	-	-	-	-	✓	583	✗	155
18	-	-	-	-	-	-	-	-	-	-	-	-	-	-	✗	37	✓	784
19	-	-	-	-	-	-	-	-	-	-	-	-	-	-	✗	76	✓	775
20	-	-	✓	8943	✓	7	✓	192	✓	604	✓	1	✓	30	✓	622	✓	780
21	-	-	✗	0	✓	7	✓	190	✓	577	✓	0	✓	30	✗	42	✓	807
22	-	-	-	-	-	-	-	-	-	-	-	-	-	-	✓	605	✗	141
23	-	-	-	-	-	-	-	-	-	-	-	-	-	-	✗	42	✓	819
24	-	-	-	-	-	-	-	-	-	-	-	-	-	-	✗	73	✓	790
25	-	-	✓	7958	✓	8	✓	209	✓	539	✓	1	✓	27	✓	530	✓	786
26	-	-	✗	1	✓	8	✓	208	✓	558	✓	1	✓	30	✗	46	✓	784
Avg. Passed	572		8121		8		214		586		1		31		627		824	

✓: satisfied, ✗: not satisfied, -: not checked, time(ms): verification time, Avg. Passed(ms): average verification time of satisfied requirements.

*Execution times are measured in milliseconds (ms).

The experimental results are shown in the third part of Table 3. The satisfied requirements are marked with checkmarks, unsatisfied ones are marked with crossmarks. When the relevant activity is not included in the case, corresponding requirement is not checked. For example, 7 of 9 cases do not implement compensation, and the relevant Requirement 17–19 and 22–24 are not checked.

As for functionality, 7 of 9 cases violate at least one requirements of workflow. Five cases violate requirement 6, which implies that these smart contract may be deadlocked. This violation is especially harmful since the funds may be locked in the Shipment contract. After a closer examination, we find that this requirement is violated when there are unexpected entrances to the activities, and strengthening the pre-conditions of these activities may help with this issue.

As for legality requirements, Case IoT violates Requirement 17 and 22, because it does not specify the limit of compensation. Due to the nature of lique- fied natural gas, Case LNG specifically assigns the limit of compensation and the time period for reselling. Although it violates 6 relevant legality requirements, its additional terms are protected by the laws, and should be deemed as legal. As for the rest of the cases, compensation and reselling are not implemented, thus relevant requirements are not checked.

Since the unsatisfied requirements take much shorter time than the satisfied ones, we evaluate the efficiency of our framework through the average time for satisfied requirements. The verification time is mainly influenced by the range and number of parameters. Number of edges in Shipment TA and code lines of the original contract may influence the verification time. For example, Case NFT1 [19] and Case NFT2 [18] are implemented by the same team, and share a similar workflow. However, the pre-condition of activity *rearrange* of Case NFT1 is too weak, so that the Shipment TA contains more edges and the verification time is about 100 times than Case NFT2.

The complexity of the requirements may also influence the verification time, as Requirement 10 takes longer time than others. It is influenced by the complexity of the TCTL formulas, and the optimization in UPPAAL.

5.2 Comparison with VeriSolid and MvSC

We compare our framework against two verification tools, VeriSolid [30] and mvSC [43], since they both deal with time and can verify customized properties. VeriSolid transforms model-based Solidity contracts as transition systems, and verifies them in nuXmv [15]. VeriSolid applies abstraction to time variables, where time-guarded transitions are specially modeled. mvSC models smart contracts as NTA and verifies them in UPPAAL. Both works consider time in modeling, and are similar to our approach.

Both tools verify a single contract only, so we simplify the Case Medical and verify the Property 1–11 with them. The result is shown in Table 4. It can be observed that our framework can deal with a wider range of properties than VeriSolid, and runs significantly faster than mvSC.

Table 4. Comparison with VeriSolid [30] and mvSC [43]

Tool	MariSmart		VeriSolid		mvSC	
Requirement	result	time(s)	result	time(s)	result	time(s)
1	✗	0.001	✗	<10	✓	1503.558
2	✗	0.000	✗	<10	✗	0.001
3	✓	0.528	✓	<10	✓	1611.017
4	✗	0.002	✗	<10	✗	0.001
5	✓	0.47	✓	<10	✓	1429.493
6	✗	0.001	✗	<10	✗	0.001
7	✓	0.656	-	-	✓	1585.734
8	✓	0.578	-	-	✓	1520.306
9	✓	0.628	-	-	✓	1605.873
10	✗	0.001	-	-	✗	0.000
11	✗	0.001	-	-	✗	0.001

VeriSolid can only verify Property 1–6, since it only takes calling a function or statement as propositional formula, while specifying the value of variables is not supported. As for verification time, VeriSolid does not record verification time originally. Since it is built as a Web application, it is unfair to calculate the entire runtime including waiting for pages to respond. Basically speaking, all verification can be finished within 10 s.

The tool mvSC models smart contracts as NTAs, where functions are modeled as TAs and statements are modeled as edges. Compared to our framework, the NTA generated by mvSC contains more transitions, so that the verification time significantly increases. As for verification results, only the result of Property 1 is different to the other two. This is because mvSC only supports single contract, and requires to call the constructor prior to any other functions. Due to the limitations of mvSC, the counterexample in the other two tools are excluded, such that the Stakeholder contracts trigger before the shipment creation.

6   Related Work

Formal methods used in smart contract verification [39] mainly include model checking based techniques [4,8,13,30,43] and theorem proving based techniques [7,24].

Hirai [24] formalized the EVM semantics in theorem prover Isabelle/HOL and manually proved safety properties. Da et al. [25] proposed a tool that automatically translates Michelson contract [20] to WhyML, and verifies them in Why3. As in [7], Amani et al. extended an existing EVM formalisation in Isabelle/HOL. Approaches based on theorem proving usually take EVM bytecode or other low level code as input. Most of them are performed manually, which requires users to be experts of theorem proving. By contrast, our framework automatically models the smart contracts so that users are free from writing formal properties.

Bai et al. [8] verified template-based smart contracts with the model checker SPIN [31]. Compared to our work, the template designed in [8] is highly abstract and only defines basic interactions between two contract parties. Users of [8] have to pay more effort to develop specific smart contracts for their business. As for verification, [8] only discusses an approach to verify smart contract with SPIN, and performs a case study manually. No automated framework is involved. Alqahtani et al. [4] modeled smart contracts as a finite state machine and checked the requirements in NuSMV [16]. [4] mainly proves the feasibility of the approach yet with little automation. FsolidM [29] is a framework for designing Ethereum smart contracts. The same team further proposed VeriSolid [30] based on FSolidM. This tool first generates augmented transition system, then transforms it into BIP [10] transition system, and finally verifies it with nuXmv [15]. Zhao et al. [43] proposed an approach to verify smart contracts with UPPAAL, especially the contracts with time constraints. Comparing to our work, [43] models each statement as one edge, which results in a larger amount of locations and transitions, and also a longer time for verification. Most of the works require users to write formal properties manually, while our framework can verify important domain-specific requirements automatically.

7   Conclusion

In this paper, we have proposed an automatic verification framework for template-based smart contracts and requirements specific to maritime transportation domain. The collected requirements can be checked with one-button verification in our framework. The effectiveness and the feasibility of our verification framework are shown by the experimental results on nine real-world smart contracts. Compared to existing works, our approach can verify a wider range of domain-specific requirements within a reasonable time.

Due to the limitation of UPPAAL, the syntax of input smart contracts is limited. For example, the random generator function *keccak* is not supported by the framework. In the future, we will expand the range of which the framework can process, and extend the framework to a broader domain.

Acknowledgments. This work is partially supported by the National Natural Science Foundation of China (No. 62072443).

Disclosure of Interests. The authors have no competing interests to declare that are relevant to the content of this article.

References

1. Maritime code of PRC (1993). https://www.gov.cn/guoqing/2020-12/24/content_5572935.htm
2. Marismart framework (2023). https://github.com/MariSmartSourceCode/MariSmart
3. Ahmad, R.W., Salah, K., Jayaraman, R., Yaqoob, I., Omar, M., Ellahham, S.: Blockchain-based forward supply chain and waste management for Covid-19 medical equipment and supplies. IEEE Access **9**, 44905–44927 (2021)
4. Alqahtani, S., He, X., Gamble, R., Mauricio, P.: Formal verification of functional requirements for smart contract compositions in supply chain management systems (2020)
5. Alur, R., Courcoubetis, C., Dill, D.: Model-checking for real-time systems. In: 1990 Proceedings of Fifth Annual IEEE Symposium on Logic in Computer Science, pp. 414–425. IEEE (1990)
6. Alur, R., Dill, D.: Automata for modeling real-time systems. In: Paterson, M.S. (ed.) ICALP 1990. LNCS, vol. 443, pp. 322–335. Springer, Heidelberg (1990). https://doi.org/10.1007/bfb0032042
7. Amani, S., Bégel, M., Bortin, M., Staples, M.: Towards verifying Ethereum smart contract bytecode in Isabelle/HOL. In: Proceedings of the 7th ACM SIGPLAN International Conference on Certified Programs and Proofs, pp. 66–77 (2018)
8. Bai, X., Cheng, Z., Duan, Z., Hu, K.: Formal modeling and verification of smart contracts. In: Proceedings of the 2018 7th International Conference on Software and Computer Applications, pp. 322–326 (2018)
9. Balci, G., Surucu-Balci, E.: Blockchain adoption in the maritime supply chain: examining barriers and salient stakeholders in containerized international trade. Transp. Res. Part E: Logist. Transp. Rev. **156**, 102539 (2021)

10. Basu, A., et al.: Rigorous component-based system design using the BIP framework. IEEE Softw. **28**(3), 41–48 (2011)
11. Bavosa, A.: Smart contracts (2018). https://github.com/ajb413/eth-shipment-tracking/tree/master
12. Behrmann, G., David, A., Larsen, K.G.: A tutorial on UPPAAL 4.0. Department of Computer Science, Aalborg University (2006)
13. Bhargavan, K., et al.: Formal verification of smart contracts: short paper. In: Proceedings of the 2016 ACM Workshop on Programming Languages and Analysis for Security, pp. 91–96 (2016)
14. Buterin, V., et al.: A next-generation smart contract and decentralized application platform. White Paper **3**(37), 2–1 (2014)
15. Cavada, R., et al.: The NUXMV symbolic model checker. In: Biere, A., Bloem, R. (eds.) CAV 2014. LNCS, vol. 8559, pp. 334–342. Springer, Cham (2014). https://doi.org/10.1007/978-3-319-08867-9_22
16. Cimatti, A., Clarke, E., Giunchiglia, F., Roveri, M.: NuSMV: a new symbolic model verifier. In: Halbwachs, N., Peled, D. (eds.) CAV 1999. LNCS, vol. 1633, pp. 495–499. Springer, Heidelberg (1999). https://doi.org/10.1007/3-540-48683-6_44
17. Clarke, E.M., Jr., Grumberg, O., Kroening, D., Peled, D., Veith, H.: Model Checking. MIT Press, Cambridge (2018)
18. Elmay, F.K., Madine, M., Salah, K., Jayaraman, R.: NFTs for trusted traceability and management of digital twins for shipping containers. In: 2023 IEEE International Conference on Pervasive Computing and Communications Workshops and other Affiliated Events (PerCom Workshops), pp. 433–438. IEEE (2023)
19. Elmay, F.K., Salah, K., Jayaraman, R., Omar, I.A.: Using NFTs and blockchain for traceability and auctioning of shipping containers and cargo in maritime industry. IEEE Access **10**, 124507–124522 (2022)
20. Foundation, T.: Michelson: the language of smart contracts in Tezos. https://tezos.gitlab.io/active/michelson.html#language-semantics
21. Ganne, E.: Can blockchain revolutionize international trade? [online] (2018). https://www.wto.org/english/res_e/booksp_e/blockchainrev18_e.pdf
22. Group, B.C.: Digital innovation in trade finance: have we reached a tipping point? (2017). https://www.swift.com/news-events/news/digital-innovation-trade-finance-have-we-reached-tipping-point
23. Hasan, H., AlHadhrami, E., AlDhaheri, A., Salah, K., Jayaraman, R.: Smart contract-based approach for efficient shipment management. Comput. Ind. Eng. **136**, 149–159 (2019)
24. Hirai, Y.: Defining the Ethereum virtual machine for interactive theorem provers. In: Brenner, M., et al. (eds.) FC 2017. LNCS, vol. 10323, pp. 520–535. Springer, Cham (2017). https://doi.org/10.1007/978-3-319-70278-0_33
25. da Horta, L.P.A., Reis, J.S., de Sousa, S.M., Pereira, M.: A tool for proving Michelson smart contracts in WHY3. In: 2020 IEEE International Conference on Blockchain (Blockchain), pp. 409–414. IEEE (2020)
26. Keith, M., Edward, S.: Master ex-ship LNG sales agreement between Cheniere Marketing, Inc. and Gaz De France International Trading S.A.S (2007). https://www.sec.gov/Archives/edgar/data/3570/000119312507106384/dex102.html
27. Kernighan, B.W., Ritchie, D.M.: The C programming language (2002)
28. Luu, L., Chu, D.H., Olickel, H., Saxena, P., Hobor, A.: Making smart contracts smarter. In: Proceedings of the 2016 ACM SIGSAC Conference on Computer and Communications Security, pp. 254–269 (2016)

29. Mavridou, A., Laszka, A.: Designing secure Ethereum smart contracts: a finite state machine based approach. In: Meiklejohn, S., Sako, K. (eds.) FC 2018. LNCS, vol. 10957, pp. 523–540. Springer, Heidelberg (2018). https://doi.org/10.1007/978-3-662-58387-6_28

30. Mavridou, A., Laszka, A., Stachtiari, E., Dubey, A.: VeriSolid: correct-by-design smart contracts for Ethereum. In: Goldberg, I., Moore, T. (eds.) FC 2019. LNCS, vol. 11598, pp. 446–465. Springer, Cham (2019). https://doi.org/10.1007/978-3-030-32101-7_27

31. Mikk, E., Lakhnech, Y., Siegel, M., Holzmann, G.J.: Implementing statecharts in PROMELA/SPIN. In: Proceedings. 2nd IEEE Workshop on Industrial Strength Formal Specification Techniques, pp. 90–101. IEEE (1998)

32. Mueller, B.: A framework for bug hunting on the Ethereum blockchain. ConsenSys/mythril (2017)

33. Nguyen, S., Chen, P.S.L., Du, Y.: Risk assessment of maritime container shipping blockchain-integrated systems: an analysis of multi-event scenarios. Transp. Res. Part E: Logist. Transp. Rev. **163**, 102764 (2022)

34. Omar, I.A., Debe, M., Jayaraman, R., Salah, K., Omar, M., Arshad, J.: Blockchain-based supply chain traceability for COVID-19 personal protective equipment. Comput. Ind. Eng. **167**, 107995 (2022)

35. Patro, P.K., Ahmad, R.W., Yaqoob, I., Salah, K., Jayaraman, R.: Blockchain-based solution for product recall management in the automotive supply chain. IEEE Access **9**, 167756–167775 (2021)

36. ShipChain: Shipchain smart contracts (2020). https://github.com/ShipChain/smart-contracts/tree/master

37. Song, D.: A literature review, container shipping supply chain: planning problems and research opportunities. Logistics **5**(2), 41 (2021)

38. Szabo, N.: Smart contracts (1994). http://www.fon.hum.uva.nl/rob/Courses/InformationInSpeech/CDROM/Literature/LOTwinterschool2006/szabo.best.vwh.net/smart.contracts.html

39. Tolmach, P., Li, Y., Lin, S.W., Liu, Y., Li, Z.: A survey of smart contract formal specification and verification. ACM Comput. Surv. (CSUR) **54**(7), 1–38 (2021)

40. Uncitral: Hamburg rules (1987). https://uncitral.un.org/zh/texts/transportgoods/conventions/hamburg_rules

41. Zhao, X., Lu, Y.: Marismart verifier webpage (2023). http://124.16.137.30:50002/#/dashboard-en

42. Zhao, X., Wei, Q., Zhu, X.Y., Zhang, W.: A smart contract development framework for maritime transportation systems. In: 2023 IEEE 23rd International Conference on Software Quality, Reliability, and Security Companion (QRS-C), pp. 310–319 (2023). https://doi.org/10.1109/QRS-C60940.2023.00091

43. Zhao, Y., Zhu, X., Li, G., Bao, Y.: Time constraint patterns of smart contracts and their formal verification. J. Softw. **33**(8), 2875–2895 (2022)

Formal Methods

QuanSafe: A DTBN-Based Framework of Quantitative Safety Analysis for AADL Models

Yiwei Zhu[1], Jing Liu[1]([✉]), Haiying Sun[1], Wei Yin[2]([✉]), and Jiexiang Kang[2]([✉])

[1] East China Normal University, Shanghai, China
`ywzhu@stu.ecnu.edu.cn`, `{jliu,hysun}@sei.ecnu.edu.cn`
[2] China Aeronautical Radio Electronics Research Institute, Shanghai, China
`{yin_wei,kang_jiexiang}@careri.com`

Abstract. The safety of modern safety-critical systems is increasingly receiving attention. AADL, as an effective modeling language, is widely used for architectural modeling of embedded safety-critical systems. Currently, the main challenges facing the safety analysis of AADL models are the system's dynamic behavior, state space explosion, rare event prediction, and the lack of explanation of unsatisfied specifications. To address these issues, we propose QuanSafe, a discrete-time Bayesian network (DTBN)-based framework of quantitative safety analysis for AADL models. The dynamic behaviors and temporal features of AADL models can be described entirely using DTBN. Moreover, DTBN can effectively avoid state space explosion and poor performance when dealing with rare events. At the same time, DTBN has the ability of diagnostic analyses, which helps improve the system. QuanSafe provides a complete algorithm to transform AADL models into DTBN models. In addition, it supports multiple automated safety analysis methods with improved metrics. We conduct a case study on the Aircraft System. The experimental results show that our approach has higher efficiency and more comprehensive analysis capabilities than existing research.

Keywords: AADL · Discrete-time Bayesian Network · Model-based Safety Analysis · Safety Analysis

1 Introduction

Modern safety-critical systems are widely used in various fields, such as automotive, aerospace, healthcare, and nuclear industries. Failures of these systems can lead to catastrophic consequences, and thus, their safety is increasingly receiving attention. However, analyzing the safety and reliability of these systems has become increasingly challenging as their scale and complexity continue to increase. In recent years, Model-Based Safety Analysis (MBSA) techniques have been developed to address the safety analysis and validation of complex safety-critical systems. MBSA techniques have facilitated the birth and development of various general-purpose modeling languages. The Architecture Analysis and

G. Bai et al. (Eds.): ICECCS 2024, LNCS 14784, pp. 201–222, 2025.
https://doi.org/10.1007/978-3-031-66456-4_11

Design Language (AADL) [28], one of the representative modeling languages, is widely used for the architecture modeling of embedded safety-critical systems. The AADL also provides the Error Model Version 2 (EMV2) Annex [25] that supports annotating system fault and failure information.

Various approaches have been proposed for safety analysis against AADL models, while most still face challenges. The system's dynamic behavior, such as the temporal behavior of systems and sequential dependencies of failures, is one of the challenges. Unfortunately, classical safety analysis techniques (including, but not limited to, Fault Tree Analysis (FTA) [17], Failure Mode and Effects Analysis (FMEA) [26], and Event Tree Analysis (ETA) [10]) cannot account for these aspects. Markov Chains (MCs) [11] and their extensions have been widely used for dynamic reliability analysis of AADL models. Many works have adopted, directly or indirectly, MCs as their formalism. In these works, high-level models, such as Dynamic Fault Trees (DFTs) [16] and Petri Nets (PNs) [13], are used to describe the safety behaviors of AADL models, which are then automatically transformed into MCs. However, MCs often suffer from state space explosion [18], and thus, they are not practical for highly complex systems [14]. Statistical Model Checking (SMC) [1] based approach is another popular technique. SMC avoids state exploration through simulation and statistical methods, but it performs poorly in predicting the satisfaction of properties with very low probability, so-called rare events [21]. In such cases, the number of samples required for model checking explodes, which makes the algorithm less efficient [33]. Additionally, few approaches are available to explain why a given safety specification is not achieved and to offer a way to improve it.

To address these challenges, we present QuanSafe, a discrete-time Bayesian network (DTBN) [19] based framework of quantitative safety analysis for AADL models. The DTBN extends the ability to model dynamic systems by explicitly incorporating time into the classical Bayesian network (BN) [24] formalism. The dynamic behaviors and temporal features of AADL models can be described entirely using DTBN. By considering local dependencies among variables, DTBN effectively avoids the state space explosion problem. Moreover, our approach does not rely on simulation and can handle systems with rare events well. At the same time, DTBN can provide a diagnostic analysis of the causes of faults, which helps improve the system.

We propose an approach to transform AADL models (with error model) into DTBN models. In our approach, the results of quantitative safety analysis for AADL models are obtained by performing different types of inference on the DTBN. In addition, we propose timed Risk Achievement Worth (tRAW), an improved sensitivity analysis measure leading to a more precise guidance of the system's maintenance activities. QuanSafe automates the processes above and has been integrated into the OSATE [12], an open-source AADL tool environment. We demonstrate the capability and efficiency of QuanSafe by conducting a case study on a real-world safety-critical system, the Aircraft System.

In summary, this paper makes the following contributions:

1. We designed a mapping algorithm to transform AADL models into DTBN models.

2. We propose QuanSafe, a DTBN-based framework of quantitative safety analysis for AADL models, which supports multiple automated safety analysis methods with improved metrics.
3. We apply the proposed approach to the Aircraft System. Experimental results demonstrate the effectiveness and efficiency of our approach.

The paper is organized as follows. Section 2 discusses some related work. Section 3 briefly introduces DTBN. Section 4 describes in detail the methodology of model transformation. Section 5 demonstrates the specific process of the QuanSafe framework. Section 6 provides a case study. Finally, Sect. 7 concludes the paper.

2   Related Work

In the research on the safety analysis of AADL models, Liu et al. [22] proposed a new FMEA methodology called RFMEA, which performs quantitative analyses of AADL-based embedded systems. In the work [23], AADL models were transformed into the equivalent HiP-HOPS models for FTA and FMEA analyses. COMPASS [7,8] is a toolset for analyzing and validating safety-critical systems, allowing FTA and FMEA analyses to be performed on a subset of the AADL. These methods are only applicable to the static systems.

Some work focuses on checking the functional correctness of the system without considering the quantitative aspects. Ahmad et al. [2] studied behavior modeling and formal verification of CTCS-3 expressed by AADL. Bae et al. [3] defined the formal semantics of Multirate Synchronous AADL in Real-Time Maude. The approach was applied to verify the distributed designs in AADL. In the work [15], Event-B was used to prove the liveness properties of AADL models. The authors in [20] used Maude and SMT solving to verify the underlying synchronous designs of the system expressed in HybridSynchAADL, an extended language.

MCs are popular for quantitative analysis of dynamic systems. The authors in [30,31] transformed AADL models into DSPNs. They obtained the probability of occurrence of system hazards by solving the DSPN using MCs. Wakankar et al. [29] mapped AADL models to DTMCs and applied them to the quantitative reliability assessment of Computer Based Systems (CBS). The work in [32] combines the probabilistic model checking approach to risk analysis of AADL models. The authors transformed AADL models into CTMCs and finally used the PRISM tool to analyze whether the reliability of the system reached a specified threshold. However, the above approaches require traversing the entire state space and are prone to performance bottlenecks when the system scales up.

On the other hand, Bruintjes et al. [9] introduced a SMC method for a subset of AADL and developed a probabilistic analysis tool based on Monte Carlo methods. The authors in [36] transformed a subset of AADL to stateful timed CSP. Then, SMC was performed in the PAT tool for the concurrent behavior of AADL models. Bao et al. [4] proposed Uncertainty-Aware Hybrid AADL. In this work, the extended AADL models were mapped to the NPTA model, and finally, the performance of the system was quantitatively evaluated using the

UPPAAL-SMC, a SMC tool. Yang et al. [34] conducted a similar research using another AADL-based language called SHML. Although SMC methods enable quantitative analysis of complex systems, their poor performance in predicting rare events limits the application of such methods in practice.

Our approach takes both dynamic and quantitative aspects of the system into account. DTBN can effectively avoid state space explosion and poor performance when dealing with rare events. Meanwhile, the BN formalism allows for a more detailed explanation of the cause of system failure.

3   Discrete-Time Bayesian Network

A Bayesian network (BN) is a directed acyclic graph in which the nodes represent variables and are connected by directed arcs [24]. The arcs denote dependencies or causal relationships between linked nodes. The leaf nodes (nodes without incoming) have marginal prior probability tables associated with them, and the other nodes have conditional probability tables (CPTs) associated with them. The CPT quantifies the effect of the parents on the child nodes [18].

Classical BNs are only suitable for modeling systems in static domains. The discrete-time Bayesian network (DTBN) was developed to account for sequential dependencies between components by explicitly incorporating time in BN formalism. It has the capability of assessing the reliability and availability of systems with complex temporal sequences and phased tasks.

In DTBN, the time line $]0, +\infty]$ is divided into $n + 1$ intervals. Each variable has a finite number $n + 1$ of states. The n first states divide the time interval $]0, T]$ (T is the mission time) into n intervals, and the last state $n + 1$ represents the time interval $]T, +\infty]$ [19]. According to the terminology introduced in [19], if the random variable A is mentioned to be in state $i(1 \leq i \leq n)$, or $A = i$, it simply means that A has failed in the ith interval or $t_A \in](i - 1)\Delta, i\Delta]$:

$$P(A = i) = P((i - 1)\Delta < t_A \leq i\Delta) = \int_{(i-1)\Delta}^{i\Delta} f_A(t)dt = F_A(i\Delta) - F_A((i - 1)\Delta)$$
(1)

where t_A is the failure time of component A, F_A is the cumulative failure distribution function, Δ is the interval length $\Delta = T/n$, and n is the time granularity.

Similarly, if A is said to be in state $n + 1$, or $A = n + 1$, this means A has survived the mission time T:

$$P(A = n + 1) = P(t_A > T) = \int_{T}^{+\infty} f_A(t)dt = 1 - F_A(T)$$
(2)

4   Model Transformation

In order to systematically transform AADL models (including the error model) into DTBN models, the transformation process is divided into two parts. First, the structure of the DTBN is constructed, including the creation of nodes and directed edges, and then the prior probability tables or CPTs are populated depending on the type of node. The following subsections describe the specific transformation methodology.

4.1 An Example of an AADL Model with Error Model

The model transformation approach is illustrated by an example of an Auxiliary Climb Control (ACC) component. The ACC is expressed in an AADL model, as shown in Fig. 1. The ACC accepts inputs from two angle-of-approach (AoA) sensors and switching commands given by an external system. When the ACC is switched on and in a normal state, it sends commands to the flight surface to control the climb of the aircraft via the *cmd* port. When the output data from the two AoAs differ, the ACC may make unsafe control decisions. At this point, the external system should shut down the ACC, and the pilot should take over control of the aircraft. If the external system does not turn off the ACC in time, incorrect decisions will be executed by the flight surface, which in turn leads to safety risks. In addition, the ACC may fail.

```
 1: --- Auxiliary Climb Control component          21: component error behavior
 2:                                                 22:   events
 3: device AuxiliaryClimbControl                    23:     ACCFail: error event;
 4: features                                        24:   transitions
 5:    aoa1:    in data port   common::sensor.aoaReading;    25:     t1: operational -[
 6:    aoa2:    in data port   common::sensor.aoaReading;    26:         1ormore(aoa1{BadValue},aoa2{BadValue})
 7:    onoff:   in feature     common::command.accOnOff;     27:         ]-> badcontrol;
 8:    cmd:     out feature    common::command.accControl;   28:   propagations
 9:                                                 29:     p1: badcontrol -[
10: annex emv2 {**                                  30:         onoff{NoACCTurnOff}]-> cmd{BadACCControl};
11: use types ErrorLibrary, acemlib;               31:     p2: all -[ACCFail]-> cmd{ServiceOmission};
12: use behavior acemlib::ACCStates;               32: end component;
13:                                                 33:
14: error propagations                             34: properties
15:    aoa1:    in propagation    {BadValue};       35:     EMV2::OccurrenceDistribution => [
16:    aoa2:    in propagation    {BadValue};       36:         Distribution => Weibull;
17:    onoff:   in propagation    {NoACCTurnOff};   37:         ShapeParameter => 2.0; ScaleParameter => 7.69E5;
18:    cmd:     out propagation   {BadACCControl,   38:     ] applies to ACCFail;
19:                               ServiceOmission}; 39: **};
20: end propagations;                               40: end AuxiliaryClimbControl;
```

Fig. 1. An Auxiliary Climb Control component modeled by AADL.

4.2 Create Event Nodes

In our approach, DTBN is event-driven. Each event node in the network represents a failure event that occurs in the system. In AADL models, we interpret the following four elements as failure events occurring in components:

- **error state**: the component enters some error state.
- **error event**: some error event has occurred within the component.
- **in propagation {error_type}**: the component received an incoming error of type *error_type*.
- **out propagation {error_type}**: the component propagated an outgoing error of type *error_type*.

We adopt the following convention: use **errState**, **errEvent**, **inProp** and **out-Prop** to represent the four elements above. We traverse each component in the system and generate a corresponding node for each of the above elements it contains, and then add it to the DTBN. For example, the ACC has three **inProp** elements, two **outProp** elements, one **errEvent** and one **errState**. These elements are mapped individually to the seven nodes shown in Fig. 2. Note that the state *operational* is the initial state of the component, not an error state. In DTBN, we only consider the fault events of the system. Therefore, we do not create a node for it.

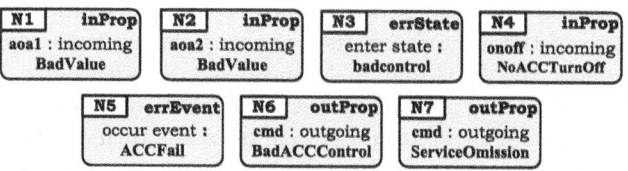

Fig. 2. The event nodes in the DTBN of ACC.

4.3   Create Logic Nodes and Edges

Error behaviors in AADL models are classified into two types: intra-component and inter-component. These error behaviors are defined through a set of rules. Each rule can be viewed as a description of dependencies (edges) between a set of event nodes in DTBN. Often, these rules may contain logical operators, and we use logic nodes to implement the corresponding logical functions. The methods for mapping these rules to nodes and edges in DTBN will be given below.

Table 1. Correspondence between logical operators in AADL and logic nodes in DTBN

Logical operator in AADL	Logic node in DTBN
or	XOR node
1ormore	OR node
*k*ormore $(k \geq 2)$	K/M node
and	AND node
all	AND node (*parents* ≥ 3)

Map Intra-component Error Behavior. Error behavior within a component is defined using two types of rules:

- **transitions**: describes migrations of component state.
- **propagations**: describes the behavior of the component to propagate errors at the port.

The expression for rule **transitions** is:

$$<src\_state> \rightarrow [trigger] \rightarrow <dst\_state>$$

where src_state is the state that the component is currently in, dst_state is the target state to which the component will migrate, $trigger$ is a logical expression containing zero or more **errEvent** and **inProp** elements. Logical operators supported by $trigger$ are and, or, $ormore$ and all. Table 1 gives the correspondence between logic nodes in DTBN and logical operators in AADL.

The rule **propagations** is similar to **transitions**. The difference between the two is that the target of **propagations**'s expression is not a *state* but an **outProp** element:

$$<src\_state> \rightarrow [trigger] \rightarrow <out\_prop>$$

The algorithm for mapping rule **propagations** is shown in Algorithm 1. The approach to mapping rule **transitations** can be analogized to it. Function $getRelatedNode()$ is used to get the corresponding node in DTBN for a given AADL element. Function $createSubGraph()$ is used to create a subgraph for the *trigger* and return the root node (node without outcoming) of it. Function $createEdge()$ adds a directed edge between two nodes. Function $createLogicNode()$ generates a logic node based on the type of logical operator.

Algorithm 1. Mapping rule **propagations**

Input: $Rule_p$: An instance of rule **propagations**;
       $DTBN$: The DTBN model;
Output: Return the DTBN model after mapping rule $Rule_p$;
 1: $N_{out} \leftarrow getRelatedNode(Rule_p.out\_prop)$;
 2: $N_{trigger} \leftarrow createSubGraph(Rule_p.trigger)$;
 3: **if** src_state is initial state or keyword all **then**
 4:    $E_{t\_o} \leftarrow createEdge(N_{trigger}, N_{out})$;
 5:    add E_{t_o} to $DTBN$;
 6: **end if**
 7: **if** src_state is an **errState** element **then**
 8:    $N_{PAND} \leftarrow createLogicNode(PAND)$;
 9:    $N_{src} \leftarrow getRelatedNode(Rule_p.src\_state)$;
10:    $E_{t\_p} \leftarrow createEdge(N_{trigger}, N_{PAND})$;
11:    $E_{s\_p} \leftarrow createEdge(N_{src}, N_{PAND})$;
12:    $E_{p\_o} \leftarrow createEdge(N_{PAND}, N_{out})$;
13:    add $E_{t\_p}, E_{s\_p}, E_{p\_o}$ to $DTBN$;
14: **end if**
15: **return** $DTBN$;

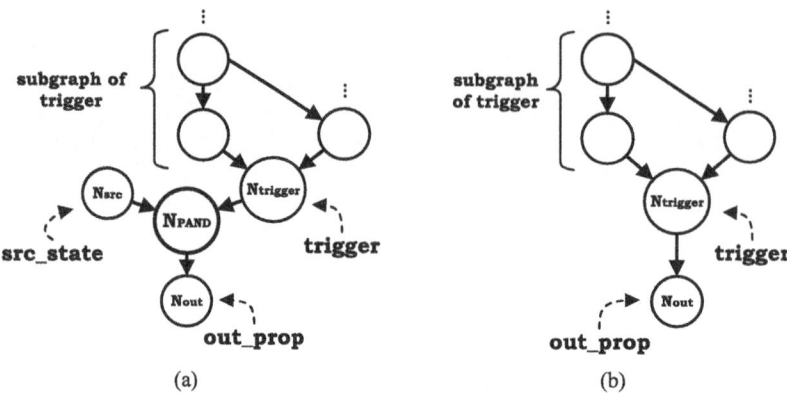

Fig. 3. Two cases for mapping rule **propagations**. (a) *src_state* is an **errState** element. (b) *src_state* is an initial state or the keyword *all*.

If *src_state* is an **errState** element, the rule is triggered by the precondition that the system has to be in the *src_state* state before the *trigger*. In DTBN, this sequential dependency can be expressed using the logic of priority AND (PAND) node. The PAND node indicates that the output occurs if and only if all input events occur in a particular order. The order of occurrence is the order in which the input events are connected to the node from left to right. In this case (Lines 7–13 of Algorithm 1, Fig. 3(a)), first, a logic node N_{PAND} is created. Then, N_{src} (the corresponding node of *src_state*) and $N_{trigger}$ (the root node of *trigger*'s subgraph) are set to be the parents of N_{PAND}. Finally, a directed edge is created from N_{PAND} to N_{out} (the corresponding node of *out_prop*).

If *src_state* is an initial state or the keyword *all*, it means only the *trigger* must be satisfied to trigger the rule. In this case (Lines 3–5 of Algorithm 1, Fig. 3(b)), adding a directed edge from $N_{trigger}$ to N_{out} is sufficient.

We describe in detail how function *createSubGraph*() constructs a subgraph for the *trigger*. If the *trigger* is a single **errEvent** or **inProp** element, at this point, its corresponding node in the DTBN is returned. Otherwise, the *trigger* is decomposed into a triple (*subexpr1, op, subexpr2*), where *op* is the highest priority logical operator in the expression, *subexpr1* and *subexpr2* are the two sub-expressions that *op* connect. In this case, first, to create a logic node N_{OP} according to the type of *op*. Then, the nodes $N_{subexpr1}$ and $N_{subexpr2}$, the root nodes of the subgraphs of *subexpr1* and *subexpr2*, respectively, are set as the parents of N_{OP}. Finally, N_{OP} is returned. The subgraphs of *subexpr1* and *subexpr2* can be obtained by recursively executing the *createSubGraph*() function.

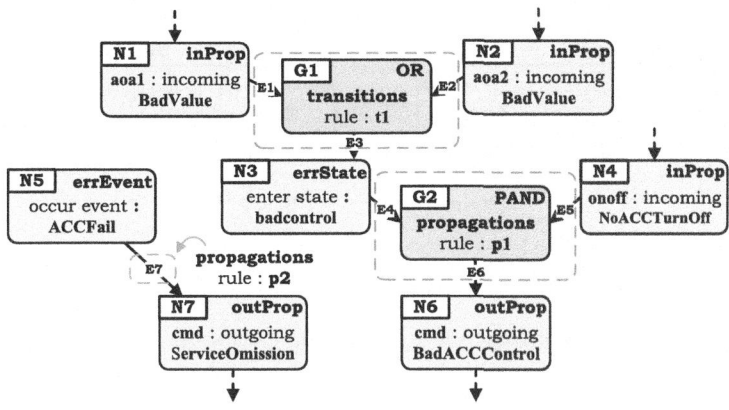

Fig. 4. The complete DTBN of ACC.

After creating all logic nodes and edges, the complete DTBN of ACC is shown in Fig. 4. The nodes and edges contained in the dashed boxes of the three colors are created according to the rules t1, p1 and p2 in ACC, respectively. The dashed arrows represent the connections between ACC and external systems.

Map Inter-component Error Behavior. If the component is a **system** component, the rule **connections** is used to describe the connections of ports between its subcomponents. Each connection has the following form:

$$<c1.out\_port> \rightarrow <c2.in\_port>$$

where *out_port* is an output port of subcomponent $c1$, and *in_port* is an input port of another subcomponent $c2$. These connections are mapped as dependencies between nodes associated with error propagations on the relevant ports. As shown in Fig. 5, an error $Error1$ propagates through the connection from the output port $out1$ of subcomponent $Subcomp1$ to the input port $in1$ of subcomponent $Subcomp2$. Assume that $E1$ is the **outProp** element describing the outgoing of $Error1$ from $out1$, while the **inProp** element $E2$ describes the incoming of $Error1$ from $in1$. Then, construct a directed edge from N_{out1} (the corresponding node of $E1$) to N_{in1} (the corresponding node of $E2$). If multiple errors propagate through a pair of ports, all pairs of nodes associated with these propagations are found in DTBN to construct directed edges.

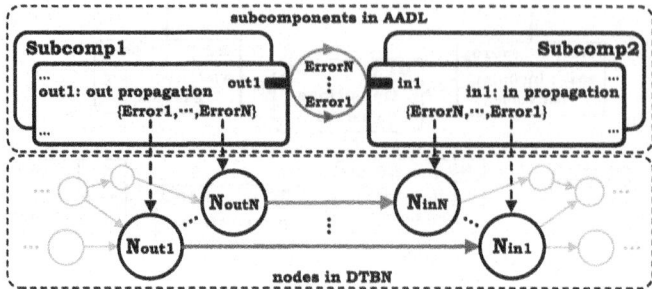

Fig. 5. Mapping a **connections** rule when multiple errors propagate through a pair of ports.

4.4   Simplify the DTBN

The DTBN generated through the above steps has many nodes that contain only one parent and one child node, called process nodes. These nodes only describe the details of error propagations in the model, and removing them will not change the inference results. Removing redundant nodes simplifies the structure of the DTBN, speeds up inference, and allows system engineers to focus on the main paths of error propagation. We use the following two rules to simplify DTBN:

- Rule 1: If the parent of a process node is a logic node, merge two nodes.
- Rule 2: Delete a process node if it does not satisfy Rule 1.

In the complete DTBN of ACC, nodes $N3$ and $N6$ conform to Rule 1. As a result, $N3$ is merged with the logic node $G1$, while $N6$ is merged with the logic node $G2$. Nodes $N1$, $N2$, $N4$, and $N7$ are deleted for conforming to Rule 2.

4.5   Populate Prior Probability Tables

All leaf nodes in the DTBN are basic events in the system. The user is required to provide the prior probability of occurrence of the basic event. Usually, the basic event is assigned a specific failure probability distribution. In AADL models, we can use the **EMV2::OccuranceDistubution** property to annotate the type and parameters of the failure distribution function for the basic event.

The failure probability distribution of the event $ACCFail$ in ACC is specified as a Weibull distribution with a cumulative failure distribution function:

$$F(t) = 1 - exp\left[-(\frac{t}{\alpha})^{\beta}\right] \tag{3}$$

where α is the scale parameter, β is the shape parameter, and t is the time variable. For $ACCFail$, it can be known from the annotations that $\alpha = 7.69 \times 10^5 h$ and $\beta = 2.0$.

We use $P(ACC = i)$ to denote the probability that ACC fails in the ith time interval. Assume $T = 10000h$ (mission time), $n = 2$ (time granularity), therefore $\Delta = 5000h$. The time intervals are $]0, \Delta]$, $]\Delta, 2\Delta = T]$, and $]T, +\infty[$. Then, according to Eq. (1), (2) and (3):

$$P(ACC = 1) = F(5000) - F(0) \approx 4.225 \times 10^{-5}$$
$$P(ACC = 2) = F(10000) - F(5000) \approx 1.267 \times 10^{-4}$$
$$P(ACC = 3) = 1 - F(10000) \approx 9.998 \times 10^{-1}$$

4.6 Populate CPT

All nodes in DTBN, except leaf nodes, contain at least two parent nodes and hence are logic nodes. Five types of logic nodes can be included in the DTBN: OR, AND, XOR, PAND and K/M nodes. Here, we provide the method to populate the CPT of the logic node according to the logic type.

Given a logic node G, $eventG$ is the event represented by G, t_g is the time interval at which $eventG$ occurs, and $Parent_i(i = 1, 2, \ldots, m)$ is the ith parent node of G, where m is the number of parents of G. Let sequence $T = <t_1, t_2, \ldots, t_m>$, where $t_i(i = 1, 2, \ldots, m)$ is the time interval at which the event represented by $Parent_i$ occurs. The order of the elements in T is the order in which the parents of G are connected to G from left to right. $1 \leq t_g, t_i \leq n+1$, where n is the temporal granularity of the mission time.

For any type of G, the conditional probability distribution of G has a similar form:

$$P(G = t_g \mid T) = \begin{cases} \begin{cases} 1 & \langle condition \rangle \\ 0 & \text{otherwise} \end{cases} & t_g \leq n \\ 1 - \sum\limits_{i=1}^{n} P(G = i \mid T) & t_g = n + 1 \end{cases} \tag{4}$$

where $\langle condition \rangle$ is the condition that makes the conditional probability one when $t_g \leq n$. Table 2 shows the corresponding $\langle condition \rangle$ for each type of logic node. When $t_g = n + 1$, for each evidence T, it is necessary to ensure that the sum of the probabilities, over all intervals, must equal one.

Table 2. Corresponding $\langle condition \rangle$ for each type of logic node in DTBN

Logic node in DTBN	$\langle condition \rangle$ in Eq. (4)
OR node	$t_g = t_{min}$
AND node	$t_g = t_{max}$
K/M node	$t_g = t_{klarge}$
XOR node	$t_g = t_{min}, c(t_{min}, T) = 1$
PAND node	$t_g = t_{max}, isAsc(T)$

- **OR node:** $t_{min} = \min(t_1, t_2, \ldots, t_m)$. $t_g = t_{min}$ indicates that the conditional probability is one when $eventG$ occurs in the same time interval as the earliest parent event.
- **AND node:** $t_{max} = \max(t_1, t_2, \ldots, t_m)$. $t_g = t_{max}$ indicates that the conditional probability is one when $eventG$ occurs in the same time interval as the latest parent event.
- **K/M node:** t_{klarge} is the kth largest element in T. $t_g = t_{klarge}$ means $eventG$ occurs in the same time interval as the kth occurring parent event, where $1 \le k \le m$. When $k = 1$, the K/M node is equivalent to the OR node.
- **XOR node:** The XOR node can be obtained by adding constraints to the OR node. The main difference between the two is that, in a time interval, $eventG$ occurs when one and only one parent event occurs. The function $c(t_{min}, T)$ counts the number of times t_{min} appears in T, $c(t_{min}, T) = 1$ represents the above constraint.
- **PAND node:** The PAND node requires T to be monotonically increasing. It can be obtained by adding constraints to the AND node. The function $isAsc(T)$ determines whether T is monotonically increasing or not.

Assume that $n = 2$ and G has only two parent nodes, A and B. The CPTs of the PAND and XOR nodes are shown in Fig. 6(a) and (b), respectively.

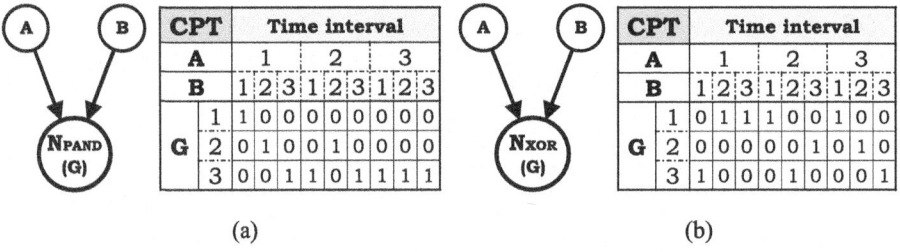

(a)   (b)

Fig. 6. (a) The CPT of the PAND node. (b) The CPT of the XOR node.

5   A DTBN-Based Framework of Quantitative Safety Analysis for AADL Models

In this section, we give the specific process of the QuanSafe. Figure 7 shows the workflow of the framework, which consists of three phases. First, AADL models with the error model are constructed based on the real system and the specifications. Second, according to the mapping rules, the AADL models are transformed into DTBN models. Finally, a qualitative analysis and three quantitative analyses, which are risk analysis, diagnostic analysis and sensitivity analysis, of the system are performed using DTBN. The results of the analysis will be used for model improvement and iteration.

The model transformation and quantitative analysis processes above have been automated by QuanSafe, which is developed into a plugin for the OSATE tool and open-source on GitHub for other researchers. The implementation details of the three quantitative analysis methods are given below.

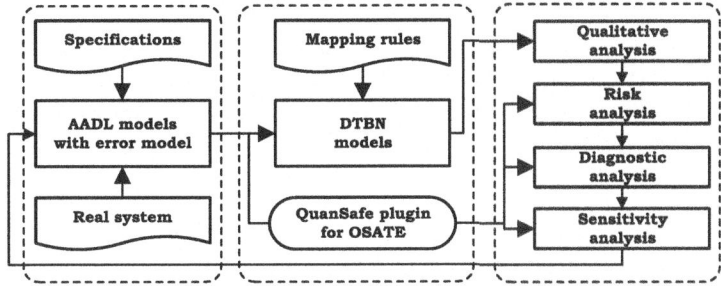

Fig. 7. The workflow of the QuanSafe.

5.1 Risk Analysis

We show the risk level of a system by quantifying its probability of failure over its life cycle, which can be realized by forward prediction in DTBN. In DTBN, a system failure event is represented by the root node, and we use root event (RE) to refer to it. The cumulative probability of RE in the first $i(1 \leq i \leq n)$ time intervals is:

$$P(RE <= i) = \sum_{1}^{i} P(RE = j) \tag{5}$$

By executing Eq. (5) for each value of i, the change in system failure probability over time can be obtained.

5.2 Diagnostic Analysis

One of the advantages of DTBN over existing methods is its ability to perform diagnostic analysis (or probability updating) given new observations (evidence). Given a specific state of the RE, the posterior probabilities of the basic events (BEs) within the system are computed, and the Most Probable Cause(s) (MPC) [6] of the RE is determined. In this way, a basis for system optimization and improvement can be obtained. Typically, we have two different types of observations for RE:

– Type 1: The RE is known to occur in the jth time interval, i.e., $RE = j$.
– Type 2: The RE is known to occur in the first j time intervals, i.e., $RE <= j$.

where $1 \leq j \leq n$.

To determine MPC, one of the above two observations can be used as evidence to perform a diagnosis. For example, if evidence of Type 1 is observed, the probability that a basic event BE_i in the system occurs before the RE occurs can be calculated as:

$$P(BE_i \leq j \mid RE = j) \tag{6}$$

5.3 Sensitivity Analysis

Sensitivity analysis can be used to measure the contribution of any event in the DTBN to the probability of the RE. The nodes with the highest impact can be identified by calculating and ranking their importance measures.

The Risk Achievement Worth (RAW) [5] is a commonly used measure, which indicates the increase in the RE probability if a given BE occurs. It shows the focus of the prevention activities in the maintenance phase of the system. In DTBN, for a basic event BE_i, its RAW during the mission time is:

$$RAW(BE_i) = P(RE \leq n \mid BE_i \leq n) - P(RE \leq n) \tag{7}$$

In QuanSafe, we use timed-RAW(tRAW), an enhanced importance measure that improves the evaluation results of RAW using the temporal features provided by DTBN. Considering that the probability of BEs varies over time, BEs may have different results in terms of prioritization according to RAW values in different time intervals. The tRAW computes the RAW for each BE in a specific time interval. For a basic event BE_i, its tRAW in time interval $j(1 \leq j \leq n)$ is:

$$tRAW_j(BE_i) = P(RE = j \mid BE_i = j) - P(RE = j) \tag{8}$$

The tRAW shows the importance of BE in different system lifecycle phases, leading to more precise resource allocation for maintenance activities.

6   Case Study

6.1   The Aircraft System

The Aircraft System is modeled by the following elements:

- **Angle of Attack (AoA) sensor**: The AoA is used to measure the angle of inclination of an aircraft. In this case, there are two kinds of AoAs, which come from different vendors and have different failure rates. The AoA may generate a bad value.
- **Auxiliary Climb Control (ACC)**: The ACC controls aircraft climb automatically. This component has been mentioned as a sample in Sect. 4.
- **AoA Discrepancy Detector**: The AoA Discrepancy Detector signals the watchdog if two AoAs produce different outputs. The detector may also fail.
- **Watchdog**: The watchdog receives inputs from the AoA Discrepancy Detector and the pilot. When the watchdog detects an ACC anomaly and is acknowledged by the pilot, it will turn off the ACC. It may also fail.
- **Pilot**: We consider the impact of human factors on system safety. When the watchdog fails, the pilot must infer the ACC anomaly from the aircraft's flight behavior, turn off the ACC and control the flight surface. While the pilot controls the airplane, he may also make flight control mistakes.
- **Flight Surface**: The flight surface directly controls the aircraft. It receives commands from the ACC or the pilot. The flight surface may experience a mechanical failure that prevents it from responding to control commands.
- **Engine**: Engines power the aircraft. In the system, there are four engines and they are redundant with each other. Only if more than two engines stop working simultaneously, than the aircraft will lose power.

According to the above description, we constructed the structural and error model of the Aircraft System using AADL. The structural model of the system is shown in Fig. 8(a). The system uses AoA of Type A.

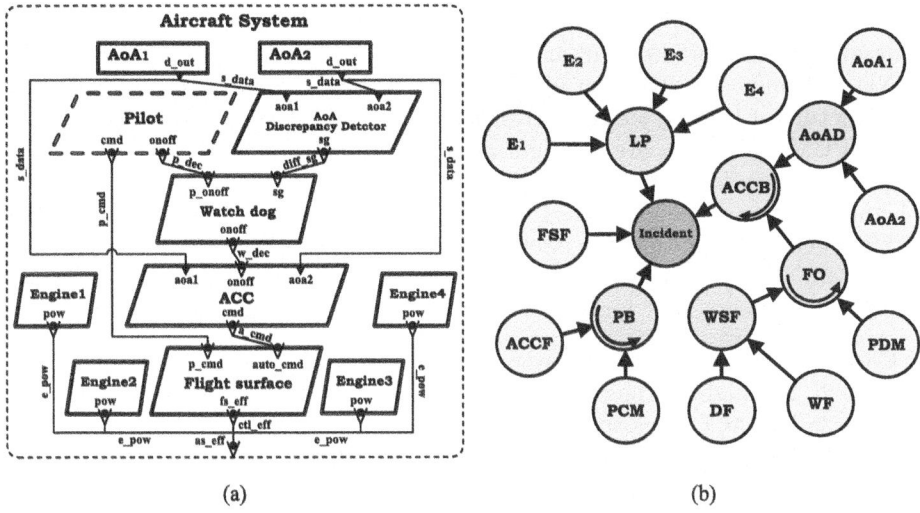

(a)   (b)

Fig. 8. (a) The AADL structural model of the Aircraft System. (b) The DTBN of the Aircraft System. (Color figure online)

Table 3. Nodes in DTBN of Fig. 8(b).

Node	Type	Event	$\alpha(h)$	β
Incident	OR, Root	Aircraft incident		
LP	K/M ($K = 3$)	System loss power		
ACCB	PAND	ACC performs bad control		
PB	PAND	Pilot performs bad control		
AoAD	OR	AoA sensors produce different outputs		
FO	PAND	Failed to turn off ACC in time		
WSF	AND	Watchdog system fails		
E_1-E_4	Leaf	Engine i fails ($1 \leq i \leq 4$)	6.67×10^5	2.0
FSF	Leaf	Flight surface fails	5.00×10^8	1.5
ACCF	Leaf	ACC fails	7.69×10^5	2.0
PCM	Leaf	Pilot makes flight control mistakes	6.67×10^5	1.0
PDM	Leaf	Pilot makes decision mistakes	5.00×10^5	1.0
WF	Leaf	Watchdog fails	6.67×10^5	0.8
DF	Leaf	AoA Discrepancy Detector fails	1.00×10^6	1.0
AoA$_{1,2}$ (Type A)	Leaf	AoA$_i$ (Type A) fails ($1 \leq i \leq 2$)	4.00×10^4	1.0
AoA$_{1,2}$ (Type B)	Leaf	AoA$_i$ (Type B) fails ($1 \leq i \leq 2$)	6.67×10^5	1.0

6.2 Model Transformation and Qualitative Analysis

Based on the method proposed in Sect. 4, the AADL model of the Aircraft System is transformed into the DTBN model by QuanSafe, as shown in Fig. 8(b). Table 3 gives a detailed description of each node in the DTBN. All nodes in yellow represent the system's basic events; the green one is the root node, which represents an aircraft incident in the system. The direction of the arrow in a PAND node indicates the order in which its parents are connected to it. Based on the structure of DTBN, we can perform qualitative analysis. For example, both nodes, PCM and PDM, are the rightmost parent nodes of a PAND node, which means that the pilot is regarded as the backup of the system. The node WSF has a higher priority than the node PCM, which implies that the watchdog, in its normal operation, prevents the pilot from posing a safety risk due to mishandling the system. In addition, the root node is an OR node, and it has four parent nodes, which means that four cases lead to an aircraft incident; they are:

- Case 1: The system loss of power.
- Case 2: The flight surface fails.
- Case 3: The ACC performs bad control.
- Case 4: The pilot performs bad control when ACC is off.

6.3   Risk Analysis

QuanSafe quantitatively analyzes the risk of the system using Eq. (5) mentioned in the previous sections. The life cycle of the Aircraft System is set to 100000 hours, i.e., mission time $T = 100000h$, with a time granularity n of 10. Then, predictions are made using DTBN. The results are shown as the orange bars in Fig. 9. The x-axis represents the time intervals, each with a length of $\Delta = T/n = 10000h$, and the y-axis represents the cumulative probability of an aircraft incident, which is the root event, occurring in the system. We observe that the probability of the incident grows over time at a progressively faster rate and reaches a maximum value in the last time interval: 2.22×10^{-3}.

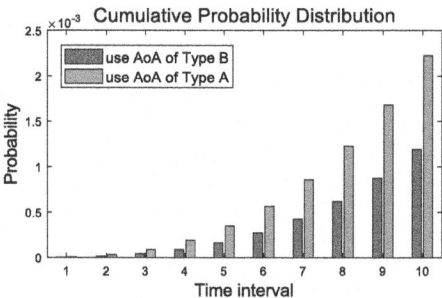

Fig. 9. The cumulative probability of an aircraft incident occurring in the system.

Here, we discuss the performance comparison of our approach with two representative works [34,35]. The AADL model of the Aircraft System is transformed into a continuous time Markov chain (CTMC) according to [35] while into a Networks of Stochastic Hybrid Automata (NSHA) model according to [34]. The

Table 4. Performance comparison of risk analysis methods for the Aircraft System.

$|AoA|$	PRISM (CTMC)			UPPAAL-SMC (NSHA)				Our Approach (DTBN)			
	States	Time (s)	Risk (10^{-3})	Samples	Time (s)	Risk (10^{-3})	Rel Err. (%)	n	Time (s)	Risk (10^{-3})	Rel Err. (%)
2	25242	201	2.181	50000	18	2.075 ± 0.177	8.105	5	<1	2.329	6.790
				100000	36	2.199 ± 0.129	5.896	10	<1	2.223	1.949
				500000	183	2.176 ± 0.057	2.623	15	2	2.194	0.606
				1000000	369	2.164 ± 0.040	1.850	20	**9**	2.177	**0.165**
3	79242	1134	2.221	50000	29	2.043 ± 0.177	7.999	5	<1	2.372	6.809
				100000	58	2.143 ± 0.128	5.758	10	<1	2.263	1.917
				500000	274	2.235 ± 0.058	2.629	15	2	2.233	0.559
				1000000	568	2.245 ± 0.041	1.863	20	**10**	2.216	**0.219**
4	241242	5199	2.233	50000	41	2.286 ± 0.187	8.368	5	<1	2.386	6.834
				100000	78	2.238 ± 0.131	5.856	10	<1	2.275	1.900
				500000	390	2.221 ± 0.058	2.609	15	3	2.245	0.528
				1000000	800	2.251 ± 0.041	1.857	20	**13**	2.227	**0.256**

system risk is then verified using the PRISM (a probabilistic model checker) and the UPPAAL-SMC (a statistical model checker), respectively. The confidence level of the UPPAAL-SMC is set to 99%. The results are shown in Table 4. In general, the risks obtained by CTMC are exact, while NSHA and our methods produce approximate results. Column *Rel Err.* shows the relative error of each approach to the results of the CTMC-based method. Column $|AoA|$ represents the number of AoA included in the system.

Taking the Aircraft System with two AoAs as an example, the larger the value of n, the higher the accuracy of DTBN. Starting from $n = 10$, the relative error produced by DTBN stays within 2%. Such an error is perfectly acceptable since a risk analysis that is accurate to the right order of magnitude is often sufficient in practice [27]. Although pursuing higher accuracy (i.e., selecting a larger value of n) also increases the execution time of the DTBN, the runtime required to obtain a satisfactory error is still much lower than the runtime of the CTMC-based method. Actually, at a time granularity n of 20, DTBN reaches a high accuracy of 99.835% (i.e., $1 - 0.165\%$) while taking only a tiny fraction (4.48%, i.e., $9s/201s$) of the time the CTMC spends. This advantage will be even more obvious when the allowed error margin is more tolerant.

The comparison with the NSHA model shows the advantages of our approach over SMC-based methods for handling systems with rare events. The Rel Err. produced by the NSHA model is significantly higher than ours due to the presence of rare event FSF (whose failure rate is about $2.0 \times 10^{-9}/h$) in the system. More samples are required to attain a high confidence ratio and a low error margin, which increases the prediction time-consuming. Take the case of $|AoA| = 2$ as an example. Even with a sample size of 1 million, it's Rel Err. (1.850%) is only close to that of the DTBN with a time granularity of 10 (1.949%), while its runtime (369s) far exceeds our approach (<1 s). In fact, the accuracy of DTBN is only related to the degree of temporal discretization (the value of n). Therefore, its performance is not affected by rare events in the system.

As the $|AoA|$ increases (i.e., system scales and complexities increase), the execution time of all three methods goes up to varying degrees. When the $|AoA|$

becomes twice as large (from 2 to 4), the state space of the system's CTMC model expands by about 10 times, while the runtime increases by about 26 times. Keeping the Rel Err. close, the runtime of the NSHA model with different sample sizes increases by a factor of 2 on average. In contrast, our method increases by only 47% on average, which implies that our approach is more promising for dealing with larger and more complex systems.

6.4   Diagnostic Analysis

Based on the risk analysis results, we can use QuanSafe to diagnose the MPC of the aircraft incident. Here we consider the evidence that the incident is known to have occurred at ith $(1 \leq i \leq 10)$ time interval and compute the posterior probabilities of the BEs in the system using Eq. (6). The cases of the incident occurring within each time interval are considered, and we focus on the top six BEs with the highest posterior probability in the diagnostic results for each time interval. The results are shown in Fig. 10(a).

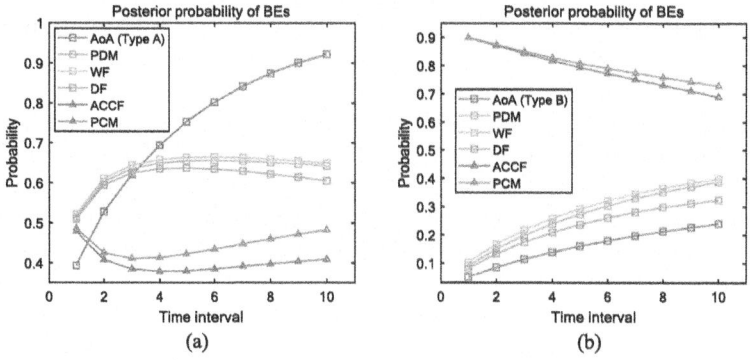

Fig. 10. (a) Diagnostic analysis for the system uses AoAs of Type A. (b) Diagnostic analysis for the system uses AoAs of Type B.

According to the structure of DTBN, we can categorize these basic event nodes into two classes. The first category of nodes uses square markers (including AoA, PDM, WF, and DF), and they can all affect node ACCB, while the second category of nodes (which use triangle markers) can all affect node PB. The higher posterior probability of the first category of nodes indicates that the incident is mainly caused by the ACC's bad control. At the same time, we observe that the posterior probability of the AoA failure grows rapidly and remains high after the third time interval. This suggests that the AoA sensor is the least reliable component within the system.

We optimize the system by replacing the AoA with a lower failure rate (AoA of Type B, mentioned in Table 3) and then re-execute the risk analysis on the modified system to evaluate the improvement. The blue bars in Fig. 9 show the cumulative probability of an incident for the modified system. The maximum probability of an aircraft incident over the system's lifetime is reduced to 1.19×10^{-3}, a reduction of 46.41%, and it grows much slower than the system using the AoA of Type A. It proves that our optimization measures are effective and that the modified system will be safer. Further, Fig. 10(b) shows the diagnostic analysis results on the modified system. At this point, the posterior probabilities of the nodes, ACCF and PCM, are higher than the others, indicating that the MPC of the incident has changed to ACC failures and pilot control mistakes.

6.5 Sensitivity Analysis

The sensitivity analysis aims to find a practical basis for prevention activities. It answers which BEs should be prioritized when performing maintenance on the system. We calculate the tRAW and RAW of all BEs in the improved system using Eq. (8) and Eq. (7), respectively. The results are recorded in Table 5. The prioritization based on the RAW measure shows that we should prevent FSF (flight surface failure) in the first place. As the FSF is a single point of failure, it will definitely lead to an aircraft incident if it occurs. For the tRAW measure, we selected its results at time intervals 1, 4, 7 and 10. We note that several BEs have significant discrepancies in priorities across time intervals. For example, from the 1st to the 10th time interval, the priority of E_1-E_4 increases from 8 to 5, while the priority of $AoA_{1,2}$ decreases from 6 to 8. This suggests that during the 1st time interval (i.e., the first 10000 h of the system life cycle), the prevention activities should prioritize avoiding AoA failures over engine failures, whereas the opposite conclusion is reached during the 10th time interval (i.e., 90000–100000 hours). It proves that the importance measure of BEs is indeed time-specific, and using the tRAW measure allows for more precise guidance of maintenance activities.

Table 5. The RAW and tRAW of basic events.

Node	$tRAW_1$	Rank	$tRAW_4$	Rank	$tRAW_7$	Rank	$tRAW_{10}$	Rank	RAW	Rank
FSF	9.99×10^{-1}	1	9.99×10^{-1}	1	9.99×10^{-1}	1	9.99×10^{-1}	1	9.99×10^{-1}	1
ACCF	1.49×10^{-2}	2	1.42×10^{-2}	2	1.36×10^{-2}	2	1.30×10^{-2}	3	1.28×10^{-2}	3
PCM	1.69×10^{-4}	3	2.70×10^{-3}	3	8.25×10^{-3}	3	1.68×10^{-2}	2	1.44×10^{-2}	2
DF	2.00×10^{-5}	4	2.11×10^{-4}	5	5.05×10^{-4}	5	8.43×10^{-4}	6	7.63×10^{-4}	6
PDM	1.00×10^{-5}	5	4.43×10^{-4}	4	1.94×10^{-3}	4	4.85×10^{-3}	4	3.97×10^{-3}	4
$AoA_{1,2}$	6.63×10^{-6}	6	6.88×10^{-5}	7	1.62×10^{-4}	8	2.66×10^{-4}	8	2.29×10^{-4}	8
WF	5.82×10^{-6}	7	8.27×10^{-5}	6	2.25×10^{-4}	7	4.08×10^{-4}	7	3.27×10^{-4}	7
E_1-E_4	1.52×10^{-7}	8	3.86×10^{-5}	8	3.57×10^{-4}	6	1.45×10^{-3}	5	1.42×10^{-3}	5

7 Conclusion and Future Work

In this paper, we introduce QuanSafe, a DTBN-based framework of quantitative safety analysis for AADL models, to address the challenges that mainstream methods face for safety analysis in AADL. DTBN provides a complete description of the dynamic and temporal behavior of AADL models and offers its unique diagnostic analysis capabilities. At the same time, DTBN effectively avoids state space explosion and handles systems with rare events well. QuanSafe provides a complete algorithm to transform AADL models into DTBN models and supports multiple automated safety analysis methods with improved metrics. We apply the QuanSafe to the Aircraft System. The experimental results show that our approach has higher efficiency and more comprehensive analysis capabilities than existing research.

In the future, we aim to refine the framework's design and performance by applying our approach to more real-world safety-critical systems. Further, we plan to combine our work with assorted AADL extension languages to enable the quantitative safety analysis of various complex systems, such as concurrent and synchronous systems.

Acknowledgments. This work was supported in part by the National Key Research and Development under Project 2022YFB3305200, the NSFC Project 61972150, Shanghai Trusted Industry Internet Software Collaborative Innovation Center; and the Fundamental Research Funds for the Central University.

References

1. Agha, G., Palmskog, K.: A survey of statistical model checking. ACM Trans. Model. Comput. Simul. (TOMACS) **28**(1), 1–39 (2018)
2. Ahmad, E., Dong, Y., Larson, B.R., Lü, J., Tang, T., Zhan, N.: Behavior modeling and verification of movement authority scenario of Chinese train control system using AADL. Sci. China Inf. Sci. **58**(11), 1–20 (2015)
3. Bae, K., Ölveczky, P.C., Meseguer, J.: Definition, semantics, and analysis of multirate synchronous AADL. In: Jones, C., Pihlajasaari, P., Sun, J. (eds.) FM 2014. LNCS, vol. 8442, pp. 94–109. Springer, Cham (2014). https://doi.org/10.1007/978-3-319-06410-9_7
4. Bao, Y., Chen, M., Zhu, Q., Wei, T., Mallet, F., Zhou, T.: Quantitative performance evaluation of uncertainty-aware hybrid AADL designs using statistical model checking. IEEE Trans. Comput. Aided Des. Integr. Circuits Syst. **36**(12), 1989–2002 (2017)
5. Borgonovo, E.: Sensitivity analysis. In: Tutorials in Operations Research: Advancing the Frontiers of OR/MS: From Methodologies to Applications, pp. 52–81 (2023)
6. Boudali, H., Dugan, J.B.: A discrete-time Bayesian network reliability modeling and analysis framework. Reliab. Eng. Syst. Saf. **87**(3), 337–349 (2005)
7. Bozzano, M., Bruintjes, H., Cimatti, A., Katoen, J.-P., Noll, T., Tonetta, S.: COMPASS 3.0. In: Vojnar, T., Zhang, L. (eds.) TACAS 2019. LNCS, vol. 11427, pp. 379–385. Springer, Cham (2019). https://doi.org/10.1007/978-3-030-17462-0_25
8. Bozzano, M., Cimatti, A., Katoen, J.P., Nguyen, V.Y., Noll, T., Roveri, M.: Safety, dependability and performance analysis of extended AADL models. Comput. J. **54**(5), 754–775 (2011)
9. Bruintjes, H., Katoen, J.P., Lesens, D.: A statistical approach for timed reachability in AADL models. In: 2015 45th Annual IEEE/IFIP International Conference on Dependable Systems and Networks, pp. 81–88. IEEE (2015)

10. Čepin, M., Čepin, M.: Event tree analysis. In: Assessment of Power System Reliability: Methods and Applications, pp. 89–99 (2011)
11. Douc, R., Moulines, E., Priouret, P., Soulier, P.: Markov Chains. Springer, Cham (2018)
12. Feiler, P.: Open source AADL tool environment (OSATE). In: AADL Workshop, Paris, pp. 1–40 (2004)
13. Giua, A., Silva, M.: Petri nets and automatic control: a historical perspective. Annu. Rev. Control. **45**, 223–239 (2018)
14. Guo, Y., Zhong, M., Gao, C., Wang, H., Liang, X., Yi, H.: A discrete-time Bayesian network approach for reliability analysis of dynamic systems with common cause failures. Reliab. Eng. Syst. Saf. **216**, 108028 (2021)
15. Hadad, A.S.A., Ma, C., Ahmed, A.A.O.: Formal verification of AADL models by Event-B. IEEE Access **8**, 72814–72834 (2020)
16. Junges, S., Guck, D., Katoen, J.P., Stoelinga, M.: Uncovering dynamic fault trees. In: 2016 46th Annual IEEE/IFIP International Conference on Dependable Systems and Networks (DSN), pp. 299–310. IEEE (2016)
17. Kabir, S.: An overview of fault tree analysis and its application in model based dependability analysis. Expert Syst. Appl. **77**, 114–135 (2017)
18. Kabir, S., Papadopoulos, Y.: Applications of Bayesian networks and petri nets in safety, reliability, and risk assessments: a review. Saf. Sci. **115**, 154–175 (2019)
19. Khakzad, N., Khan, F., Amyotte, P.: Risk-based design of process systems using discrete-time Bayesian networks. Reliab. Eng. Syst. Saf. **109**, 5–17 (2013)
20. Lee, J., Bae, K., Ölveczky, P.C., Kim, S., Kang, M.: Modeling and formal analysis of virtually synchronous cyber-physical systems in AADL. Int. J. Softw. Tools Technol. Transfer **24**(6), 911–948 (2022)
21. Legay, A., Lukina, A., Traonouez, L.M., Yang, J., Smolka, S.A., Grosu, R.: Statistical model checking. In: Steffen, B., Woeginger, G. (eds.) Computing and Software Science. LNCS, vol. 10000, pp. 478–504. Springer, Cham (2019). https://doi.org/10.1007/978-3-319-91908-9_23
22. Liu, Y., Shen, G., Huang, Z., Yang, Z.: Quantitative risk analysis of safety-critical embedded systems. Softw. Qual. J. **25**, 503–527 (2017)
23. Mian, Z., Bottaci, L., Papadopoulos, Y., Mahmud, N.: Model transformation for analyzing dependability of AADL model by using hip-hops. J. Syst. Softw. **151**, 258–282 (2019)
24. Pearl, J.: Bayesian networks (2011)
25. Procter, S., Feiler, P.: The AADL error library: an operationalized taxonomy of system errors. ACM SIGAda Ada Lett. **39**(1), 63–70 (2020)
26. Sharma, K.D., Srivastava, S.: Failure mode and effect analysis (FMEA) implementation: a literature review. J. Adv. Res. Aeronaut. Space Sci. **5**(1–2), 1–17 (2018)
27. Sinha, S., Goyal, N.K., Mall, R.: Survey of combined hardware-software reliability prediction approaches from architectural and system failure viewpoint. Int. J. Syst. Assurance Eng. Manag. **10**, 453–474 (2019)
28. Tan, Y., Zhao, Y., Ma, D., Zhang, X.: A comprehensive formalization of AADL with behavior annex. Sci. Program. **2022**, 1–26 (2022)
29. Wakankar, A., Kabra, A., Bhattacharjee, A., Karmakar, G.: Architectural model driven dependability analysis of computer based safety system in nuclear power plant. Nucl. Eng. Technol. **51**(2), 463–478 (2019)
30. Wei, X., Dong, Y., Li, X., Wong, W.E.: Architecture-level hazard analysis using AADL. J. Syst. Softw. **137**, 580–604 (2018)

31. Wei, X., Dong, Y., Yang, M., Hu, N., Ye, H.: Hazard analysis for AADL model. In: 2014 IEEE 20th International Conference on Embedded and Real-Time Computing Systems and Applications, pp. 1–10. IEEE (2014)

32. Wei, X., Dong, Y., Ye, H.: Qasten: integrating quantitative verification with safety analysis for AADL model. In: 2015 International Symposium on Theoretical Aspects of Software Engineering, pp. 103–110. IEEE (2015)

33. Xie, J., Tan, W., Fang, B., Huang, Z.: Towards a statistical model checking method for safety-critical cyber-physical system verification. Secur. Commun. Netw. **2021**, 1–12 (2021)

34. Yang, C., et al.: Uncertainty modeling and quantitative evaluation of cyber-physical systems. In: 2021 IEEE 45th Annual Computers, Software, and Applications Conference (COMPSAC), pp. 874–883. IEEE (2021)

35. Yuan, C., Wu, K., Chen, G., Mo, Y.: An automatic transformation method from AADL reliability model to CTMC. In: 2021 IEEE International Conference on Information Communication and Software Engineering (ICICSE), pp. 322–326. IEEE (2021)

36. Zhang, F., Zhao, Y., Ma, D., Niu, W.: Formal verification of behavioral AADL models by stateful timed CSP. IEEE Access **5**, 27421–27438 (2017)

A Event-B-Based Approach for Schedulability Analysis For Real-Time Scheduling Algorithms through Deadlock Detection

Jiale Quan and Qin Li$^{(\boxtimes)}$ (iD)

Shanghai Key Laboratory of Trustworthy Computing, East China Normal University,
Shanghai, China
qli@sei.ecnu.edu.cn

Abstract. Event-B is a refinement-based formal method that enables incremental modeling of complex systems and supports verifying system properties. Real-time systems adhere to strict timing constraints by the tasks within the system. The real-time scheduling algorithm serves as the cornerstone to guarantee the timely completion of tasks. Therefore, modeling real-time scheduling algorithms and verifying schedulability represent prominent areas of focus within the realm of real-time systems. While existing approaches often employ model checking, the scalability of the model and the problem of state explosion during verification remain challenges. Relying on theorem proving, Event-B allows for rigorous verification of system properties and circumvents state explosion. Benefiting from model refinement, the abstract model can be extended and refined to implement various scheduling algorithms.

This paper introduces an Event-B-based framework for modeling real-time scheduling algorithms and verifying properties, including schedulability. The framework provides a common refinement pattern for modeling the schedulability of the Event-B model. It facilitates the transformation of the schedulability analysis on the obtained model into the deadlock detection problem within the model. Deadlock detection can be effectively addressed through either theorem proving or model checker. We utilized Event-B to model and refine several real-time scheduling algorithms. Following the formal verification of functional and environmental requirements, we analyzed and verified the model's schedulability within the proposed framework.

Keywords: Event-B · Real-Time scheduling · Schedulability analysis · Deadlock detection · Model checking

1 Introduction

Event-B [1] is a formal modeling method based on set theory and first-order logic, which is usually used for system-level modeling and verification. It allows

© The Author(s), under exclusive license to Springer Nature Switzerland AG 2025
G. Bai et al. (Eds.): ICECCS 2024, LNCS 14784, pp. 223–244, 2025.
https://doi.org/10.1007/978-3-031-66456-4_12

for the specification of systems using mathematical notations and supports step-wise refinement. It can be used to verify various aspects of a system, including functional and environmental properties. Timing properties are critical properties in many systems. A real-time operating system (RTOS) is a specialized operating system designed for real-time applications with requirements concerning strict time constraints. Failure to meet these time constraints can lead to system failures or produce results that are considered unacceptable. Therefore, real-time scheduling necessitates adherence to stringent timing constraints, and the schedulability analysis is highly prominent within this realm. The system model of real-time scheduling is inherently intricate, necessitating the fulfillment of timing properties and other functional and environmental properties. Event-B, therefore, proves to be a viable approach for modeling and verifying real-time scheduling algorithms.

Related Works. Utilizing formal methods to model and employ the schedulability analysis on real-time scheduling algorithms has been a well-established practice for a considerable duration. Certain studies [6–8,18] employ UPPAAL to construct timed automaton for representing various real-time scheduling algorithms and schedulers. Formal property specification, including schedulability, is expressed with Timed Computational Tree Logic (TCTL). UPPAAL explores the model's state space and verifies whether it adheres to the properties specified by TCTL. Haul et al. [9] focus on formal schedulability analysis in the context of multi-core RTOS. They utilize a High-level Petri net as the modeling formalism to capture the behavior and properties of an RTOS called Trampoline and use the Romeo model checker for verification. The above work commonly entails directly modeling the scheduler or scheduling algorithm and verifying the system model using model checking. Approaches directly tailored to specific scheduling algorithms may suffer from insufficient scalability, while employing model checking to verify properties can lead to state explosion. Event-B's formal verification method, which relies on theorem proving, effectively circumvented state explosion. While Event-B is not specifically designed for modeling time, some work [4,5,12,13,17] have introduced approaches to articulate and model timing properties within Event-B, thereby enhancing its suitability for modeling and verifying real-time scheduling algorithms.

Our work introduces a framework to facilitate schedulability analysis within the Event-B model. Initially, the formal method Event-B is employed to construct a model of the real-time scheduling algorithm. The model undergoes successive refinement to encompass the functional and environmental requirements essential for the real-time scheduling algorithm. The early abstract model has a certain degree of scalability and can be further refined to accommodate various real-time scheduling algorithms. As a timing constraint, the schedulability is formalized by leveraging the trigger-response pattern in the final refinement. We prove that any detected deadlock state within the final model must inherently arise from the refinement of schedulability. Consequently, we can analyze the model's schedulability based on the outcomes of the deadlock detection by the proB model checker. Therefore, the deadlock state detected by model checking indicates a violation of the system model's schedulability.

The contributions of this paper are as follows:

1. We use Event-B to model and verify the real-time scheduling algorithms. The refinement-based approach reduces the modeling complexity and extends refinement to different algorithms.
2. Our framework provides a method to transform the difficult-to-analyze timing properties, such as schedulability in Event-B, into a solvable deadlock detection problem that can be automated through model checkers.

The rest of the paper is organized as follows: Sect. 2 introduces real-time scheduling, Event-B, and modeling timing properties in Event-B. Section 3 provides an overview of the proposed framework. Section 4 describes the approach to the modeling and refinement of the case study. Section 5 presents the verification outcome, and Sect. 6 summarizes the paper and outlines future work.

2 Preliminaries

2.1 Real Time Scheduling

Each activity in real-time scheduling is referred to as a *task* τ. A task τ_i, comprising a series of jobs $J_{i,j}$, can be periodic or aperiodic. Each job is characterized by an arrival time $a_{i,j}$, an execution time $c_{i,j}$, and an absolute deadline $d_{i,j}$. Additionally, each task τ_i has a worst case execution time (WCET), $C_i = max\{c_{i,j}\}$, and a minimal inter-arrival time, $T_i = min\{a_{i,j+1} - a_{i,j}\}$. A task is periodic if $a_{i,j+1} = a_{i,j} + T_i$ for any job $J_{i,j}$; otherwise, it is called aperiodic. When a task's completion by its deadline is essential to the system's operation, it is said to have a hard deadline. A task's deadline is deemed soft if it is preferable to achieve it, but it can still be tolerated if missed. The completion time of any given task $f_{i,j}$ for any hard real-time task in the system must be less than or equal to the task's absolute deadline, or $f_{i,j} \leq d_{i,j}$. Usually, the hard tasks in the system are periodic. Unless specified, the absolute deadline of a job in the periodic task is the arrival time of the next coming job, as well as $d_{i,j} = a_{i,j+1}$. Aperiodic tasks typically have a soft deadline.

The real-time scheduling algorithm can be divided into fixed and dynamic according to its priority strategy. Liu and Layland's rate monotonic algorithm [11] is a widely recognized fixed priority strategy that ensures adherence to strict deadlines for periodic tasks. Priority is determined based on the task's period, with shorter periods receiving higher priority. The rate monotonic algorithm is a fundamental framework for evolving subsequent real-time scheduling algorithms. Within dynamic priority scheduling, one of the most common algorithms is the Earliest Deadline First (EDF) algorithm [15], according to which the priority is inversely proportional to the absolute deadline of the task.

Real-time systems typically include a combination of periodic and aperiodic tasks. To accommodate this mixed workload, the Polling Server [14] and the

Deferrable Server [16] were developed as extensions to the rate monotonic algorithm. The mentioned technique introduces a dedicated periodic task called the *Server*. The *Server* is designed to allocate its capacity to handle requests from aperiodic tasks and replenish periodically. A polling server starts periodically and services pending requests for aperiodic tasks. Otherwise, the polling server will be suspended until the next period. The feature of polling is that if new aperiodic task requests occur when the server is suspended, they must wait until the server starts in the next period before they can be responded to and processed. A deferrable server retains its budget for a period, even if there are no requests for aperiodic tasks. As long as this budget has not been exhausted, the system is ready to service requests for aperiodic tasks.

Figure 1 shows an example of the polling server and the deferrable server algorithm. The periodic task set of this example is shown in the table at the top of the figure. Task C_1 to C_4 are aperiodic tasks. As depicted in the figure, it is evident that the C_1 task within the polling server example misses the point when the server is started at t = 0, consequently delaying its execution until t = 5. In contrast to the polling server, the deferrable server reserves capacity and preempts task B's execution to address the needs of task C_1 at t = 2.

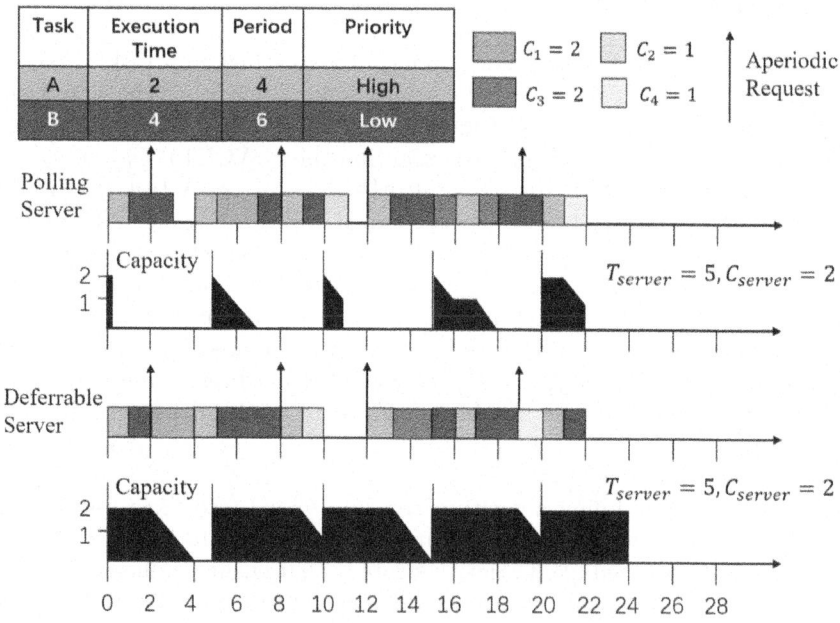

Fig. 1. Examples of Polling Server and Deferrable Server

2.2 Event-B

Event-B is a formal method for modeling and verifying systems at the system level, an improvement to the B method [3]. The model undergoes incremental refinement and development, resulting in a more concrete model. An Event-B model is made of two elements: context and machine. A context describes the static part of a model, which contains constants and sets (user-defined types) together with axioms that specify their properties. A machine represents the dynamic part of the model, which defines variables, invariants, and a set of events (Fig. 2).

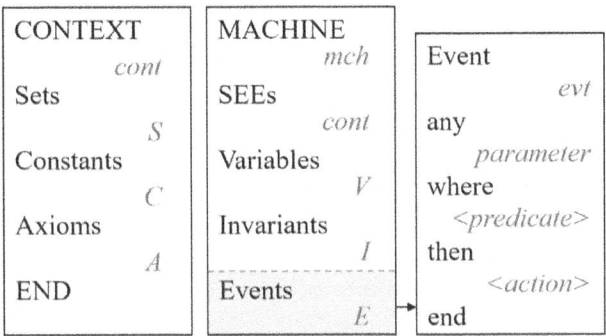

Fig. 2. Event-B Component

The Rodin [2] platform is an industrial-level tool designed for Event-B modeling and verification. It facilitates the automatic generation of proof obligations associated with the constructed models and offers support for both interactive and automated theorem-proving techniques. ProB [10] is a B-method-based tool that offers an automated verification approach capable of conducting static analysis on B method models. It verifies the model's properties and is equipped to detect errors and ambiguities within the model, which assists developers in identifying and rectifying inconsistencies in system design and implementation. It can be integrated into the Rodin platform as a plug-in or be used independently.

2.3   Time Modelling in Event-B

Event-B is a modeling language that does not have explicit support for expressing and verifying time constraints. Prior research has proposed various approaches for incorporating time into Event-B models.

Butler et al. [4] utilize natural numbers to represent the current time and model the progress of time through tick events in classical B. The timing constraint is modeled as a guard in tick events. Sarshogh [13] categorizes timing properties into delay, expiration, and deadline. A typical timing constraint is related to a trigger event and a response event. Those properties are denoted as:

$$Deadline(Trigger, Response, Deadline) \tag{1}$$

$$Delay(Trigger, Response, DelayTime) \tag{2}$$

$$Expiry(Trigger, Response, ExpiryTime) \qquad (3)$$

The deadline property states that if the *Trigger* event occurs, the events *Response* must occur before *deadline* passes. The delay property means that the response event cannot occur before *DelayTime* of the trigger event occurring. Similarly, the expiry property means that the response event cannot occur after *ExpiryTime* of the trigger event occurring.

Figure 3 shows how to model a deadline property in Event-B. A global clock variable *clock* is introduced to model the discrete time. Additionally, two boolean variables, *trigger* and *response*, signify the occurrence of the *Trigger* and *Response* event, respectively. Furthermore, a variable *triggerTime* is utilized to record the timestamp of the *Trigger* event. The *Trigger* event sets *trigger* to TRUE, indicating that the trigger event has occurred, and employs *triggerTime* to record the time point. The *Response* event necessitates fulfilling the condition that the *trigger* is TRUE and subsequently sets *response* to TRUE. The event guard of the event *Tick* that advances the time strictly prevents the clock from advancing to a value that may violate the deadline property when the trigger event occurs, and the response event does not occur.

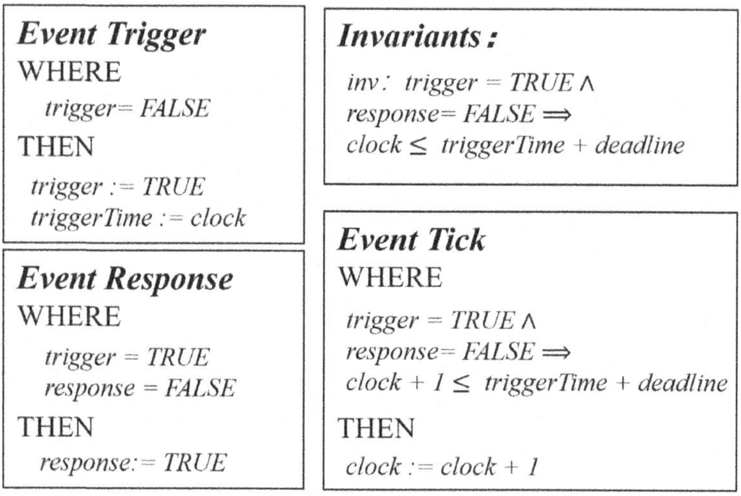

Event Trigger
WHERE
   trigger= FALSE
THEN
   trigger := TRUE
   triggerTime := clock

Event Response
WHERE
   trigger = TRUE
   response = FALSE
THEN
   response:= TRUE

Invariants :
inv: trigger = TRUE $\wedge$
response= FALSE $\Longrightarrow$
clock $\leq$ *triggerTime + deadline*

Event Tick
WHERE
trigger = TRUE $\wedge$
response= FALSE $\Longrightarrow$
clock + 1 $\leq$ *triggerTime + deadline*
THEN
clock := clock + 1

Fig. 3. Modeling A Deadline Property in Event-B

Zhu [19] proposes a framework that adopts the above-mentioned patterns to model the timing constraints of the abstract model and distinguishes the timing properties of different system design phases. Our approach also employs such a trigger-response pattern to model timing constraints and extends it for subsequent refinement.

3   Framework

In this section, we outline our framework-based methodology. The methodology framework is shown in Fig. 4.

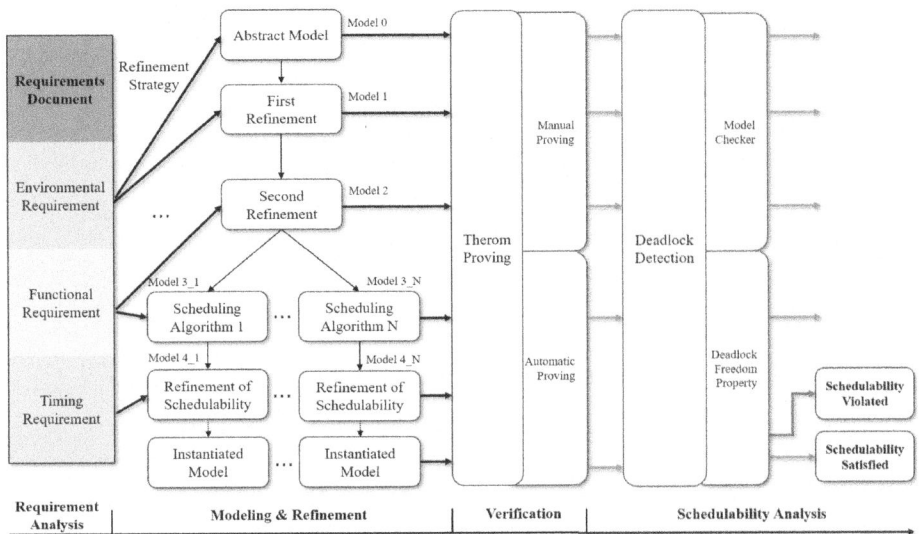

Fig. 4. Framework

3.1   Requirement Analysis

System analysis and requirements document is required before modeling a system using the Event-B formal method. The system analysis encompasses the initial studies of the system we intend to construct. The requirements document, mostly written in natural language, explicitly states the functions and constraints of the system. Upon analyzing the real-time scheduling algorithm, we derive and categorize the requirements, as depicted in Table 1.

3.2   Modeling and Refinement

The refinement strategy employed by our model is illustrated in the Table 1. The modeling approach commences with establishing a fundamental abstract scheduling model, Model 0. Model 0 encompasses essential scheduling requirements that most scheduling algorithms satisfy. In the "critical properties" paragraphs of each subsection of Sect. 4, we describe the requirements formulated through invariants modeled at each model layer.

Table 1. Requirements & Refinement Strategy

Models	Requirements	Type
Model 0	A task can have different states such as created, ready, running, and finished.	ENV-1
	The operating system manages a ready set, which contains all tasks in the ready state.	ENV-2
	At any given time, at most one task can be executed on the core.	FUN-1
	The core is kept as busy as possible, meaning that the core remains idle only when there are no ready tasks to execute.	FUN-2
Model 1 ~ Model 3	The system encompasses periodic tasks, aperiodic tasks, and a Server task.	ENV-3
	The priority assignment of periodic tasks, including the server, follows the Rate Monotonic strategy.	ENV-4
	The operating system manages a ready queue for each priority level, which contains all tasks in the ready state at that priority.	ENV-2
	The system selects the eligible task with the highest priority to execute on the core.	FUN-3
Model 4	The system satisfies schedulability, implying that each periodic task must finish its current job before the next job is created.	TIMING

Through a systematic series of refinements, common environmental and functional requirements related to the real-time algorithm and real-time scheduling environment are progressively integrated.

The general model can then be refined to accommodate distinct real-time scheduling algorithms. After modeling various algorithms, a general refinement is employed for Model 4 to formalize schedulability. The model constants can be refined to instantiate the obtained model, supporting the schedulability analysis in different initial situations.

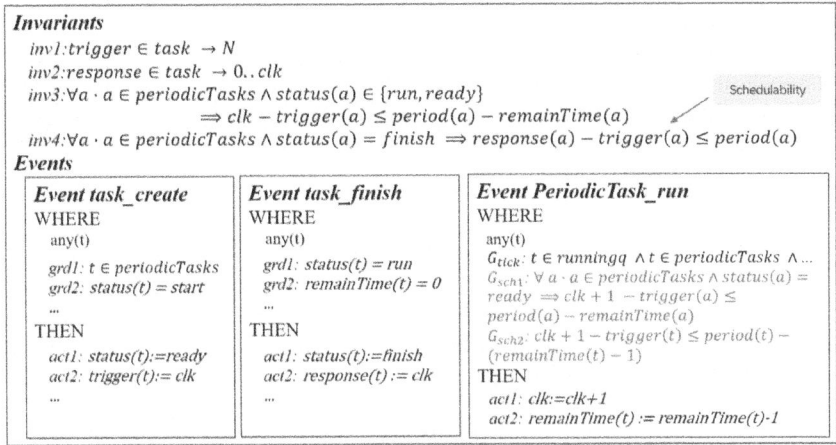

Fig. 5. Formalize Schedulability in Machine 4

3.3 Verification

An invariant indicates that all states of the model should satisfy this invariant. Rodin examines all events, and events that involve changes in relevant variables within the definition of invariants will generate proof obligations of the invariant preservation type. Proving these proof obligations ensures that all events preserve the invariant.

As elucidated in the preceding subsection, we employ invariants to model the system's environmental and functional requirements. Consequently, we can demonstrate that all model states adhere to these requirements by proving the proof obligations of invariant preservation type. Additionally, the model will generate other proof obligations that need to be proved to guarantee the correctness of the model's definition and the refinement during modeling.

Meanwhile, the model must be deadlock-free to guarantee that the system will not enter a deadlock state due to specific execution paths. Verifying the model's deadlock-free property can be achieved by employing either a model checker or proving a theorem that establishes the property of deadlock-free. The theorem follows the format in Fig. 6. Its validation implies that the event guards of at least one event evaluate to true, thus indicating the existence of at least one executable event within the model.

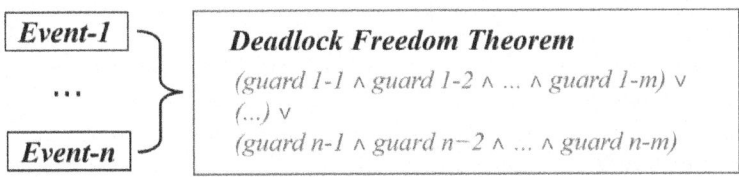

Fig. 6. Deadlock Freedom Theorem

As shown in the framework, before the refinement of schedulability, the model should not only meet the correctness of all proof obligations but also avoid deadlock. The conjunction of those two aspects serves as a comprehensive verification, affirming the satisfaction of the model's critical properties.

3.4 Schedulability Analysis

The trigger-response pattern is employed within the final refinement to model the schedulability in Fig. 5. The run event in the figure uses the run event of a periodic task as an example, introducing two new guards, G_{sch1} and G_{sch2}. The run event for aperiodic tasks only introduces one guard G_{sch1}. We begin by presenting the following definition:

Definition 1. M_{before} *is denoted as the model before the refinement of schedulability. It has a total of n events, each with a corresponding guard condition G_n. The guard condition of the run event is specially denoted as G_{tick}.*

Definition 2. M_{after} *is denoted as the model after the refinement of schedulability. It has a total of n events, each with a corresponding guard condition G_n. The guard condition of the run event is denoted as $(G_{tick} \land G_{sch})$, where G_{sch} represents the newly added guard.*

According to the definition, the model depicted in Fig. 5 can be represented as M_{after}. G_{sch} is therefore equal to $G_{sch} = G_{sch1} \land G_{sch2}$. We take the run event of a periodic task as an example and give the following theorem:

Theorem 1. *If guard G_{sch} is unsatisfied, the model M_{after} does not satisfy schedulability.*

Proof. If guard G_{sch} is unsatisfied, then there could be three cases:

1. G_{sch1} is unsatisfied. If G_{sch1} is unsatisfied, it implies that there exists a periodic task a that satisfies $clk + 1 - trigger(a) > period(a) - remainTime(a)$. The inequality can be rewritten to $clk + 1 + remainTime(a) > period(a) + trigger(a)$. Figure 7 illustrates the timeline. The one-unit interval from time clk to $clk + 1$ is occupied by task t. The task t is selected by the scheduling policy modeled in the model. Even in the ideal scenario, task a cannot utilize this time unit. Otherwise, it would violate the scheduling policy. The earliest time task a can start executing is time $clk + 1$, and a surpasses the deadline.
2. G_{sch2} is unsatisfied. If G_{sch2} is unsatisfied, it implies $clk + 1 - trigger(t) < period(t) - (remainTime(t) - 1)$. The inequality can be rewritten to $clk + remainTime(t) < period(t) + trigger(t)$. Figure 7 illustrates the timeline, and the task t surpasses the deadline.
3. Both G_{sch1} and G_{sch2} are unsatisfied. The tasks a and t surpass their respective deadline, similar to prior cases.

Thus, in each case, the model fails to meet the schedulability.

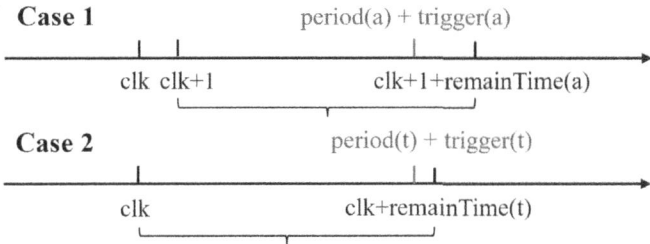

Fig. 7. Timeline of different situations

After explaining the relationship between G_{sch} and schedulability through Theorem 1, we proceed to introduce the subsequent theorem:

Theorem 2. *Let M_{before} model be proved to be deadlock-free and all proof obligations of both M_{before} and M_{after} hold. If M_{after} is deadlock-free, then M_{after} satisfies schedulability. Otherwise, M_{after} does not satisfy schedulability.*

Proof. The deadlock-free property of M_{before} can be expressed as $DLF_{before} = (G_1 \vee .... \vee G_{tick} \vee ... G_n)$. The M_{before} model is proved deadlock-free, indicating that the value of DLF_{before} is true. The deadlock-free property of M_{after} can be expressed as $DLF_{after} = (G_1 \vee .... \vee (G_{tick} \wedge G_{sch}) \vee ... G_n)$. If the M_{after} model is deadlock-free, the model therefore satisfies schedulability. If the M_{after} model is deadlocked, then it indicates that the value of DLF_{after} is false. The disparity between the two theorem lies solely in the presence of G_{sch}. This implies that the value of G_{sch} is false. Under Theorem 1, it can be deduced that the dissatisfaction of G_{sch} indicates that the model fails to satisfy the schedulability.

Theorem 2 elucidates the connection between the occurrence of deadlock and the verification of schedulability under our approach. Therefore, we can analyze the model's schedulability by transforming it into a deadlock detection problem.

4   Modeling

This section details the specific steps of modeling and refining the abstract model to a few real-time scheduling algorithms while covering new requirements.

4.1   Abstract Model

The abstract model fulfills the fundamental requirements outlined in the requirement table. It is a foundational model that can be further refined to accommodate specific scheduling environments and algorithms while preserving its adherence to the basic scheduling requirements.

Context. The set TASK represents the task type that runs in the model, and the set STATUS represents the four possible cases in a task's state. The constant *task* represents all the potential tasks in the model.

CONTEXT Context 0
  SETS
    TASK STATUS
  CONSTANTS
    start ready run finish task
  AXIOMS
    @axm1: $partition(STATUS, start, ready, run, finish)$
    @axm2: $task \subseteq TASK$

Machine. The variable $status$ represents the task's status, and the $readyq$ represents the ready set, which contains tasks in a ready state. The variable $ctrl$ is a boolean variable that indicates whether scheduling is completed.

MACHINE Machine 0
  invariants
    @inv1: $status \in task \rightarrow STATUS$
    @inv2: $readyq = \{a \cdot a \in task \land status(a) = ready|a\}$
    @inv3: $runningq = \{a \cdot a \in task \land status(a) = run|a\}$
    @inv4: $ctrl \in BOOL$

---

event t_schedule_busy
  any(t)
  where
    @grd1: $t \in readyq$
    @grd2: $runningq = \varnothing$
    @grd3: $ctrl = FALSE$
  then
    @act1: $readyq := readyq \setminus \{t\}$
    @act2: $runningq := \{t\}$
    @act3: $status(t) := run$
    @act4: $ctrl := TRUE$

event run2ready
  any(t)
  where
    @grd1: $t \in runningq$
    @grd2: $ctrl = FALSE$
  then
    @act1: $runningq := \varnothing$
    @act2: $readyq := readyq \cup \{t\}$
    @act3: $status(t) := ready$

Fig. 8. Abstract Event in Model 0

Figure 8 illustrates two abstract events within the model. The former indicates that the system selects a task from $readyq$ to run on the core, and the latter indicates that the running task is replaced and re-enters the $readyq$.

Critical Properties. The invariants of defining requirements FUN-1 and FUN-2 are listed as follows.

$$\forall a, b \cdot a \in runningq \land b \in runningq \implies a = b \tag{4}$$

According to (4), any two tasks running in the running set are the same.

$$running q = \varnothing \wedge ctrl = TRUE \implies ready q = \varnothing \qquad (5)$$

This invariant states that the system is idle only if the $readyq$ is an empty set.

4.2 Series of Refinements

Considering space constraints, we consolidate the description of refinement steps before refinement to specific algorithms. These hierarchical layers aim to progressively enhance the granularity of the abstract model to encompass the common requirements of the following specific algorithms. Our case study considers real-time algorithms based on a fixed priority strategy. These models primarily address the ensuing requirements ENV-3,4, the rewritten ENV-2, and FUN-3.

Context. The variable $task$ is partitioned into three types representing different tasks. We define the structure and property of the task queue in the context and define the operation of the queue. As context illustrates, constant $priority$ is assigned according to axm2, which formalizes the rate monotonic strategy. The constant $QUEUE$ represents the structure of the task queue, and each task is bound to an index using a total injection function. The $APPEND$ operation is responsible for adding a task to the tail of the task queue.

CONTEXT Context 1& Context 2
 CONSTANTS
  server periodicTasks aperiodicTasks period priority QUEUE APPEND
 AXIOMS
  @axm1: $partition(task, \{server\}, periodicTasks, aperiodicTasks)$
  @axm2: $\forall t1, t2 \cdot t1 \in dom(period) \wedge t2 \in dom(period) \wedge t1 \neq$
 $t2 \wedge period(t1) > period(t2) \Rightarrow priority(t1) < priority(t2)$
  @axm3:
 $QUEUE = \{q | q \in \mathbb{N}_1 \rightarrowtail task \wedge finite(q) \wedge dom(q) = 1 .. card(q)\} \cup \{\varnothing\}$
  @axm4: $APPEND \in QUEUE \times task \rightarrow QUEUE$
  @axm5: $\forall q, t, i \cdot q \in QUEUE \wedge t \in task \wedge t \notin ran(q) \Rightarrow (i \in dom(q) \Rightarrow APP$
 $END(q \mapsto t)(i) = q(i)) \wedge (i = card(q) + 1 \Rightarrow APPEND(q \mapsto t)(i) = t)$

Machine. We define clk to formalize the time clock and use a time variable $lastTimeRequest$ to record the timestamp when the periodic task was last created. To enhance the representation of the task queue concerning priority, we introduce a new variable called $prioQueue$. This variable replaces the former $readyq$ and represents the task queue with different priorities.

MACHINE Machine 1 & Machine 2
  invariants
    @inv1: $clk \in \mathbb{N}_1$
    @inv2: $lastTimeRequest \in periodicTasks \rightarrow \mathbb{N}$
    @inv3: $prioQueue \in prios \rightarrow QUEUE$
    @inv4: $serverQueue = prioQueue(priority(server))$

Gluing Invariants and Critical Properties. Given that we have refined and replaced the original variable $readyq$ with the new variable $prioQueue$, the model requires gluing invariants. These invariants are essential for clarifying the relationship between the newly introduced variable and the original one.

$$\forall p \cdot p \in prios \implies t | t \in ran(prioQueue(p)) \subseteq readyq \qquad (6)$$

$$\forall t \cdot t \in readyq \land t \in periodicTasks \implies t \in ran(prioQueue(priority(t))) \qquad (7)$$

$$\forall t \cdot t \in readyq \land t \in aperiodicTasks \implies t \in ran(prioQueue(priority(server))) \qquad (8)$$

Each task t that is an element of $prioQueue$ is also an element of $readyq$. Similarly, each task t that is an element of $readyq$ is also an element of the correct priority queue in $prioQueue$. These gluing invariants demonstrate that the two variables, albeit with a modified structure, encompass identical elements.

The requirement FUN-3 can be defined through the following invariant:

$$\forall t, other, p \cdot t \in runningq \land t \in periodicTasks \land ctrl = TRUE$$

$$\land other \in periodicTasks \land p \in prios \land other \in ran(prioQueue(p)) \land other \neq t \qquad (9)$$

$$\implies priority(t) > priority(other)$$

This invariant guarantees that whenever the system selects a periodic task from the task queue, it has the highest priority among all tasks. The invariant for selecting aperiodic tasks is structured similarly.

---

event schedule_aperi_atPeriod_notBackground
  where **any(t)**
    @grd1: $t \in \{a, p \cdot p \in prios \land a \in task \land a \in ran(prioQueue(p)) | a\}$
    @grd2: $runningq = \varnothing \land ctrl = FALSE \land t = serverQueue(1)$
    @grd3: $\forall th \cdot th \in periodicTasks \land status(th) = finish \implies$
$clk - lastTimeRequest(th) \neq period(th)$
    @grd4: $\exists p \cdot p \in periodicSet \implies status(p) = ready$
    @grd5: $((clk - 1) \bmod period(server) = 0$
    @grd6: $\forall other, p \cdot other \in periodicTasks \land p \in prios \land other \in$
$ran(prioQueue(n)(p)) \implies priority(server) \geq priority(other)$
    @grd7: $serverCapacity > 0$
  then
    @act1: $prioQueue := prioQueue \mathbin{\lhd\mkern-9mu-} \{priority(server) \mapsto$
$(DEQUEUE(serverQueue))\}$
    @act2: $runningq := \{t\}$
    @act3: $status(t) := run$
    @act4: $serverAvailable := isNotEmpty(DEQUEUE(severQueue))$
    @act5: $serverQueue := DEQUEUE(serverQueue)$

Fig. 9. The Scheduling Event in Machine 3_polling

4.3   Implementing the Scheduling Algorithm

We then refine the previous model to a few specific scheduling algorithms in this layer. We take the polling and deferrable servers as examples because they are server-based and adopt a rate-monotonic priority allocation strategy.

Context. The polling and deferrable server models share a context that defines a constant $capacity \in \mathbb{N}_1$ to represent the server's budget.

Polling Server. The variable $serverCapacity$ represents the remaining budget for the server to consume, and the variable $serverAvaible$ is defined to indicate whether the server is available. In the context of a polling server, the server is available only when the system clock aligns with the server's activation time, concurrently with aperiodic tasks within the server queue.

MACHINE Machine 3_polling
    invariants
      @inv1: $serverCapcaity \in \mathbb{N}$
      @inv2: $serverAvailable \in BOOL$
      @inv3: $((clk - 1)\ mod\ period(server) = 0 \wedge serverQueue \neq \varnothing) \implies$
    $serverAvailable = TRUE$

The scheduling event for the aperiodic task within the algorithm is shown in Fig. 9. Event guard grd4 signifies the presence of a currently available periodic task, while event guard grd5 indicates that the current system clock aligns with the activation time of the server task. Additionally, event guard grd6 denotes that the server task has higher priority than other ready periodic tasks. Consequently, the event selects the aperiodic task for immediate execution.

Deferrable Server. In a deferrable server, the server is available only when the server has an available budget and pending aperiodic requests.

MACHINE Machine 3_deferrable
    invariants
      @inv1: $serverCapacity \in \mathbb{N}$
      @inv2:
    $(serverCapacity > 0 \wedge serverQueue \neq \varnothing) \implies serverAvailable = TRUE$

Figure 10 depicts the preempted event within the algorithm. Event guard grd5 ensures the completion of recreating all necessary periodic tasks. Event guard grd6 indicates that the highest priority among the periodic tasks in the *prioQueue* surpasses the priority of the current server task. Therefore, the system preempts the aperiodic task to allow for the selection of periodic tasks with higher priority in future scheduling.

```
event run2ready_aperi_higherPeriTask
    where any(t)
        @grd1: t ∈ runningq ∧ ctrl = FALSE ∧ t ∈ aperiodicSet
        @grd2: ∀th · th ∈ periodicTasks ∧ status(th) = finish ⟹
clk − lastTimeRequest(th) ≠ period(th)
        @grd3: priority(server) < max({other, p · other ∈ periodicTasks
        ∧p ∈ prios ∧ other ∈ ran(prioQueue(p))|priority(other)} ∪ 0)
    then
        @act1: runningq := ∅
        @act2:
prioQueue := prioQueue ◁− {priority(server) ↦ (APPEND(q ↦ t))}
        @act3: status(t) := ready
        @act4: serverQueue := APPEND(serverQueue ↦ t)
        @act5: serverAvailable := ¬isZero(serverCapacity)
```

Fig. 10. The Preempted Event in Machine 3_Deferrable

4.4   Final Refinement: Formalization of Schedulability

Schedulability refers to the ability to complete periodic tasks before their dead-line when given a specific task set, which is a typical timing property. In our model, we apply the trigger-response pattern to formalize the schedulability. We have made certain modifications to this pattern based on our model. We employ time variables instead of boolean variables to capture the interval during which the task must fulfill the schedulability. Furthermore, in the original pattern, the machine invariant specifies the timing constraints that the clock must adhere to during the interval when the trigger event has been executed, and the response event has not been executed, implicitly ensuring compliance with schedulability. Utilizing time variables to record the timestamp of the trigger and response events enables a more intuitive definition of schedulability through invariants. All events in the pattern are preserved. In our model shown in 11, the trigger event corresponds to the *peri_first_create* event, and the response event corresponds to the *run2finish_peri* event. The tick event corresponds to the *t_run_busy_peri* event.

```
event peri_first_create
    where any(t)
        @grd1: t ∈ periodicTasks
        @grd2: status(t) = start
        ...
    then
        @act1: status(t) := ready
        @act2: trigger(t) := clk

event run2finish_peri
    where any(t)
        @grd1: status(t) = run
        @grd2: remainTime(t) = 0
        ...
    then
        @act1: status(t) := finish
        @act2: response(t) := clk
```

```
event t_run_busy_peri
    where any(t)
        @G_tick: t ∈ runningq ∧ t ∈ periodicTasks ∧ ...
        @G_sch1:
∀a · a ∈ periodicTasks ∧ status(a) = ready ⟹
clk + 1 − trigger(a) ≤ period(a) − remainTime(a)
        @G_sch2:
clk + 1 − trigger(t) ≤ period(a) − (remainTime(t) − 1)
    then
        @act1: clk := clk + 1
        @act2: remainTime(t) := remainTime(t) − 1
```

Fig. 11. Trigger-Response Pattern Event in Machine 4

Context. We use $executeTime \in task \to \mathbb{N}_1$ to represent the execution time required by the task.

Machine. We define variables $trigger \in task \to \mathbb{N}$ and $response \in task \to 0..clk$ to record the timestamp of task creation and completion.

Critical Property. The absolute deadline for a periodic task is the arrival time of the next creation time. Therefore, we employ the following invariant to formalize the schedulability of periodic tasks.

$$\forall a \cdot a \in periodicTasks \wedge status(a) = finish \\ \implies response(a) - trigger(a) \leq period(a) \tag{10}$$

The timestamp for any periodic task to complete, subtracted from its timestamp to creation, must be less than its period. It ensures the periodic task can finish its current job before the next job is created.

With only this invariant, the proof obligation generated by the $run2finish\_peri$ event cannot be proven, as no existing theorems and invariants to elucidate the relationship between clk and $response$. We require the following invariant.

$$\forall a \cdot a \in periodicTasks \wedge status(a) \notin \{start, finish\} \\ \implies clk - trigger(a) \leq period(a) - remainTime(a) \tag{11}$$

The invariant, along with the event guard in $run2finish\_peri$, elucidates the relationship between clk and $response$, providing essential support for proving the proof obligations generated by the $run2finish\_peri$ event.

5   Verification

5.1   Proof Obligations

The Rodin Platform automatically generates the proof obligations (PO). The proof obligation is a theorem that needs to be proved to verify the correctness of the model. Although Rodin has a built-in automatic prover, there may still be some proof obligations that the automated tools cannot discharge. In such cases, these undischarged POs are left manually for the developers to prove.

It is imperative to prove the model that does not formalize schedulability, is deadlock-free before schedulability analysis. We choose to verify it by proving the deadlock freedom theorem. The theorem generates a proof obligation, which the model automatically adds to the proof obligation list.

Table 2. Proof statistics

Machine	Total	Automatic	Manual
Mch 0(DLF theorem included)	54	54(100%)	0(0%)
Mch 1(DLF theorem included)	82	82(100%)	0(0%)
Mch 2(DLF theorem included)	129	115(89%)	14(11%)
Mch 3_polling(DLF theorem included)	96	77(80%)	19(20%)
Mch 3_deferable(DLF theorem included)	93	81(87%)	12(13%)
Mch 4_polling	94	91(97%)	3(3%)
Mch 4_deferrable	94	91(97%)	3(3%)
Total	642	591(92%)	51(8%)

Table 2 provides an overview of the proof obligations statistics in the model. It presents statistics on the number of POs and their status in automatic and manual proving. Over 92% of the POs have been discharged automatically, indicating a significant level of automation.

5.2   Schedulability Analysis

In the obtained model, under the definition in Sect. 3.4, M_{before} constitutes a series of models ranging from $M0$ to $M3$, while M_{after} comprises the latter two models that formalize schedulability. In the preceding subsection, we demonstrated the satisfaction of proof obligations for all the models, and the models from $M0$ to $M3$ are deadlock-free. Therefore, following Theorem 2, we can analyze the schedulability of the latter two models by detecting the deadlock state. Given that $M4$ introduces event guards comparing time, the conventional theorem proving method becomes unsuitable for verifying deadlock-freeness. Due to the complexity of managing the comparison of time variables within theorem proving, we employ the ProB model checker on the latter two models to detect whether the model enters a deadlock state. The proB model checker detects deadlock in the Event-B model by exploring its finite state space. However, since we use natural numbers to represent the clock in the model and the clock's value continues to increase, the instantiated Event-B model has infinite states. To address this issue, we introduce an event in the model that resets all variables, including the clock, to their initial values once the clock reaches a certain threshold. This event changes the model's state to a finite number, allowing the proB model checker to detect deadlock in a finite number of states.

Table 3. Task Set

Task	Exec time(Capacity)	Period	Priority
Polling Server	(2)	5	2
Deferrable Server	(2)	5	2
P1	1	4	3
P2	2	6	1
AP1	2	/	/
AP2	1	/	/
AP3	2	/	/

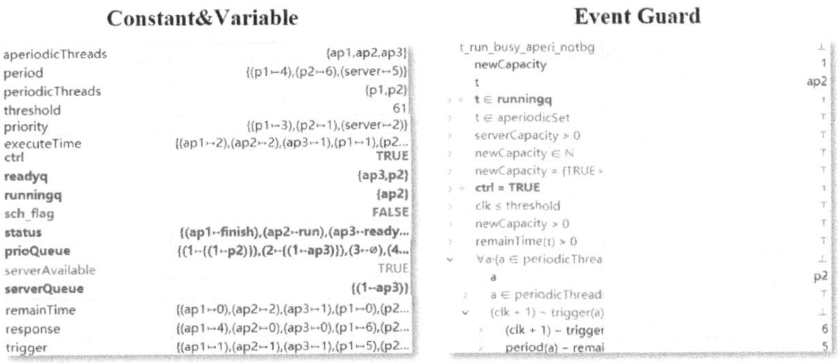

Constant&Variable		Event Guard	
aperiodicThreads	{ap1,ap2,ap3}	t_run_busy_aperi_notbg	⊥
period	{(p1↦4),(p2↦6),(server↦5)}	newCapacity	1
periodicThreads	{p1,p2}	t	ap2
threshold	61	t ∈ runningq	⊥
priority	{(p1↦3),(p2↦1),(server↦2)}	t ∈ aperiodicSet	⊤
executeTime	{(ap1↦2),(ap2↦2),(ap3↦1),(p1↦1),(p2...	serverCapacity > 0	⊤
ctrl	TRUE	newCapacity ∈ N	⊤
readyq	{ap3,p2}	newCapacity = (TRUE ↦	⊤
runningq	{ap2}	ctrl = TRUE	⊥
sch_flag	FALSE	clk ≤ threshold	⊤
status	{(ap1↦finish),(ap2↦run),(ap3↦ready...	newCapacity > 0	⊤
prioQueue	{(1↦{(1↦p2)}),(2↦{(1↦ap3)}),(3↦ø),(4...	remainTime(t) > 0	⊤
serverAvailable	TRUE	∀a·(a ∈ periodicThrea	⊥
serverQueue	{(1↦ap3)}	a	p2
remainTime	{(ap1↦0),(ap2↦2),(ap3↦1),(p1↦0),(p2...	a ∈ periodicThread	⊤
response	{(ap1↦4),(ap2↦0),(ap3↦0),(p1↦6),(p2...	(clk + 1) ↦ trigger(a)	⊥
trigger	{(ap1↦1),(ap2↦1),(ap3↦1),(p1↦5),(p2...	(clk + 1) ↦ trigger	6
		period(a) ↦ remai	5

Fig. 12. Deadlock Counterexample

We instantiate Machine 4_polling and Machine 4_deferrable based on the task set specified in the table 3. The reset threshold is configured to match the least common multiple of the periodic task periods in this task set, specifically set at 60. The models can subsequently undergo schedulability analysis using the proB model checker to detect potential deadlock scenarios. Machine 4_polling successfully passed the proB model checker, with no instances of deadlock states detected. Therefore, we can conclude that this task set is schedulable under the deferrable server algorithm in this model. However, in the case of Machine 4_deferrable, proB enters a deadlock state following the event trace. According to Theorem 2, the model does not satisfy schedulability.

From Fig. 12, it can be observed that the guard related to schedulability in the task run event is false, indicating that task P2 cannot complete its execution before its deadline. Figure 13 illustrates the task execution flow based on the trace provided by proB. At time $t = 6$, the server has a higher priority than P2, leading to the execution of task AP2. Consequently, P2 fails to complete before $t = 7$. By assigning initial values to the constants in the context of each layer, developers can accommodate different task sets for schedulability analysis.

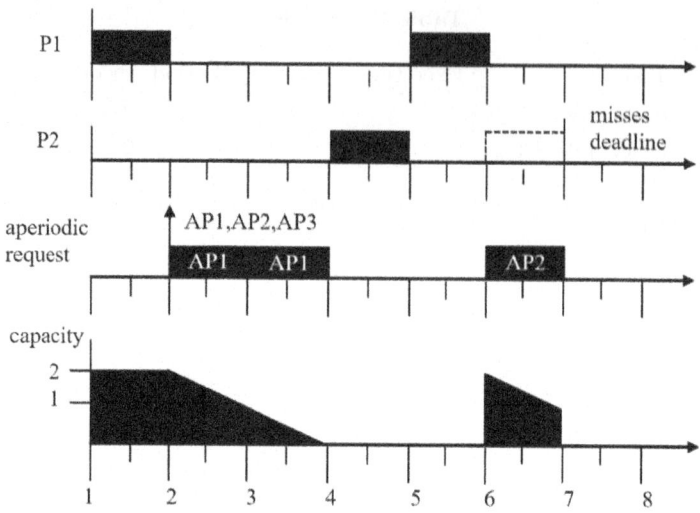

Fig. 13. Task Execution Flow

6 Conclusion

Our work proposes an approach that integrates model refinement and model checking for the formal modeling and schedulability analysis of a real-time scheduling model. The abstract model allows the refinement of various scheduling algorithms. The initialized models are subjected to model checking, and their schedulability can be verified by detecting deadlock states. Our approach enables schedulability analysis under a specified task set in the Event-B model and provides counterexamples when schedulability is violated.

The scheduling algorithm refined through this method remains a preliminary, simplistic algorithm. It does not encompass the modeling and implementing of more complicated algorithms, nor does it account for factors such as task jitter time. Also, the high time consumption of deadlock detection in ProB poses challenges for supporting schedulability analysis in scenarios involving large-scale task sets and relatively high reset thresholds. As a potential direction in the future, we intend to refine more complicated algorithms and investigate a more efficient deadlock detection method for Event-B models.

Acknowledgments. This work is supported by the National Key Research and Development Program (2022YFB3305200), the National Natural Science Foundation of China NSFC (No. 92370201, No. 62272165), the "Digital Silk Road" Shanghai International Joint Lab of Trustworthy Intelligent Software (No. 22510750100) and Shanghai Trusted Industry Internet Software Collaborative Innovation Center.

References

1. Abrial, J.R.: Modeling in Event-B: System and Software Engineering. Cambridge University Press, Cambridge (2010)
2. Abrial, J.R., Butler, M., Hallerstede, S., Hoang, T.S., Mehta, F., Voisin, L.: Rodin: an open toolset for modelling and reasoning in event-b. Int. J. Softw. Tools Technol. Transf. **12**, 447–466 (2010)
3. Abrial, J.R., Hoare, A., Chapron, P.: The b-book. (No Title) (1996)
4. Butler, M., Falampin, J.: An approach to modelling and refining timing properties in b (2002)
5. Cansell, D., Méry, D., Rehm, J.: Time constraint patterns for event B development. In: Julliand, J., Kouchnarenko, O. (eds.) B 2007. LNCS, vol. 4355, pp. 140–154. Springer, Heidelberg (2006). https://doi.org/10.1007/11955757_13
6. David, A., Illum, J., Larsen, K.G., Skou, A.: Model-based framework for schedulability analysis using UPPAAL 4.1. In: Model-Based Design for Embedded Systems, pp. 117–144. CRC Press (2018)
7. Foughali, M., Hladik, P.E.: Bridging the gap between formal verification and schedulability analysis: the case of robotics. J. Syst. Archit. **111**, 101817 (2020)
8. Guan, N., Gu, Z., Deng, Q., Gao, S., Yu, G.: Exact schedulability analysis for static-priority global multiprocessor scheduling using model-checking. In: Obermaisser, R., Nah, Y., Puschner, P., Rammig, F.J. (eds.) SEUS 2007. LNCS, vol. 4761, pp. 263–272. Springer, Heidelberg (2007). https://doi.org/10.1007/978-3-540-75664-4_26
9. Haur, I., Béchennec, J.L., Roux, O.H.: Formal schedulability analysis based on multi-core RTOS model. In: 29th International Conference on Real-Time Networks and Systems, pp. 216–225 (2021)
10. Leuschel, M., Butler, M.: ProB: a model checker for b. In: FME 2003: Formal Methods: International Symposium of Formal Methods Europe, Pisa, Italy, 8–14 September 2003. Proceedings, LNCS, pp. 855–874. Springer, Berlin, Heidelberg (2003). https://doi.org/10.1007/978-3-540-45236-2_46
11. Liu, C.L., Layland, J.W.: Scheduling algorithms for multiprogramming in a hard-real-time environment. J. ACM (JACM) **20**(1), 46–61 (1973)
12. Rehm, J.: From absolute-timer to relative-countdown: patterns for model-checking (2008)
13. Sarshogh, M.R., Butler, M.: Specification and refinement of discrete timing properties in event-b (2011)
14. Sha, L., Lehoczky, J.P., Rajkumar, R.: Solutions for some practical problems in prioritized preemptive scheduling, pp. 181–191. IEEE (1986). In: Unknown Host Publication Title
15. Sprunt, B., Sha, L., Lehoczky, J.: Aperiodic task scheduling for hard-real-time systems. Real-Time Syst. **1**, 27–60 (1989)
16. Strosnider, J.K., Lehoczky, J.P., Sha, L.: The deferrable server algorithm for enhanced aperiodic responsiveness in hard real-time environments. IEEE Trans. Comput. **44**(1), 73–91 (1995)
17. Sulskus, G., Poppleton, M., Rezazadeh, A.: An interval-based approach to modelling time in Event-B. In: Dastani, M., Sirjani, M. (eds.) FSEN 2015. LNCS, vol. 9392, pp. 292–307. Springer, Cham (2015). https://doi.org/10.1007/978-3-319-24644-4_20

18. Waszniowski, L., Hanzálek, Z.: Formal verification of multitasking applications based on timed automata model. Real-Time Syst. **38**, 39–65 (2008)
19. Zhu, C., Butler, M., Cirstea, C.: Refinement of timing constraints for concurrent tasks with scheduling. In: Butler, M., Raschke, A., Hoang, T.S., Reichl, K. (eds.) ABZ 2018. LNCS, vol. 10817, pp. 219–233. Springer, Cham (2018). https://doi.org/10.1007/978-3-319-91271-4_15

Validation of railML Using ProB

Jan Gruteser$^{(\boxtimes)}$ ⓘ and Michael Leuschel ⓘ

Faculty of Mathematics and Natural Sciences,
Heinrich Heine University Düsseldorf, Universitätsstr. 1, 40225 Düsseldorf, Germany
{jan.gruteser,michael.leuschel}@hhu.de

Abstract. The aim of this work is to translate railML 3 files to the formal B-method to enable formal verification and validation. railML is an XML-based format designed to facilitate the exchange of information about railway systems. Our approach allows syntactic and semantic validation against predefined and custom rules, using PROB and its integrated B-Rules DSL. In addition, a B-model can be generated to animate the dynamic behaviour of the specification, which can also be used for simulations and statistical tests using SIMB. A technique for creating visualisations of the topology using Graphviz is presented. Finally, real-life railML case studies show that the implemented validation process can be effectively applied to complex models and that errors in these models can be successfully detected.

1 Introduction

The Railway Markup Language (railML®) [2,14,28] is an XML-based language for railway system specifications that enables a standardised data exchange between railway applications. This work targets the current version railML 3.2 [3, 19,23] and aims to convert railML specifications to a representation in the B-method [5] for formal verification and validation of the specified railway system. This is done using PROB [26], an animator, constraint solver, and model checker for the B-method. The B-method has been used for a variety of industrial applications [11,15]. Based on set theory and first-order logic, it allows formal specification of state machines that represent the properties of a system. This enables the use of model checking, proofs or simulations to verify safety and consistency of the specification. For safety-critical railway applications the use of formal methods is recommended by the Cenelec norm EN50128 [13]. By tra-

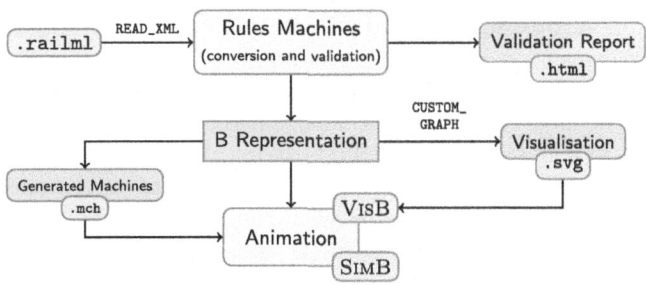

Fig. 1. PROB-based railML Toolchain

The original version of the chapter has been revised. The casing "railML" been corrected. A correction to this chapter can be found at
https://doi.org/10.1007/978-3-031-66456-4_23

© The Author(s), under exclusive license to Springer Nature Switzerland AG 2025, corrected publication 2025
G. Bai et al. (Eds.): ICECCS 2024, LNCS 14784,, pp. 245–256, 2025.
https://doi.org/10.1007/978-3-031-66456-4_13

nslating from railML to B, it is possible to apply these techniques to both the static parameters and the dynamic behaviour of the model.

The core contributions of this work are:

- a representation of the railML core in the B formal method with focus on topology, movable elements and key elements for modelling interlocking logic,
- the validation of predefined and custom semantic (static) consistency rules,
- a B specification of dynamic behaviour for a variety of railML concepts, along with tooling to animate, visualise and simulate the behaviour in PROB,
- automatic visual rendering of the railML topology, for use with PROB/VISB,
- the application of the toolchain (Fig. 1) on several industrial specifications (such as Norway's railway network), uncovering new issues and errors.

2   The Import Process

Our implementation makes use of a domain specific language built on top of B[1] [17,18], which was designed for data validation [8,21,25]. By formulating so-called rules and computations, it is possible to import data from external sources such as XML or CSV files, check validation rules and compute derived data. If a rule is violated, counterexamples are provided, accompanied by meaningful messages to help the user understand the violation. The B-Rules DSL takes care of ordering the rules and computations, taking success and failure of rules into account. This is used to sequentially import and validate railML data in PROB. In this way, the entire conversion process can be handled on the B-side using PROB without any additional dependencies. It also allows direct use of PROB's constraint-solving capabilities to compute logical relationships directly from the data. The import and validation process is based on multiple B machines with specific rules and computations, each linked to one of the railML subschemas.

The *data conversion* from strings to standardised data types is crucial for correct interpretation of data values [31]. Hence, after the raw import of XML data by PROB's function READ_XML, the data is converted into a B representation using records for tags and matching B types for attributes. PROB automatically performs type checking. The handling of attributes in B is non-trivial, since they do not always have to be present in XML files. Therefore, we treat attributes as sets of their type, allowing unspecified attributes to be represented by the empty set. Attributes with predefined string values are modelled as enumerated sets.

The imported data is then subjected to an extended *syntactic validation* using the railML 3 schema with the help of additional rules. This is necessary to ensure well-definedness in B and also because the schema is quite flexible for certain elements. Ordinary switches, for example, need to have exactly two branches for correct animation in B, whereas the schema only requires them to have zero to two branches. Using B-Rules DSL computations, we can *derive relationships* from syntactically valid data, such as the path of a route or the set of aspect relations of a signal, which can then be validated semantically.

---

[1] https://prob.hhu.de/w/index.php?title=Rules-DSL.

The phase of *semantic validation* is possibly the most interesting for the application of the B-method, as it allows easy checking of the relationships and properties of the specified elements in B. As the officially available semantic constraints by railML.org are currently very limited, the constraints validated here are mostly based on our own considerations and experience. A complete list of our implemented rules can be obtained from [16]. For instance, one of our rules checks that no buffer stop shares the same location with a border marked as an open end. The rules can be easily extended as more constraints become available.

After completing the steps outlined above, the validated data is available in classical B data structures and can be used in other B machines.

3   Modelling of the Data in B

A common problem with formal modelling of non-formal schemas is that they may be interpreted differently by users, leading to inconsistencies in usage that the formal model has to deal with [31]. Another related problem is the lack of data essential for the correct functioning of the model. To overcome this, our import process can infer missing data from other data, such as intrinsic coordinates for locations, track ends, activation blocks for routes, and signal plans.

A crucial concept is the modelling of *locations* within the rail topology and their connectivity. In railML, the locations for all infrastructure elements are specified in terms of net elements and intrinsic coordinates within them ranging from 0.0 to 1.0. In B, we represent a location by a triple, containing the ID of the net element, the intrinsic coordinate and the application direction of the location (*normal* or *reverse*). This approach is also proposed by [27]. The use of real numbers is possible due to an extension of classical B with the REAL type (see also [30]). This allows the decimal values to be imported directly from the XML file, without discretising them as integers.

To derive a fine-grained "topology relation" we obviously cannot consider all possible (continuous) intrinsic coordinates, but we have to focus on locations of interest. These are precisely those on which infrastructure elements are located. For this, all available spot and linear locations together with the endpoints of each net element are considered as part of the relation. This enables precise modelling of train movements, but also the computation of paths between locations. To simplify computations of route and section paths, it is assumed that the topology does not contain cycles. As the intrinsic coordinates are mandatory for specifying locations in B, these are derived, where possible, for locations without explicit intrinsic coordinates as the fraction of their absolute position and their length.

The directions of locations are determined by the net relations, whereby the direction changes for the same connected intrinsic coordinates and is retained for different intrinsic coordinates, such as $1.0 \rightarrow 0.0$ (here *normal*). Within a net element, *normal* indicates that the relation is traversed from 0.0 to 1.0, for *reverse* the same applies in the opposite direction.

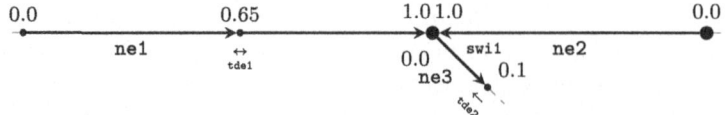

Fig. 2. Simple Example of a Topology With Locations

Example. Figure 2 shows a topology together with a few locations of infrastructure elements. This results in the following set of spot locations in B:

{"swi1" ↦ {("ne1",1.0,direction_normal)},

"tde1" ↦ {("ne1",0.65,direction_normal), ("ne1",0.65,direction_reverse)},

"tde2" ↦ {("ne3",0.1,direction_reverse)}}.

Note that the *both* directed location of tde1 is split into a *normal* and a *reverse* directed location to make it applicable to the relation.

Movable elements of railML that are imported into B are switches, switch crossings, movable crossings, and derailers. In general, each movable element can be in an undefined position or in one of the positions predefined by the schema for its interlocking part. A tricky aspect is that multiple interlocking elements can refer to the same infrastructure element, and we may have to pre-compute all possible traversing paths over a movable element. We refer to [16] for details.

Interlocking Objects: Track vacancy detection (TVD) sections are used to maintain safety distances between operating trains. They are defined by *demarcating elements*, such as train detection elements. In B, TVD sections are modelled as a subset of the topology relation that contains all possible paths between its demarcating elements. railML defines a *route* path by its entry and exit, and additional switch positions to exclude ambiguous paths. Based on this, a unique route path can be derived in B as a function that is a subset of the topology relation. The semantic validation ensures that the path continuously connects entry and exit. The railML signalling is self-contained and managed by *signal plans*. We use B records with entries for their *master, slave* and *distant* signals containing the specified signal aspects. For signals without explicit signalling data, additional aspect relations are inferred based on the specified routes.

4   Animation

To study the dynamic behaviour of a railML system, we implemented a B machine, which contains several operations for animation of the model. An overview of the operations can also be obtained from Fig. 3. The implementation is inspired by the interlocking model by Abrial [6], which aims to verify that only safe train movements take place in the system, i.e. that collisions or derailments can never occur. Invariants ensure that safety-critical properties are always fulfilled. We provide natural language descriptions for invariants and operation guards, which can be displayed within our tooling IDE (ProB2-UI [9]). This should give domain experts a better understanding of what a formula actually

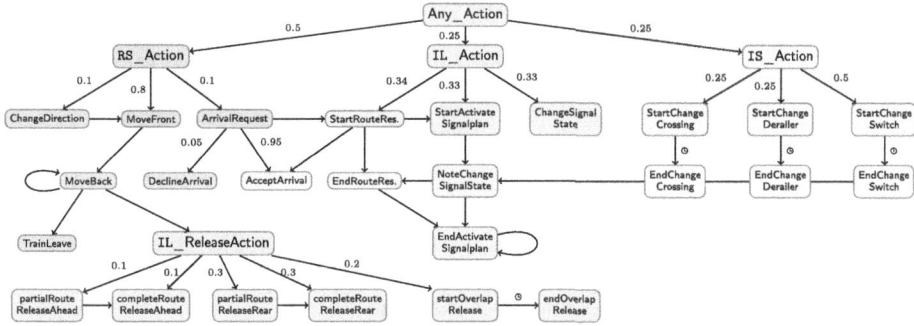

Fig. 3. Concept of the SIMB Activations for Simulation of railML Models

expresses and why it may not be satisfied. The general idea of the model is described below.

The model allows the control of movable *infrastructure* elements. For each type, there are two operations to start and finish the movement. For switches, the specified position restrictions for other elements must be satisfied for the new state. Switch crossings are controlled as a single switch. The movement of coupled switches can be triggered simultaneously.

The *interlocking* system is crucial for safe train movements and thus constitutes the core of the model, including routes, TVD sections, overlaps, and signalling. A route reservation can be started when a train is on one of its activation sections. During the process, the route's associated movable elements can be changed to their target positions and any associated signal plans can be activated. The operation guard ensures that no conflicting routes or overlaps intersect the route's path, and that the associated TVD sections are vacant. As soon as all preconditions, e.g. flank protection, are fulfilled, the reservation can be completed. The route path is maintained by locking all movable elements involved. Invariants are used to monitor safety requirements, such as ensuring all reserved routes are disjoint or no other train is on a reserved route. A route can be released either partially or completely if the train has passed all associated TVD sections or if the sections have not yet been occupied by the train.

Train movements to the next location of the topology relation can occur unless the train's front has the same location as a closed signal. Passing over a train detection element causes the corresponding TVD section to be occupied or freed. A train can change its direction if it is on exactly one TVD section that is a berthing track. New trains can arrive at open ends of the track and request occupancy. The interlocking can then accept or decline the arriving train. Trains can leave the controlled area by releasing its reserved routes and freeing its remaining occupied locations.

5   Generation of Standalone B Machines

The tooling to inline a railML model into a standalone B model was again developed in B itself, calling the railML import rules and using external functions

Table 1. Machine Statistics for Rule-Based and Generated Animation Machine

	Constants	Variables	Properties	Invariants	Operations
Rule-Based	8	580	13	618	257
Generated	201	30	204	70	26

of PROB to write the imported data to designated files. The motivation is to improve the performance, as the operations of the import process can be omitted, and the variables generated by the rules machines can be converted to constants.

Data Machine. In the data machine all records of imported railML types and all derived data structures are stored. In contrast to the rule-based machine, the data in the generated machines can be accessed and changed directly via the data machine file. This can be beneficial if the user wants to try out minor modifications without a complete re-import.

Animation Machine. This machine contains the logic for the dynamic behaviour (Sect. 4). By converting the variables to constants and omitting the import operations of the rules machines, the generated animation machine has significantly fewer invariants, as can be seen from the comparison of the machine statistics in Table 1. This benefits model checking and the overall performance.

Validation Machine. Having the generated data machine, it is possible to reference it from within any rules machine and to formulate custom rules using the B-Rules DSL that are not covered by the validation rules during import, which can be useful for case-specific requirements, such as regulatory purposes or standards. We provide a few example rules, one of which checks that the specified maximum allowed speed for switch branches does not exceed the prescribed value.

6   Simulation

Explicit state model checking suffers from the problem of state space explosion. Hence, it can be used to detect errors, but is not generally suitable for exhaustive verification of industrial railML models. In the future we may look at bounded model checking. Another alternative is SIMB [34], which performs timed probabilistic simulations in PROB. Statistical tests, as proposed by Cappart et al. [12], are used to validate properties by several Monte Carlo (MC) simulations.

For a simulation we need to specify an activation diagram, i.e. the order in which operations are activated. The configuration presented in Fig. 3 is used in our simulations with the animation machine. From the root activation `Any_Action`, the next activation is chosen by probabilistic branching into rolling stock, interlocking, and infrastructure actions. The arrows from each inner activation back to the root activation have been omitted for clarity. When an activation has been queued, it activates the operation and any subsequent activations, such

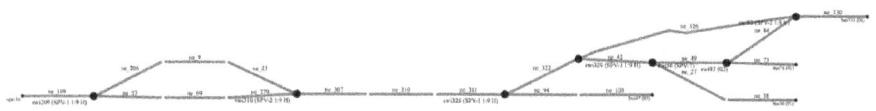

Fig. 4. Flåmsbana Visualised without Positioning Data

as ending a route reservation after it has been started. The parameters of activations without fixed variables are chosen uniformly. The default value for the time elapsed between two events is 1 s. For certain activations, the defined time span is used (☉ in Fig. 3), e.g. the *typicalThrowTime* of movable elements as the time between its start and end event. Our configuration provides support for interactive simulation [33] and allows for real-time simulations to check whether the model behaves as expected, while MC simulations can be conducted on multiple traces for statistical statements about certain properties. A common check is to run multiple simulations for a given number of steps or time, using a left-tailed hypothesis test to examine the probability of an invariant violation. Although this approach cannot replace an exhaustive model check, it can help with validation and reveal problems if the hypothesis is rejected (i.e. the probability that all invariants are true is lower than assumed). Besides the invariant check, specific properties can be examined, including that a route becomes eventually reserved or that a train has passed a certain position after a certain time.

7   Visualisation

Visualisations can help identify errors that might not be noticed simply by looking at the data. VISB [35], a visualisation technique based on Scalable Vector Graphics (SVG), is available in PROB for this purpose. It allows an individual visualisation of a machine's current state and, in PROB2-UI [9], the execution of events triggered by user interaction with a specific element within the SVG.

Create SVG Using Graphviz. We use PROB's interface to Graphviz, a tool for visualising graphs and therefore well suited to a topology-based model such as railML 3, to create SVGs. There have been attempts by railML.org to visualise pure relational topology data in graph form with Graphviz [22]. In contrast, the aim here is to automatically create a view that is more similar to an interlocking view, in order to better represent train positions and further properties in VISB. For the visualisation of railML topologies, we developed three graph definitions.

   The first attempts to visualise the topology using only its relational data and the placement algorithm of a Graphviz engine, without using any positional data. It uses the relationship between net elements and intermediate or delimiting infrastructure elements. This results, for example, in the visualisation shown in Fig. 4, which is a satisfactory representation of the actual topology.

   For railML files that come with explicit *visualization* data, we implemented two graph definitions, namely for files exported by the D4R Track Planner [1] and by railOscope [4]. The strategy to be used depends on the file, which means that applying a different strategy is unlikely to produce a correct visualisation.

Visualisation in VisB. First, we need to post-process the SVG output by Graphviz for VisB by adding identifiers and paths for visualising properties like occupancy or reservations. The converted SVG can then be used in the animation machine. The appearance of SVG elements is controlled by VisB definitions, specifying sets of updates for attributes of existing SVG objects. Tooltips with detailed state information on movable elements, signals and occupied sections are also provided. Figure 5 shows a selection of the properties visualised in VisB.

Without animation, the visualisation can help to gain a better understanding of the topology. Along with animation and real-time simulation, it makes the current state more comprehensible and possible errors easier to identify.

8   Case Studies

Unfortunately, there is a lack of publicly available examples for the relatively new railML 3 format. The only official examples are the Simple and Advanced Example by railML.org [2]. Fortunately, the entire Norwegian railway network has been made publicly available by the Norwegian Railway Directorate[2] [10].

For each example, we applied the import process, i.e. syntactic and semantic validation, and then performed simulations and (not complete) model checking for a selection of those. We measured the duration of the rule-based import and the loading times of the generated animation machine (until its initialisation) for selected models in Table 2. Each benchmark was run once without measurement to parse the generic machine, and then ten times in succession. The median for the total run time and wall time is calculated. It turned out that even for the two largest topologies currently available (BB and OSL), the run times are satisfactory, with just over five minutes for import and validation.

The XML conversion times in PROB vary from negligible 30 ms (Simple Ex., FLB) to 1050 ms (BB, OSL). Abstract constants in combination with *memoization*, a technique used by PROB for storing the results of function applications, are used to significantly improve the performance of repeated queries on XML elements.

Simple Example by railML.org. This is the smallest example available. For version 12, our semantic rules identified three overlapping TVD sections due to a misconfiguration of the location of a train detection element. The animation model was simulated for a scenario with two initial trains, once in real time and 20 times with a MC simulation using a hypothesis test for the fulfilment of all invariants for traces with 300 steps. An invariant violation occurred in every trace due to a train running over an activated derailer, as its protection signal lacks an interlocking definition. The execution of all simulations took approximately 400 s. The model was checked for deadlocks until 60% of the current

---

[2] https://www.jernbanedirektoratet.no/en/infrastructure-model-a-digital-description-of-the-railway.

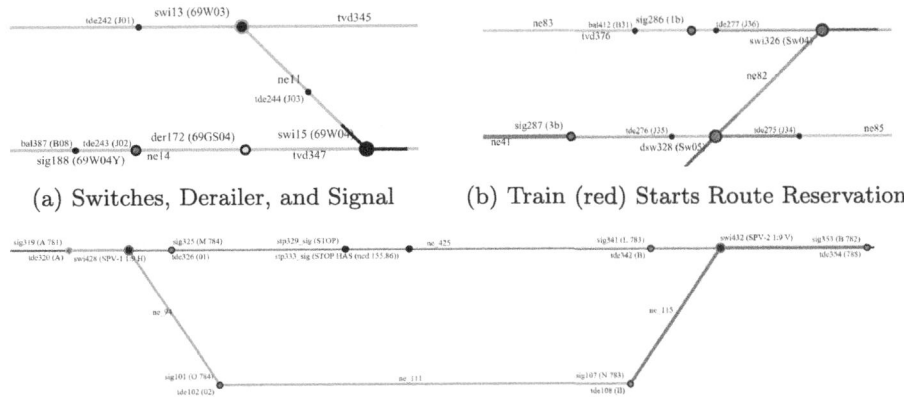

(a) Switches, Derailer, and Signal   (b) Train (red) Starts Route Reservation

(c) Reserved Route (orange) with Overlap (violet) and Passable Signal

Fig. 5. Visualisation in VISB

queue had been processed (432 816/720 012 states). Until then, eight deadlocks with invariant violations were found, which occurred for the same reasons as in the simulations.

Norwegian Railway Network. The files contain detailed infrastructure and basic interlocking data with the topologies varying considerably in size. There are small lines, such as Flåmsbana, only a few kilometres long, and long lines spanning several hundred kilometres, such as Bergensbanen, and spatially small topologies with many branches, such as Oslo main station. The (not yet railML 3-certified) models are created with railOscope, which then generates a railML file.

With the help of semantic validation, errors could be detected in the exported models. Some routes had identical entry and exit signals, while others could not be computed because contradictory switch positions were specified. The investigation revealed that the branches of all switch crossings on the interlocking side had been swapped during export, so that the forced switch positions did not match the branches of the intended route. This bug has been reported and fixed, so the updated models pass our semantic validations.

Due to the complexity of the models, there is an explosion of the state space, and even MC simulations are hardly applicable. Hence, behaviour has mainly been investigated using real-time simulations in combination with VISB. When applied with two trains, at least two erroneous traces with invariant violations were found for the Flåmsbana model. These can also be exported as a standalone HTML file, which allows playback without knowledge of PROB for easy inspection by domain experts [32]. One problem identified in this way is an incompletely configured TVD section at Flåm, which can result in an occupied section that is no longer used by a train. Some HTML exports of simulation traces are available on our webpage [16] and can be viewed in a standard browser.

Table 2. Import and Loading Times of Selected Models [ms]

Model	Rule-based import		Generated machine	
	Run time	Wall time	Run time	Wall time
Simple Example	5957	9234	2053	4471
Advanced Example	9345	12 851	3002	5979
Bergensbanen (BB)	278 986	315 169	10 971	23 787
Flåmsbana (FLB)	7499	10 856	2460	5174
Oslo S (OSL)	138 859	174 770	10 620	24 080

9   Conclusion, Related Work and Outlook

In summary, we developed an approach to transform railML data to a formal classical B model. In addition to syntactic and semantic validation, it is possible to animate and simulate the models. Three visualisation strategies for the railway topologies have been implemented, one of which works independently of the given visualisation data in railML. This can be used during animation and simulation, but also to recognise possible errors in the topology. Finally, a number of case studies were conducted, which showed that import and validation is feasible even for large models in an acceptable time. The animation and simulation of large models is still a challenge, as the computation of guards and VisB definitions can become expensive. Model checking is generally only partially applicable due to the state space explosion problem, but can also help to uncover inconsistencies in the specifications. Indeed, our approach helped to uncover and correct errors in existing railML models, in particular from the Norwegian railway network.

There is considerable academic and industrial interest in data validation using B, see, e.g., [11]. As far as we know, no work has tackled the railML schema yet along with automated visualisation and animation of dynamic aspects. ProB's XML file usage has been used in other works [29,31]. An interesting related work is [20,36], which uses a DSL integrated with visualisation. In future it would be interesting to combine this approach with ours, in particular since [20,36] also build on ProB. It would also be interesting to incorporate continuous aspects into a hybrid B model [7]. We plan to provide the various parts of the toolchain in a single application based on the ProB Java API [24] and to integrate it into the ProB2-UI [9]. This enables integration with other tools and application by users unfamiliar with B.

Acknowledgement. We would like to thank the Norwegian Railway Directorate, railML.org, trafIT solutions, and NEAT for their cooperation. Many thanks also to Fabian Vu for his SimB support and Philipp Körner for his feedback on the manuscript. Part of the work was funded by KI-LOK, grant # 19/21007E of BMWi.

References

1. D4R::Horizon – D4R. https://design4rail.com/service/d4rhorizon
2. Home – railML.org. https://www.railml.org/
3. railML 3 Wiki. https://wiki3.railml.org/
4. railOscope. https://railoscope.com
5. Abrial, J.R.: The B-Book: Assigning Programs to Meanings. Cambridge University Press, New York (1996)
6. Abrial, J.R.: Modeling in Event-B: System and Software Engineering. Cambridge University Press, New York (2010)
7. Aït-Ameur, Y., Bogomolov, S., Dupont, G., Iliasov, A., Romanovsky, A.B., Stankaitis, P.: A refinement-based formal development of cyber-physical railway signalling systems. Formal Aspects Comput. **35**(1) (2023)
8. Badeau, F., Chappelin, J., Lamare, J.: Generating and verifying configuration data with OVADO. In: Collart-Dutilleul, S., Haxthausen, A.E., Lecomte, T. (eds.) RSS-Rail 2022. LNCS, vol. 13294, pp. 143–148. Springer, Cham (2022). https://doi.org/10.1007/978-3-031-05814-1_10
9. Bendisposto, J., et al.: PROB2-UI: a java-based user interface for ProB. In: Lluch Lafuente, A., Mavridou, A. (eds.) FMICS 2021. LNCS, vol. 12863, pp. 193–201. Springer, Cham (2021). https://doi.org/10.1007/978-3-030-85248-1_12
10. Brand, T.: ISO RailDax timeline and railML usage in Norway. In: 43rd railML Conference, Berlin (2023)
11. Butler, M., et al.: The first twenty-five years of industrial use of the B-Method. In: ter Beek, M.H., Ničković, D. (eds.) FMICS 2020. LNCS, vol. 12327, pp. 189–209. Springer, Cham (2020). https://doi.org/10.1007/978-3-030-58298-2_8
12. Cappart, Q., Limbrée, C., Schaus, P., Quilbeuf, J., Traonouez, L.M., Legay, A.: Verification of interlocking systems using statistical model checking. In: Proceedings HASE, pp. 61–68. IEEE (2017)
13. CENELEC: Railway Applications – Communication, signalling and processing systems – Software for railway control and protection systems. Technical report EN50128, European Standard (2011)
14. Ciszewski, T., Nowakowski, W., Chrzan, M.: RailTopoModel and railML – data exchange standards in railway sector. Archives Transp. Syst. Telematics **10** (2017)
15. Ferrari, A., Beek, M.H.T.: Formal methods in railways: a systematic mapping study. ACM Comput. Surv. **55**(4), 1–37 (2022)
16. Gruteser, J.: Modeling and Verification of Railway Systems: Translation of railML Into the B-Method. Master's thesis, Heinrich Heine University Düsseldorf (2023), https://stups.hhu-hosting.de/models/railml
17. Hansen, D., Schneider, D., Leuschel, M.: Using B and ProB for data validation projects. In: Butler, M., Schewe, K.-D., Mashkoor, A., Biro, M. (eds.) ABZ 2016. LNCS, vol. 9675, pp. 167–182. Springer, Cham (2016). https://doi.org/10.1007/978-3-319-33600-8_10
18. Heinzen, C.: A user-interface Plugin for the Rule Validation Language in ProB. Master's thesis, Heinrich Heine University Düsseldorf (2018)
19. Hlubuček, A.: RailTopoModel and railML 3 in overall context. Acta Polytechnica CTU Proc. **11**, 16–21 (2017)
20. Idani, A., Ledru, Y., Ait Wakrime, A., Ben Ayed, R., Collart-Dutilleul, S.: Incremental development of a safety critical system combining formal methods and DSMLs. In: Larsen, K.G., Willemse, T. (eds.) FMICS 2019. LNCS, vol. 11687, pp. 93–109. Springer, Cham (2019). https://doi.org/10.1007/978-3-030-27008-7_6

21. Iliasov, A., Taylor, D., Laibinis, L., Romanovsky, A.B.: The SafeCap trajectory: industry-driven improvement of an interlocking verification tool. In: Milius, B., Collart-Dutilleul, S., Lecomte, T. (eds.) RSSRail 2023. LNCS, vol. 14198, pp. 117–127. Springer, Cham (2023). https://doi.org/10.1007/978-3-031-43366-5_7
22. Kolmorgen, V.P.: Governance & News. In: 43rd railML Conference, Berlin (2023)
23. Kolmorgen, V.P., Rahmig, C., von Lingen, J., Wölke, M.: The federal ministry of transport's digitalisation strategy for regional railways. Signal. Datacommun. (1+2), 6–13 (2023). https://elib.dlr.de/196688/1/06_13_Kolmorgen_etal.pdf
24. Körner, P., Bendisposto, J., Dunkelau, J., Krings, S., Leuschel, M.: Integrating formal specifications into applications: the ProB Java API. Formal Methods Syst. Des. **58**(1–2), 160–187 (2021)
25. Lecomte, T., Burdy, L., Leuschel, M.: Formally checking large data sets in the railways. CoRR **abs/1210.6815** (2012)
26. Leuschel, M., Butler, M.: ProB: A model checker for B. In: FME 2003: Formal Methods. LNCS, vol. 2805, pp. 855–874. Springer, Berlin, Heidelberg (2003)
27. Martins, J., et al.: Verification of railway network models with EVEREST. In: Proceedings MODELS, pp. 345–355 (2022)
28. Nash, A., Huerlimann, D., Schütte, J., Krauss, V.P.: railML – a standard data interface for railroad applications. In: Computers in Railways IX. WIT Transactions on The Built Environment, vol. 74, pp. 233–240. WIT Press (2004)
29. Peng, C., Keming, W.: Applying B and ProB to a Real-world Data Validation Project. In: Proceedings ISKE, pp. 521–524. IEEE (2021)
30. Rutenkolk, K.: Extending Modelchecking with ProB to floating-point numbers and hybrid systems. In: Glässer, U., Creissac Campos, J., Méry, D., Palanque, P. (eds.) ABZ 2023. LNCS, vol. 14010, pp. 366–370. Springer, Cham (2023). https://doi.org/10.1007/978-3-031-33163-3_27
31. St-Denis, R.: A comparison of three solver-aided programming languages: αRby, ProB, and Rosette. J. Comput. Lang. **77** (2023)
32. Vu, F., Happe, C., Leuschel, M.: Generating domain-specific interactive validation documents. In: Groote, J.F., Huisman, M. (eds.) FMICS 2022. LNCS, vol. 13487, pp. 32–49. Springer, Cham (2022). https://doi.org/10.1007/978-3-031-15008-1_4
33. Vu, F., Leuschel, M.: Validation of formal models by interactive simulation. In: Glässer, U., Creissac Campos, J., Méry, D., Palanque, P. (eds.) ABZ 2023. LNCS, vol. 14010, pp. 59–69. Springer, Cham (2023). https://doi.org/10.1007/978-3-031-33163-3_5
34. Vu, F., Leuschel, M., Mashkoor, A.: Validation of formal models by timed probabilistic simulation. In: Raschke, A., Méry, D. (eds.) ABZ 2021. LNCS, vol. 12709, pp. 81–96. Springer, Cham (2021). https://doi.org/10.1007/978-3-030-77543-8_6
35. Werth, M., Leuschel, M.: VisB: a lightweight tool to visualize formal models with SVG graphics. In: Raschke, A., Méry, D., Houdek, F. (eds.) ABZ 2020. LNCS, vol. 12071, pp. 260–265. Springer, Cham (2020). https://doi.org/10.1007/978-3-030-48077-6_21
36. Yar, A., Idani, A., Ledru, Y., Dutilleul, S.C.: Visual animation of B specifications using executable DSLs. In: Proceedings MODELS, pp. 617–626 (2022)

Reachability Analysis of Concurrent Self-modifying Code

Walid Messahel[1] and Tayssir Touili[2(✉)]

[1] LIPN, CNRS and University Sorbonne Paris Nord, Villetaneuse, France
[2] IRIF, CNRS and University Paris Cité, Paris, France
touili@irif.fr

Abstract. We introduce a new effective way to analyze multi-threaded programs that contain self modifying code, i.e., code that modifies itself during its execution. This kind of code is widely used in malware to hide the malicious portions of their code. We introduce a new model called Self Modifying Dynamic Pushdown Network (SM-DPN) to model such programs. A SM-DPN is a network of Self-Modifying Pushdown Systems, i.e., Pushdown Systems that can modify their instructions on the fly during execution. We use finite automata to model regular infinite sets of configurations of SM-DPNs, and propose an algorithm to compute a finite automaton corresponding to the pre* reachable configurations of SM-DPNs. This allows to perform the backward reachability analysis of self modifying multi-threaded programs. We implemented our techniques in a tool and obtained encouraging results. In particular, our tool was able to detect a self-modifying multithreaded malware.

1 Introduction

Self-modifying code is code that changes its own instructions during execution time, without any external intervention. This technique is used for different purposes: some software engineers use it to protect their products from being reverse engineered (code obfuscation), while malware writers use it to evade detection by anti-malware systems.

On the other hand, concurrent programming allows programs to perform multiple tasks simultaneously to be more efficient. Handling this class of programs is known to be delicate and bug-prone, making the process of automatically analysing such category of programs challenging.

In this work, we tackle the analysis problem of *self-modifying* and *concurrent* programs. Indeed, this kind of programs is nowadays widely used e.g. by malware writers to make their malware more efficient when executed (this is the role of concurrency) and hard to analyse and to detect by antiviruses (this is the role of self-modifying code). This makes the issue more complicated and more

This work was partially funded by the ERGANEO grant MALWARE and the french ANR grant Defmal "ANR-22-PECY-0007".

G. Bai et al. (Eds.): ICECCS 2024, LNCS 14784, pp. 257–271, 2025.
https://doi.org/10.1007/978-3-031-66456-4_14

challenging, since it involves two sources of difficulties: *concurrency* and *self-modifying code*.

There are several ways to implement a self modifying code. In this work we consider *direct memory modification*, which consists in modifying the code directly in the execution memory zone, using low-level assembly instructions such as **mov** instructions. Indeed, malware with self modifying code tend to use assembly **mov** instructions that have the ability to access and modify memory enabling the code to alter itself automatically by changing the instruction's binary code stored in memory. Let us consider the example shown in Listing 1.1.

```
1 Address     Bytecode      Assembly
2
3 0x13        b807          Mov eax,0x7
4 0x15        89c0          Mov eax,eax
5 0x17        c602          Mov [0x14],0x2
6 0x19        c618cd        Mov [0x15],0xcd
7 0x1c        c61980        Mov [0x16],0x80
8 0x1f        ff2513        Jmp 0x13
```

Listing 1.1. Binary code with self modifying instructions and thread creation

```
1 0x13    b807    Mov eax,0x7    -executing -->    Mov eax,0x2
2                                Mov [0x14],0x2
3
4 0x15    89c0    Mov eax,eax    -executing -->    int 0x80
5                                Mov [0x15],0xcd
6                                and
7                                Mov [0x16],0x80
8
9
10 0x17   c602    Mov [0x14],0x2
11 0x19   c618cd  Mov [0x15],0xcd
12 0x1c   c61980  Mov [0x16],0x80
13 0x1f   ff2513  Jmp 0x13
```

Listing 1.2. The code after applying the self-modifying instructions

The first column gives the addresses of the instructions. The binary code is given in the second column, while its corresponding assembly code is given in the right-most column. For example, the binary code **b8 07** stored at address **0x13** corresponds to the instruction **mov eax,0x7**. This piece of code contains *self-modifying code* and *concurrency* (thread creation). The code obtained after executing the self-modifying instructions is shown in Listing 1.2. Let us explain how this code is both self-modifying and multi-threaded. The instruction **mov [0x14],0x2** will replace the content in the address **0x14** with **0x2** resulting in the new instruction starting from the address **0x13**: mov eax,0x2. Thus, **mov [0x14],0x2** is self-modifying. It changes the instruction **mov eax,0x7** at address **0x13** into **mov eax,0x2**. The instruction **mov [0x15],0xcd** will replace the content in the address **0x15** with **0xcd**, and instruction **mov [0x16],0x80** will replace the content in the address **0x16** with **0x80** resulting in the new instruction starting from the address **0x15**: **Int 0x80** (the binary code **cd 80** corresponds to the assembly instruction **int 0x80**). Thus, these instructions **mov [0x15],0xcd** and **mov [0x16],0x80** are both self-modifying. They change the instruction **Mov eax,eax** at address **0x15** into **int 0x80**. This kernel function call **int 0x80** uses the content of the **eax** register as a parameter to identify which function to call. In this case, since **int 0x80** is called while **eax** contains **0x2**, the function **Fork** that creates a new process will be called. Thus, **int 0x80** will create a new thread. Therefore, this piece of code contains *self-modifying code* and *concurrency* (thread creation).

In this work, we consider the reachability analysis of this class of programs (self-modifying programs with thread creation). For this purpose, we need an adequate formalism to model such programs. PushDown Systems (PDS) are known to be a natural model for sequential programs [1], since they allow to mimic the program's stack, and thus to model procedure calls faithfully. Dynamic Pushdown Networks (DPNs) [2] were then introduced to model concurrent programs with dynamic thread creation in a precise way. A DPN can be seen as a network of pushdown systems, where each process (PDS) can create new processes (PDSs) during its execution. However, DPNs cannot model self-modifying code. On the other hand, Self-Modifying PushDown Systems (SM-PDS) were introduced in [3–5] to model sequential programs with self-modifying code. A SM-PDS can be seen as a PDS that can modify its own set of transitions during execution. Following these approaches, in this work we go one step further and propose a new model, called Self-Modifying Dynamic Pushdown Network (SM-DPN), to formally represent concurrent programs that involve self-modifying code as well as dynamic thread creation. Roughly speaking, a SM-DPN is a DPN where each process of the network is a SM-PDS, i.e., a pushdown system with self-modifying instructions.

We show how SM-DPNs can be used to naturally represent self-modifying programs with dynamic thread creation. It turns out that SM-DPNs are equivalent to standard DPNs. We show how to translate a SM-DPN to a standard DPN. This translation is exponential, making the reachability analysis on the equivalent DPN not efficient. Therefore, we propose a *direct* algorithm to compute the backward (Pre^*) reachability sets for SM-DPNs. This allows to efficiently perform reachability analysis for self-modifying concurrent programs. Our algorithm is based on (1) representing regular (potentially infinite) sets of configurations of SM-DPNs using finite state automata, and (2) applying a saturation procedure on the finite state automata in order to take into account the effect of applying the rules of the SM-DPN. We implemented our algorithms in a tool that takes as input either a SM-DPN or a self-modifying binary program with thread creation. Our experiments show that our *direct* techniques are much more efficient than translating the SM-DPN to an equivalent DPN and then applying the standard reachability algorithm for DPNs [2]. To show the efficiency of our approach, we successfully applied our tool for malware detection.

Outline. The rest of the paper is structured as follows: The related work is described in Sect. 2. Section 3 introduces our new model and shows how to translate a SM-DPN to an equivalent DPN. In Sect. 4, we give the translation from a self-modifying binary code with thread creation to a SM-DPN. In Sect. 5, we define finite automata to represent regular (potentially infinite) sets of configurations of SM-DPNs. Section 6 gives our algorithm to compute the backward reachability set of SM-DPNs. Section 7 describes our experiments.

2   Related Work

Analyzing binary code has always been an interesting field of study by a variety of computer scientists especially for security purposes either to disclose vulnerabilities or to detect hidden malware. Model checking and static analysis approaches have been extensively used to analyze binary programs in [6–10]. However, these works do not consider self-modifying code.

Analyzing self-modifying code is challenging due to the changing nature of the code. Tha majority of the works that deal with self-modifying code use dynamic analysis like [11–13]. However, dynamic analysis cannot cover all possible execution traces of the program. There are several works that perform static analysis of self-modifying code. [14] use separation logic to describe self-modifying code. However, [14] requires programs to be manually annotated with invariants. [15] propose a formal semantics for self-modifying codes. This work deals only with packing and unpacking behaviours. Bonfante et al. [16] provide an operational semantics for self-modifying programs and show that they can be constructively rewritten to a non-modifying program. However, all these specifications [14–16] are too abstract to be used in practice. [17] propose a new representation of self-modifying code named State Enhanced-Control Flow Graph (SE-CFG). SE-CFG extends standard control flow graphs with a new data structure that keeps track of the possible states that programs can reach, and with edges that can be conditional on the state of the target memory location. This SE-CFG representation does not allow to take into account the stack of the program. [18] propose abstract interpretation techniques to compute an over-approximation of the set of reachable states of a self-modifying program, where for each control point of the program, an over-approximation of the memory state at this control point is provided. [19] combine static and dynamic analysis techniques to analyse self-modifying programs. These techniques [18,19] cannot handle the program's stack.

Self Modifying PushDown Systems (SM-PDSs) were introduced in [3] and were successfully applied for the analysis of *sequential* self-modifying code [3–5]. Our work can be seen as an extension of this approach to *concurrent* self-modifying code.

On the other hand, Dynamic Pushdown Networks (DPNs) [2,20,21] and its extensions [22–26] were broadly used to analyse concurrent programs with thread creation. All these works cannot deal with self-modifying code. Other models based on networks of pushdown systems [27–31] were extensively used to deal with concurrent programs. However, these works cannot handle thread creation, nor self-modifying code.

3   Self Modifying Dynamic Pushdown Network

3.1   Definition

We introduce in this section our new model: Self Modifying Dynamic Pushdown Network. A Self Modifying Pushdown System (SM-PDS) [3] is a pushdown system that has the ability to modify its own set of rules during the execution. A

Self Modifying Dynamic Pushdown Network (SM-DPN) consists of a network of SM-PDSs that is capable of modeling a collection of pushdown processes running in parallel, where each of these pushdown processes can dynamically change its current set of rules and create new processes during its execution. Intuitively, a SM-DPN is a DPN [2] (a network of pushdown processes) where each process of the network has the ability to modify its own set of transitions. Formally:

Definition 1. *A Self Modifying Dynamic Pushdown Network (SM-DPN) is a tuple $\Re = (P, \Gamma, \Delta, \Delta_c)$, where P is a finite set of control points, Γ is a finite set of stack symbols, $\Delta \subseteq (P \times \Gamma) \times (P \times \Gamma^*) \cup (P \times \Gamma) \times (P \times \Gamma^*) \times ((P \times 2^{\Delta \cup \Delta_c}) \times \Gamma^*)$ is a finite set of transitions, and $\Delta_c \subseteq P \times (\Delta \cup \Delta_c) \times (\Delta \cup \Delta_c) \times P$ is a finite set of modifying transitions rules. A Dynamic Pushdown Network (DPN) is a SM-DPN with $\Delta_c = \emptyset$.*

Every process in the network has its current set of transition rules θ called the phase, such that $\theta \subseteq \Delta \cup \Delta_c$. Rules in Δ_c can alter a process phase. We can distinguish three different types of transition rules used by the SM-DPN:

- $((p, \gamma), (p_0, w_0)) \in \Delta$ where $p, p_0 \in P$, $\gamma \in \Gamma$, $w_0 \in \Gamma^*$. This rule can also be written as $p\gamma \hookrightarrow p_0 w_0 \in \Delta$.
  This rule means that if a process of the network is in control point p with γ as its top element of the stack then it can move to control point p_0, pop γ and push w_0.
- $((p, \gamma), (p_1, w_1), ((p_0, \theta), w_0)) \in \Delta$ where $p, p_0, p_1 \in P$, $\gamma \in \Gamma$, $w_0, w_1 \in \Gamma^*$, $\theta \subseteq \Delta \cup \Delta_c$. This rule can also be written as $p\gamma \hookrightarrow p_1 w_1 \triangleright (p_0, \theta) w_0 \in \Delta$.
  This rule means that if a process of the network is in control point p with γ as its topmost stack element, then it can move to control point p_1, pop γ, push w_1 and create a new process in the network having p_0 as its initial control point, w_0 as its initial stack content and θ as its initial current set of rules (phase).
- $(p, r_1, r_2, p_0) \in \Delta_c$ where $p, p_0 \in P$, $r_1, r_2 \in \Delta \cup \Delta_c$. This rule can also be written as $p \xrightarrow{(r_1, r_2)} p_0 \in \Delta_c$.
  This rule means that if a process of the network is in control point p and r_1 is in its current set of rules, then it can move to control point p_0 and update its current set of transition rules by replacing the rule r_1 with the rule r_2.

A local configuration of a process of the network can be represented by $(p, \theta)w$, where $p \in P$ is the control point of the process, $\theta \subseteq \Delta \cup \Delta_c$ is its current set of rules (phase), $w \in \Gamma^*$ is its stack content. For simplicity, from now on, we will sometimes write p^θ instead of (p, θ).

A global SM-DPN configuration is a word of the form $p_0^{\theta_0} w_0 p_1^{\theta_1} w_1 \ldots p_n^{\theta_n} w_n$ where $p_0, p_1, \ldots, p_n \in P$, $w_0, w_1, \ldots, w_n \in \Gamma^*$ and $\theta_0, \theta_1, \ldots, \theta_n \subseteq \Delta \cup \Delta_c$. This word means that there are n running processes in the network and for every i such as $0 \leq i \leq n$, the process i is in control point p_i, with w_i as its stack content and have θ_i as its current set of rules.

Let $Conf_\Re$ be the set of all global configurations of the SM-DPN $\Re$. We define the transition relation $\Rightarrow_\Re$ to be the smallest relation between two configurations in $Conf_\Re \times Conf_\Re$ as follows :

- Let $c = up^\theta wv$, $c' = up'^{\theta'}wv$ with $u, v \in Conf_\Re$, $w \in \Gamma^*$, if $r = p \xrightarrow{(r_1, r_2)} p' \in \Delta_c \cap \theta$, $r_1 \in \theta$ and $\theta' = (\theta \backslash \{r_1\}) \cup \{r_2\}$ then c is a predecessor of c' (also written as $c \Rightarrow_\Re c'$). The rule r moves the process from the control point p to the control point p' and changes the current phase (current set of transition rules) by removing r_1 and replacing it with r_2 without altering the content of the stack.
- Let $c = up^\theta \gamma wv$, $c' = up'^\theta w'wv$ with $\gamma \in \Gamma$, $w, w' \in \Gamma^*$, $u, v \in Conf_\Re$, and $r = p\gamma \hookrightarrow p'w' \in \Delta \cap \theta$, then $c \Rightarrow_\Re c'$. The rule r moves the SM-DPN process from the control point p to the control point p', pops γ from the stack and pushes w' into the stack. This rule maintains the current phase (θ) untouched.
- Let $c = up^\theta \gamma wv$, $c' = up_1^{\theta_1} w_1 p'^\theta w'wv$ with $\gamma \in \Gamma$, $w \in \Gamma^*$, $u, v \in Conf_\Re$, if $r = p\gamma \hookrightarrow p'w' \rhd p_1^{\theta_1} w_1 \in \Delta \cap \theta$, then $c \Rightarrow_\Re c'$. Here the rule r will move the SM-DPN process from the control point p to the control point p', pop γ from the stack, push w' into the stack and create a new process on the control point p_1, with w_1 as stack content, θ_1 as the initial phase and maintains the current phase (θ) untouched.

We define $\Rightarrow_\Re^*$ as the transitive reflexive closure of $\Rightarrow_\Re$. If a configuration c' is reachable from c_0 in i steps by applying $\Rightarrow_\Re$ i times, we write $c_0 \Rightarrow_\Re^i c'$.

We denote the set of immediate predecessors (resp. successors) of a configuration c as $Pre_\Re(c) = \{c_1 \in Conf_\Re : c_1 \Rightarrow_\Re c\}$ (resp. $Post_\Re(c) = \{c_1 \in Conf_\Re : c \Rightarrow_\Re c_1\}$). Let $Pre_\Re^*$ (resp. $Post_\Re^*$) denote the reflexive transitive closure of $Pre_\Re$ (resp. $Post_\Re$). These notations can be generalized to sets of configurations in the obvious ways.

3.2   From SM-DPN to DPN

Since the number of phases is finite, a SM-DPN can be simulated by a DPN [2] by encoding the phases along with the DPN control points.

Let $\Re = (P, \Gamma, \Delta, \Delta_c)$ be a SM-DPN, we compute the corresponding DPN $M = (P', \Gamma, \Delta')$ where P' is the DPN set of control points, Δ' is the DPN set of transition rules such that $P' = P \times 2^{\Delta \cup \Delta_c}$, and Δ' is computed as follows:

Initially $\Delta' = \emptyset$. For every $\theta \in 2^{\Delta \cup \Delta_c}$ and $r \in \theta$.

1. if $r = p\gamma \hookrightarrow p'w' \in \Delta \cap \theta$, then $(p, \theta)\gamma \hookrightarrow (p', \theta)w' \in \Delta'$.
2. if $r = p\gamma \hookrightarrow p'w' \rhd (p_2, \theta_2)w_2 \in \Delta \cap \theta$, then $(p, \theta)\gamma \hookrightarrow (p', \theta)w' \rhd (p_2, \theta_2)w_2 \in \Delta'$.
3. if $r = p \xrightarrow{(r_1, r_2)} p' \in \Delta_c \cap \theta$, then for every $\gamma \in \Gamma$, $(p, \theta)\gamma \hookrightarrow (p', \theta')\gamma \in \Delta'$, where $\theta' = (\theta \backslash \{r_1\}) \cup \{r_2\}$.

We can show that:

Proposition 1. *Let $u, v \in Conf_\Re$, $u \Rightarrow_\Re v$ iff $u \Rightarrow_M v$.*

Thus, we get:

Theorem 1. *Let $\Re = (P, \Gamma, \Delta, \Delta_c)$ be a SM-DPN, we can compute a corresponding DPN $M = (P', \Gamma, \Delta')$ such that $|P'| = |P| \cdot 2^{\mathcal{O}(|\Delta| + |\Delta_c|)}$ and $|\Delta'| = (|\Delta| + |\Delta_c| \cdot |\Gamma|) \cdot 2^{\mathcal{O}(|\Delta| + |\Delta_c|)}$.*

4 Modeling Self Modifying Concurrent Code with SM-DPN

We show in this section how to build a SM-DPN from a binary program with self-modifying code and dynamic thread creation. We suppose we are given an oracle $\mathcal{O}$ that extracts from the binary code a corresponding assembly program, together with informations about the values of the registers and the memory locations at each control point of the program. We could use Jakstab [32], IDA Pro [33] or radare2 [34] to get this oracle. We translate the assembly program into a SM-DPN whose control points are the control points of the binary program and such that the stacks of the different pushdown processes mimic the program's processes stacks. The non self-modifying instructions of the program define the rules Δ of the SM-DPN (which are standard DPN rules), and can be obtained following the translation of [10] that models non self-modifying sequential instructions of the program by a PDS. Thread creation instructions can then be represented by rules of the form $p\gamma \hookrightarrow p_1 w_1 \triangleright (p_0, \theta) w_0 \in \Delta$, where $(p_0, \theta) w_0$ corresponds to the initial configuration of the newly created thread.

As for the self-modifying instructions of the program, we use the translation of [3]. These instructions define the set of changing rules Δ_c. As mentioned in the introduction, in this work we tackle self-modifying instructions of the form $mov\ l\ v$, where l is a location of the program that stores executable data. This instruction replaces the value at location l (in the binary code) with the value v. Let i_1 be the initial instruction involving the location l, and let i_2 be the new instruction involving the location l, after applying the $mov\ l\ v$ instruction. As in [3], we assume that i_1 and i_2 have the same number of operands (to ensure that only one instruction is modified). Let r_1 (resp. r_2) be the SM-DPN rule corresponding to the instruction i_1 (resp. i_2). Suppose from control point n to n', we have this $mov\ l\ v$ instruction, then we add $n \xrightarrow{(r_1, r_2)} n'$ to Δ_c. This is the SM-DPN rule corresponding to the instruction $mov\ l\ v$ at control point n.

5 Representing Infinite Sets of Configurations of SM-DPN

Following [2], to finitely represent a regular infinite set of SM-DPN configurations, we use a special kind of automata called $\Re$-automata.

Let $\Re = (P, \Gamma, \Delta, \Delta_c)$ be a SM-DPN. a $\Re$-automaton is a tuple $A = (S, \Omega, T, s^0, F)$ with the following conditions:

1. $\Omega = (P \times 2^{\Delta \cup \Delta_c}) \cup \Gamma$ is the automaton alphabet.
2. S is a finite set containing the automaton states partitioned into two subsets S_c and S_s s.t $S = S_c \cup S_s$, $S_c \cap S_s = \emptyset$ and for every $s \in S_c$ and every p^θ such that $p \in P$, $\theta \subseteq \Delta \cup \Delta_c$, there is a unique state called $s_{p^\theta} \in S_s$.
3. There is a relation $T' \subseteq S_s \times \Gamma \times (S_s \backslash \{s_{p^\theta} : s \in S_c, p \in P, \theta \subseteq \Delta \cup \Delta_c\}) \cup S_s \times \{\epsilon\} \times S_c$ such that $T = T' \cup \{(s, p^\theta, s_{p^\theta}) : s \in S, p \in P, \theta \subseteq \Delta \cup \Delta_c\}$.
4. $s^0 \in S_c$ is the initial state.
5. $F \subseteq S$ is the set of final states.

Note that condition (3) implies the following properties:

- For each $p \in P$, $\theta \subseteq \Delta \cup \Delta_c$, $s \in S_c$, s is the only predecessor of s_{p^θ}.
- States s in S_c do not have Γ-transitions.
- Only ϵ-moves from states in S_s lead to states in S_c.
- States s in S_s do not have p-successors, for $p \in P$

For $y \in \Gamma \cup \{\epsilon\}$ and $s, s' \in S$, if $(s, y, s') \in T$ we write $s \xrightarrow{y}_T s'$. This notation can be extended in the obvious manner to sequences of symbols as follows : $\forall s \in S$, $s \xrightarrow{\epsilon}_T s$ and $\forall s, s' \in S$, $\forall y \in \Gamma \cup \{\epsilon\}$, $\forall w \in \Gamma^*$, $s \xrightarrow{yw}_T s'$ iff $\exists s'' \in S$ such that $s \xrightarrow{y}_T s'' \xrightarrow{w}_T s'$. We will remove the subscript T if it is understood in the context.

Intuitively, the conditions above make sure that every path in the $\Re$-automaton is the concatenation of paths of the form $s^0 \xrightarrow[p_0^0]{p_0^{\theta_0}} s^0_{p_0^{\theta_0}} \xrightarrow{w_0} q_0 \xrightarrow{\epsilon}$ $s^1 \xrightarrow[p_1^1]{p_1^{\theta_1}} s^1_{p_1^{\theta_1}} \xrightarrow{w_1} q_1 \xrightarrow{\epsilon} \ldots \xrightarrow{\epsilon} s^n \xrightarrow[p_n^n]{p_n^{\theta_n}} s^n_{p_n^{\theta_n}} \xrightarrow{w_n} q_n$ such that $s^0, s^1 \ldots s^n \in S_c$, $s^0_{p_0^{\theta_0}}, s^1_{p_1^{\theta_1}} \ldots s^n_{p_n^{\theta_n}} \in S_s$, $q_0, q_1 \ldots q_n \in S_s$, $p_0, p_1 \ldots p_n \in P$, $\theta_0, \theta_1 \ldots \theta_n \subseteq \Delta \cup \Delta_c$, $w_0, w_1 \ldots w_n \in \Gamma^*$.

A configuration $p_0^{\theta_0} w_0 p_1^{\theta_1} w_1 \ldots p_n^{\theta_n} w_n$ is accepted by an automaton A if there exists a path of the form $s^0 \xrightarrow[p_0^0]{p_0^{\theta_0}} s^0_{p_0^{\theta_0}} \xrightarrow{w_0} q_0 \xrightarrow{\epsilon} s^1 \xrightarrow[p_1^{\theta_1}]{p_1^{\theta_1}} s^1_{p_1^{\theta_1}} \xrightarrow{w_1} q_1 \ldots s^n \xrightarrow{p_n^{\theta_n}}$ $s^n_{p_n^{\theta_n}} \xrightarrow{w_n} q_f$ such that $q_f \in F$. We denote by $L(A)$ the set of configurations accepted by A. A set of global SM-DPN configurations $\mathcal{L}$ is regular if there exists an $\Re$-automaton A such that $\mathcal{L} = L(A)$.

5.1   An Example

Consider the regular set of configurations $C = (p_0^{\theta_0} \gamma_1^+ \gamma_2 p_1^{\theta_1} \gamma_1)^*$ of a SM-DPN. This set can be represented by the automaton $A = (S, \Omega, T, s^0, F)$ presented in Fig. 1. As can be seen, the structure and the different constraints imposed to the automaton ensure that the configurations of different processes of the network are linked by ϵ transitions in the automaton.

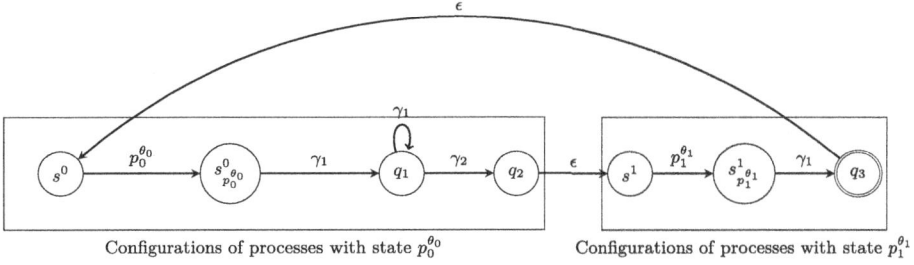

Configurations of processes with state $p_0^{\theta_0}$ Configurations of processes with state $p_1^{\theta_1}$

Fig. 1. Example of a SM-DPN automaton

6   SM-DPN Reachability Analysis

It has been shown in [2] that $Post^*$ images of regular sets of configurations are not regular for DPNs. Since SM-DPNs are equivalent to DPNs, this holds for SM-DPNs as well. Thus, we concentrate on Pre^* images in this section. Let then $\Re = (P, \Gamma, \Delta, \Delta_c)$ be a SM-DPN. Since this SM-DPN can be converted to a DPN, we can use the Pre^* algorithm for DPNs, presented in [2], to perform the reachability analysis of the SM-DPN. However, this approach is very expensive in both memory consumption and time of execution as we will show in Sect. 7. Thus we present here a *direct* algorithm to perform Pre^* images for SM-DPN.

Let $A = (S, \Omega, T, s^0, F)$ be an $\Re$-automaton. We build $A_{Pre^*} = (S, \Omega, T', s^0, F)$ by adding new transitions to the automaton A using the following saturation rules. Initially $T' = T$:

- β_1 : if $r = p_0\gamma \hookrightarrow p_1w_1 \in \Delta$, where $w_1 \in \Gamma^*, \gamma \in \Gamma$, then for every $\theta \subseteq \Delta \cup \Delta_c$ such that $r \in \theta$, if there exists a path in the form $s \xrightarrow{p_1^\theta w_1}_{T'} s'$, for $s \in S_c$, $s' \in S$, then $(s_{p_0^\theta}, \gamma, s') \in T'$.

- β_2 : if $r = p_0\gamma \hookrightarrow p_1w_1 \triangleright p_2^{\theta_2}w_2 \in \Delta$ where $w_1, w_2 \in \Gamma^*, \gamma \in \Gamma$, then for every $\theta \subseteq \Delta \cup \Delta_c$ such that $r \in \theta$, if there exists a path in the form $s \xrightarrow{p_2^{\theta_2} w_2 p_1^\theta w_1}_{T'} s'$, for $s \in S_c$, $s' \in S$, then $(s_{p_0^\theta}, \gamma, s') \in T'$.

- β_3 : if $r = p_0 \xrightarrow{(r_1,r_2)} p_1 \in \Delta_c$, then for every $\theta \subseteq \Delta \cup \Delta_c$ such that $r_2 \in \theta$, for every $\gamma \in \Gamma$, if there exists a path $s \xrightarrow{p_1^\theta \gamma}_{T'} s'$ for $s \in S_c$, $s' \in S$, then $(s_{p_0^{\theta_0}}, \gamma, s') \in T'$ with $\theta_0 = (\theta \backslash \{r_2\}) \cup \{r_1\}$.

The procedure above terminates since there is a finite number of states and phases. Let us explain the intuition behind these rules:

- Let $c = up_0^\theta\gamma v$ and $c' = up_1^\theta w_1 v$ be two global configurations and $r = p_0\gamma \hookrightarrow p_1w_1 \in \Delta$ be a SM-DPN transition rule. If $r \in \theta$, then c is a predecessor of c'. The intuition behind β_1 is that if c' is accepted by the automaton A_{Pre^*}, then c should also be accepted by A_{Pre^*}, i.e. if there exists a path that accepts c' such that $s^0 \xrightarrow{u}_{T'} s \xrightarrow{p_1^\theta}_{T'} s_{p_1^\theta} \xrightarrow{w_1}_{T'} s' \xrightarrow{v}_{T'} q_f, q_f \in F$, then, β_1 adds the new transition $(s_{p_0^\theta}, \gamma, s')$ to T'. Thus we will get a new path in the automaton such that $s^0 \xrightarrow{u}_{T'} s \xrightarrow{p_0^\theta}_{T'} s_{p_0^\theta} \xrightarrow{\gamma}_{T'} s' \xrightarrow{v} q_f, q_f \in F$. This path accepts the configuration c then, c is accepted by the automaton A_{Pre^*}.

- Let $r = p_0\gamma_0 \hookrightarrow p_1 w_1 \triangleright p_2^{\theta_0} w_2 \in \Delta$, consider now the configurations $c = u p_0^\theta \gamma_0 v$ and $c' = u p_2^{\theta_0} w_2 p_1^\theta w_1 v$, c is a predecessor of c'. The intuition behind β_2 is that if c' is accepted by the automaton A_{Pre^*}, then c should also be accepted by A_{Pre^*}, i.e. if there exists a path that accepts c' such that $s^0 \xrightarrow{u} s^1 \xrightarrow{p_2^{\theta_0}} s^1_{p_2^{\theta_0}} \xrightarrow{w_2} q_1 \xrightarrow{\epsilon} s^2 \xrightarrow{p_1^\theta} s^2_{p_1^\theta} \xrightarrow{w_1} q' \xrightarrow{v} q_f, q_f \in F$. By the saturation rule β_2, we add the new transition $(s^1_{p_0^\theta}, \gamma_0, q')$ to T'. Thus we will get a new path in the automaton in such a way $s^0 \xrightarrow{u} s^1 \xrightarrow{p_0^\theta} s^1_{p_0^\theta} \xrightarrow{\gamma_0} q' \xrightarrow{v} q_f, q_f \in F$. This path accepts the configuration c then, c is accepted by the automaton A_{Pre^*}.

- Let $r = p_0 \xrightarrow{(r_1, r_2)} p_1 \in \Delta_c$ and $c = u p_0^\theta \gamma_0 v$, $c' = u p_1^{\theta_0} \gamma_0 v$. The intuition behind the saturation rule β_3 is that if $r, r_2 \in \theta_0$ and $\theta = (\theta_0 \backslash \{r_2\}) \cup \{r_1\}$ then c is a predecessor of c'. Thus, if c' is accepted by the automaton A_{Pre^*} then so should be c. Thus, if there there exists a path in the automaton of the form $s^0 \xrightarrow{u} s \xrightarrow{p_1^{\theta_0}} s_{p_1^{\theta_0}} \xrightarrow{\gamma_0} q' \xrightarrow{v} q_f, q_f \in F$. By applying β_3 we add a new transition $(s_{p_0^\theta}, \gamma_0, q')$ to T'. Thus we will get a new path in the automaton in such a way $s^0 \xrightarrow{u} s \xrightarrow{p_0^\theta} s_{p_0^\theta} \xrightarrow{\gamma_0} q' \xrightarrow{v} q_f, q_f \in F$ which means c is accepted by the automaton A_{Pre^*}.

Thus, we can show that:

Theorem 2.
$$L(A_{Pre^*}) = Pre^*(L(A)).$$

To prove Theorem 2 we show the following two directions:

Lemma 1. $Pre^*(L(A)) \subseteq L(A_{Pre^*})$

Lemma 2. $L(A_{Pre^*}) \subseteq Pre^*(L(A))$

7   Experiments

We implemented our algorithms in a tool. Our experiments were conducted on a machine with 11th Gen Intel® Core™ i7-11850H @ 2.50 GHz × 16 CPU and 30,6 GiB of Memory. Our experiments contain two parts: (1) First, we show the efficiency of our approach vs. the approach that consists in translating the SM-DPN to a DPN and then applying the standard Pre^* algorithm of [2]. (2) Then, we show how our approach can be used for malware detection.

7.1   Our Algorithm vs. The Standard Pre^* Algorithm of DPNs

To practically prove the efficiency of our Pre^* computation compared to the approach that consists in translating the SM-DPN to an equivalent DPN and then apply the standard Pre^* algorithm of [2], we have conducted a number of experiments. To realize such experimental study we developed a script that randomly generates a number of SM-DPNs then converts them to equivalent DPNs as described in Sect. 3. The results of the comparision using the Pre^* algorithm are reported in Table 1. **Column** $|\Delta| + |\Delta_c|$ gives the total number of rules of the SM-DPN. **Column** *Our SM-DPN Algo* gives the execution time and the memory consumption of our Pre^* SM-DPN algorithm. **Column** SM-DPN $\Rightarrow$ DPN gives the conversion time it takes to translate the SM-DPN to a corresponding DPN. **Column** *DPN Results* gives the execution time of the standard Pre^* DPN algorithm of [2]. **Column** *DPN Total* gives the total time it takes to convert the SM-DPN to an equivalent DPN and then apply the Pre^* DPN algorithm of [2] (conversion + Pre^* execution time). As can be seen, our *direct* algorithm has shown a remarkable improvement in the time of execution and the memory consumption. For example, with the configuration $|\Delta| + |\Delta_c| = 3568 + 14$ we can notice the huge difference between the time taken by our SM-DPN *direct Pre^** algorithm which is 4446.49 s and the time taken by the DPN standard Pre^* algorithm of [2] (15540 s: more than 4.3 h). Also we notice a significant memory consumption improvement like the results of the example with the configuration $|\Delta| + |\Delta_c| = 522 + 14$ where our SM-DPN algorithm used 18.05 MB, while the DPN algorithm has required 1280.08 MB of memory.

7.2   Application to Malware Detection

Malware writers extensively use (1) self-modifying code as an obfuscation technique and (2) thread creation to make their malware more efficient and thus harder to be discovered. Thus, we apply our tool for malware detection. We consider a concurrent self modifying binary file infected by the BadRabbit ransomware which is a notorious malware that performs a drive-by attack to install itself through fake adobe flash installer or updates, encrypts all data and uses EternalRomance exploit to spread within the corporate network [35]. The binary "BadRabbit.exe" is obtained from **MalwareCollection** github repo [36]. The malicious behavior is unreachable if one does not take into account the *self-modifying* nature of the *concurrent* code: after executing the self-modifying code, the control point will jump to the part containing the malicious behavior. We use radare2 [34] to decompile the binary and retrieve its assembly code. Then, we model the corresponding assembly program by a SM-DPN as explained in Sect. 4. Finally, we apply our Pre^* algorithm to show that the malicious part of the code is reachable from its entry point. Thus, applying our tool, we deduce that the code is malicious.

Table 1. SM-DPN vs DPN comparison table.

$|\Delta| + |\Delta_c|$	Our SM-DPN Algo	SM-DPN $\Rightarrow$ DPN	DPN Results	DPN Total
400 + 14	117.63 s & 18.08 MB	0.19 s & 36.09 MB	231.61 s	231.80 s
600 + 12	73.51 s & 18.08 MB	1.35 s & 72.08 MB	318.41 s	319.76 s
2500 + 9	76.57 s & 0.45 MB	1.80 s & 0.45 MB	12047.41 s	12049.22 s
44 + 7	0.09 s & 0.45 MB	0.046 s & 0.45 MB	0.56 s	0.60 s
220 + 8	243.21 s & 0.45 MB	0.24 s & 0.45 MB	702.74 s	702.99 s
10 + 9	0.003 s & 0.45 MB	0.0722 s & 0.45 MB	0.1574 s	0.22 s
850 + 12	729.13 s & 36.09 MB	9.59 s & 288.08 MB	1264.47 s	1274.07 s
50 + 14	1.8492 s & 2.21 MB	0.2324 s & 36.09 MB	3108.86 s	3109.1007 s
140 + 8	224.02 s & 0.45 MB	0.08 s & 0.45 MB	1847.18 s	1847.2763 s
90 + 10	12.28 s & 4.58 MB	0.33 s & 72.08 MB	81.19 s	81.5271 s
1200 + 12	13.21 s & 36.09 MB	0.58 s & 36.09 MB	5787.19 s	5787.77 s
520 + 9	8.44 s & 18.08 MB	0.07 s & 18.08 MB	95.89 s	95.967 s
1000 + 10	48.41 s & 36.09 MB	5.974 s & 72.08 MB	146.16 s	152.13 s
900 + 20	75.5486 s & 36.09 MB	4794.896 s & 2560.09 MB	3859.33 s	8654.23 s
1600 + 13	31.17 s & 72.08 MB	799.46 s & 2560.09 MB	1392.90 s	2192.36 s
1400 + 13	19.42 s & 72.08 MB	116.17 s & 576.09 MB	223.31 s	339.48 s
1100 + 13	32.89 s & 36.09 MB	95.05 s & 576.09 MB	126.5723 s	221.6304 s
800 + 12	334.89 s & 36.09 MB	14615.336 s & 10240.09 MB	7223.04 s	21838.38 s
650 + 12	40.98 s & 18.08 MB	2734.209 s & 2560.09 MB	2255.87 s	4990.08 s
350 + 18	15.49 s & 18.08 MB	2576.876 s & 2560.09 MB	17427.96 s	20004.83 s
1300 + 15	133.52 s & 36.09 MB	4592.42 s & 5120.08 MB	2317.41 s	6909.83 s
700 + 20	408.94 s & 36.09 MB	259.71 s & 1280.08 MB	17334.025 s	17593.73 s
950 + 12	20.27 s & 36.09 MB	54.37 s & 576.09 MB	298.16 s	352.53 s
450 + 13	152.64 s & 18.08 MB	8410.54 s & 5120.08 MB	9719.08 s	18129.63 s
300 + 15	11.01 s & 9.09 MB	592.95 s & 2560.09 MB	4174.48 s	4767.4428 s
2000 + 16	6.43 s & 72.08 MB	177.087 s & 1280.08 MB	140.43 s	317.52 s
4244 + 8	887.33 s & 144.09 MB	47.45 s & 576.09 MB	1332.1188 s	1379.5702 s
2348 + 14	1219.36 s & 72.08 MB	341.98 s & 1280.08 MB	3921.81 s	4263.80 s
2846 + 10	973.38 s & 144.09 MB	105.33 s & 576.09 MB	1717.64 s	1822.98 s
2466 + 6	1712.61 s & 72.08 MB	203.66 s & 1280.08 MB	5776.01 s	5979.67 s
2014 + 14	547.006 s & 72.08 MB	43.50 s & 576.09 MB	1082.05 s	1125.559 s
3986 + 13	11054.64 s & 144.09 MB	3743.176 s & 5120.08 MB	85162.2779 s	88905.45 s
714 + 12	696.03 s & 36.09 MB	18.26 s & 288.08 MB	1814.77 s	1833.03 s
982 + 12	5206.48 s & 36.09 MB	1008.14 s & 1280.08 MB	28784.3065 s	29792.4513 s
2496 + 11	2481.03 s & 72.08 MB	504.82 s & 2560.09 MB	8365.26 s	8870.09 s
2974 + 14	848.19 s & 144.09 MB	91.57 s & 576.09 MB	1863.164 s	1954.73 s
2148 + 13	605.627 s & 72.08 MB	63.36 s & 576.09 MB	1700.17 s	1763.53 s
1998 + 7	1063.92 s & 72.08 MB	88.14 s & 576.09 MB	3216.34 s	3304.48 s
522 + 14	6321.78 s & 18.05 MB	816.55 s & 1280.08 MB	21265.877 s	22082.42 s
3100 + 13	7044.20 s & 144.09 MB	2575.38 s & 5120.08 MBs	74293.04 s	76868.43
1788 + 12	699.28 s & 72.08 MB	350.74 s & 1280.08 MB	3164.87 s	3515.61 s
874 + 16	5618.36 s & 36.09 MB	2036.18 s & 2560.09 MB	46751.55 s	48787.74 s
2746 + 15	7043.74 s & 144.09 MB	2010.69 s & 2560.09 MB	34790.60 s	36801.30 s
966 + 13	1092.55 s & 36.09 MB	196.13 s & 576.09 MB	3862.98 s	4059.12 s
2916 + 14	1202.16 s & 144.09 MB	469.23 s & 1280.08 MB	5241.58 s	5710.827 s
3568 + 14	4446.49 s & 144.09 MB	287.94 s & 1280.08 MB	15252.92 s	15540.86 s
954 + 14	70629.60 s & 36.09 MB	17.58 s & 144.09 MB	5927.26 s	5944.84 s
2500 + 11	4993.79 s & 72.08 MB	506.52 s & 2560.09 MB	87936.67 s	88443.19 s
2632 + 13	2809.73 s & 72.08 MB	153.15 s & 1280.08 MB	12487.38 s	12640.54 s

References

1. Esparza, J., Schwoon, S.: A BDD-based model checker for recursive programs. In: Berry, G., Comon, H., Finkel, A. (eds.) CAV 2001. LNCS, vol. 2102, pp. 324–336. Springer, Heidelberg (2001). https://doi.org/10.1007/3-540-44585-4_30

2. Bouajjani, A., Müller-Olm, M., Touili, T.: Regular symbolic analysis of dynamic networks of pushdown systems. In: Abadi, M., de Alfaro, L. (eds.) CONCUR 2005. LNCS, vol. 3653, pp. 473–487. Springer, Heidelberg (2005). https://doi.org/10.1007/11539452_36

3. Touili, T., Ye, X.: Reachability analysis of self modifying code. In: 22nd International Conference on Engineering of Complex Computer Systems, ICECCS 2017, Fukuoka, Japan, November 5-8, 2017, pp. 120–127. IEEE Computer Society (2017)

4. Touili, T., Ye, X.: LTL model checking of self modifying code. In: Pang, J., Sun, J. (eds.) 24th International Conference on Engineering of Complex Computer Systems, ICECCS 2019, Guangzhou, China, November 10-13, 2019, pp. 1–10. IEEE (2019)

5. Touili, T., Ye, X.: CTL model checking of self modifying code. In: Li, Y., Liew, A.W., (eds.) 25th International Conference on Engineering of Complex Computer Systems, ICECCS 2020, Singapore, October 28-31, 2020, pp. 11–20. IEEE (2020)

6. Bergeron, J., Debbabi, M., Desharnais, J., Erhioui, M.M., Lavoie, Y., Tawbi, N., et al.: Static detection of malicious code in executable programs. Int. J. of Req. Eng **2001**(184–189), 79 (2001)

7. Balakrishnan, G., et al.: Model checking x86 executables with CodeSurfer/x86 and WPDS++. In: Etessami, K., Rajamani, S.K. (eds.) CAV 2005. LNCS, vol. 3576, pp. 158–163. Springer, Heidelberg (2005). https://doi.org/10.1007/11513988_17

8. Singh, P.K., Lakhotia, A.: Static verification of worm and virus behavior in binary executables using model checking. In: Information Assurance Workshop, 2003. IEEE Systems, Man and Cybernetics Society (2003)

9. Kinder, J., Katzenbeisser, S., Schallhart, C., Veith, H.: Detecting malicious code by model checking. In: Julisch, K., Kruegel, C. (eds.) DIMVA 2005. LNCS, vol. 3548, pp. 174–187. Springer, Heidelberg (2005). https://doi.org/10.1007/11506881_11

10. Song, F., Touili, T.: Efficient malware detection using model-checking. In: Giannakopoulou, D., Méry, D. (eds.) FM 2012. LNCS, vol. 7436, pp. 418–433. Springer, Heidelberg (2012). https://doi.org/10.1007/978-3-642-32759-9_34

11. Dawei, S., Delong, L., Zhibin, Y.: Dynamic self-modifying code detection based on backward analysis. In: Proceedings of the 2018 10th International Conference on Computer and Automation Engineering, ser. ICCAE 2018 (2018)

12. Ugarte-Pedrero, X., Balzarotti, D., Santos, I., Bringas, P.G.: Sok: deep packer inspection: a longitudinal study of the complexity of run-time packers. In: 2015 IEEE Symposium on Security and Privacy, pp. 659–673 (2015)

13. Guizani, W., Marion, J.-Y., Reynaud-Plantey, D.: Server-side dynamic code analysis. In: 2009 4th International Conference on Malicious and Unwanted Software (MALWARE), pp. 55–62 (2009)

14. Cai, H., Shao, Z., Vaynberg, A.: Certified self-modifying code. In: ACM SIGPLAN Notices (2007)

15. Debray, S.K., Coogan, K.P., Townsend, G.M.: On the semantics of self-unpacking malware code. In: Citeseer (2008)

16. Bonfante, G., Marion, J.-Y., Reynaud-Plantey, D.: A computability perspective on self-modifying programs. In: 2009 Seventh IEEE International Conference on Software Engineering and Formal Methods (2009)

17. Anckaert, B., Madou, M., De Bosschere, K.: A model for self-modifying code. In: International Workshop on Information Hiding (2006)
18. Blazy, S., Laporte, V., Pichardie, D.: Verified abstract interpretation techniques for disassembling low-level self-modifying code. J. Autom. Reason. **56**(3), 283–308 (2016). https://doi.org/10.1007/s10817-015-9359-8
19. Roundy, K.A., Miller, B.P.: Hybrid analysis and control of malware. In: Jha, S., Sommer, R., Kreibich, C. (eds.) RAID 2010. LNCS, vol. 6307, pp. 317–338. Springer, Heidelberg (2010). https://doi.org/10.1007/978-3-642-15512-3_17
20. Nguyen, H.-V., Touili, T.: CARET analysis of multithreaded programs. In: Fioravanti, F., Gallagher, J.P. (eds.) LOPSTR 2017. LNCS, vol. 10855, pp. 73–90. Springer, Cham (2018). https://doi.org/10.1007/978-3-319-94460-9_5
21. Song, F., Touili, T.: Model checking dynamic pushdown networks. Formal Aspects Comput. **27**(2), 397–421 (2015)
22. Wenner, A.: Weighted dynamic pushdown networks. In: Gordon, A.D. (ed.) ESOP 2010. LNCS, vol. 6012, pp. 590–609. Springer, Heidelberg (2010). https://doi.org/10.1007/978-3-642-11957-6_31
23. Bouajjani, A., Esparza, J., Schwoon, S., Strejček, J.: Reachability analysis of multithreaded software with asynchronous communication. In: Sarukkai, S., Sen, S. (eds.) FSTTCS 2005. LNCS, vol. 3821, pp. 348–359. Springer, Heidelberg (2005). https://doi.org/10.1007/11590156_28
24. Diaz, M., Touili, T.: Model checking dynamic pushdown networks with locks and priorities. In: Podelski, A., Taïani, F. (eds.) NETYS 2018. LNCS, vol. 11028, pp. 240–251. Springer, Cham (2019). https://doi.org/10.1007/978-3-030-05529-5_16
25. Diaz, M., Touili, T.: Dealing with priorities and locks for concurrent programs. In: D'Souza, D., Narayan Kumar, K. (eds.) ATVA 2017. LNCS, vol. 10482, pp. 208–224. Springer, Cham (2017). https://doi.org/10.1007/978-3-319-68167-2_15
26. Diaz, M., Touili, T.: Reachability analysis of dynamic pushdown networks with priorities. In: El Abbadi, A., Garbinato, B. (eds.) NETYS 2017. LNCS, vol. 10299, pp. 288–303. Springer, Cham (2017). https://doi.org/10.1007/978-3-319-59647-1_22
27. Qadeer, S., Rehof, J.: Context-bounded model checking of concurrent software. In: Halbwachs, N., Zuck, L.D. (eds.) TACAS 2005. LNCS, vol. 3440, pp. 93–107. Springer, Heidelberg (2005). https://doi.org/10.1007/978-3-540-31980-1_7
28. Pommellet, A., Touili, T.: LTL model-checking for communicating concurrent programs. In: Atig, M.F., Bensalem, S., Bliudze, S., Monsuez, B. (eds.) VECoS 2018. LNCS, vol. 11181, pp. 150–165. Springer, Cham (2018). https://doi.org/10.1007/978-3-030-00359-3_10
29. Pommellet, A., Touili, T.: Static analysis of multithreaded recursive programs communicating via Rendez-Vous. In: Chang, B.-Y.E. (ed.) APLAS 2017. LNCS, vol. 10695, pp. 235–254. Springer, Cham (2017). https://doi.org/10.1007/978-3-319-71237-6_12
30. Kidd, N., Lammich, P., Touili, T., Reps, T.W.: A decision procedure for detecting atomicity violations for communicating processes with locks. Int. J. Softw. Tools Technol. Transf. **13**(1), 37–60 (2011)
31. Bouajjani, A., Esparza, J., Touili, T.: A generic approach to the static analysis of concurrent programs with procedures. In: Aiken, A., Morrisett, G., (eds.) Conference Record of POPL 2003: The 30th SIGPLAN-SIGACT Symposium on Principles of Programming Languages (2003)
32. Kinder, J., Veith, H.: Jakstab: a static analysis platform for binaries. In: Gupta, A., Malik, S. (eds.) CAV 2008. LNCS, vol. 5123, pp. 423–427. Springer, Heidelberg (2008). https://doi.org/10.1007/978-3-540-70545-1_40

33. "IDA Pro". http://www.hex-rays.com/idapro/
34. Adrian Studer, A.K.R., El-MAwgood, A.M.A.: The official radare2 book (2023). https://book.rada.re/credits/credits.html
35. Perekalin, A.: Bad rabbit: A new ransomware epidemic is on the rise (2017). https://www.kaspersky.com/blog/bad-rabbit-ransomware/19887/
36. E. xcp3r, Malwarecollection (2022). https://github.com/xcp3r/MalwareCollection

An Iterative Formal Model-Driven Approach to Railway Systems Validation

Asfand Yar[1], Akram Idani[1]($\boxtimes$), Yves Ledru[1], Simon Collart-Dutilleul[2], Amel Mammar[3], and German Vega[1]

[1] Univ. Grenoble Alpes, CNRS, Grenoble INP, LIG, 38000 Grenoble, France
{asfand.yar,akram.idani,yves.ledru,german.vega}@univ-grenoble-alpes.fr
[2] COSYS-ESTAS, Univ. Gustave Eiffel, IFSTTAR, Univ. de Lille, 59650 Villeneuve d'Ascq, France
simon.collart-dutilleul@univ-eiffel.fr
[3] TELECOM SudParis, SAMOVAR, CNRS, Institut Polytechnique de Paris, 91011 Evry, France
amel.mammar@telecom-sudparis.eu

Abstract. European Rail Traffic Management System (ERTMS) is a standard for the train control and signalling system whose application is spreading throughout Europe. The ETCS (European Train Control System) level 3 is attracting experts because it is still in the design phase. Many works provide formal models to the verification of ERTMS/ETCS using formal methods, but they did not investigate the validation problem. To deal with this challenge we propose an iterative formal model-driven approach that helps validating step-by-step a real formal specification of ERTMS/ETCS hybrid level 3. Our approach introduces Domain-Specific Languages (DSLs) to help system experts understand existing specifications that are already proven. To this purpose we extend and apply Meeduse, the only existing language workbench today that allows embedding formal semantics within DSLs. The paper presents this application and discusses the lessons learned from the railway and formal method experts' points of view.

Keywords: Domain-specific languages · B Method · Validation · Refinement · ERTMS/ETCS

1 Introduction

In a train system, functions are distributed and performed by the train or by track-side devices. Introducing new technology can lead to reallocating a function from the track to the train. This is typically the case for the railway ERTMS/ETCS[1] level 3 solution [1], where positions that were detected by the track are now provided by the train using its own localization means.

---

[1] ERTMS/ETCS: European Rail Traffic Management System/ European Train Control System.

G. Bai et al. (Eds.): ICECCS 2024, LNCS 14784, pp. 272–289, 2025.
https://doi.org/10.1007/978-3-031-66456-4_15

The ERTMS/ETCS has gained significant attention in industry and academia, and because of the underlying safety requirements the formal methods community has investigated numerous techniques to its modeling and verification. For instance, the call of solutions of the ABZ'2018 conference [2] has resulted in several realistic applications of formal methods. Most of these applications deal with verification concerns, without providing insights showing whether their formal model is valid and conforms to user requirements. Some approaches propose translations from graphical models (*e.g.,* UML, KAOS) into formal specifications, but as the transformation is not proven correct, the resulting formal models still need to be validated. And, even if the transformation is proven, validation is still necessary. Indeed, Verification cannot replace validation, and vice versa.

More generally, the use of a model provides an abstraction of reality, assuming a specific point of view [3]. However, in this abstraction, the presence of each concept and correlated requirement has to be justified. While building a Domain Specific Language (DSL) that can be executed, Meeduse [6,7] provides a solution to confront the understanding of the system architect and of the domain expert. This motivates the current work, which is an application of the tool to a real case study. Meeduse is a language workbench built on the B method allowing one to formally define the semantics of DSLs, and animate them using the ProB model-checker. In [15], we presented an approach for visual animation of B specifications using DSLs. In this paper, we extend our approach with an iterative model-driven technique. Considering an existing B specification of an ERTMS hybrid level 3 system provided by a formal methods expert [12], the paper proposes to build incrementally a meta-model of the system which defines the abstract syntax of a DSL. At each level of abstraction, the system is analyzed and the corresponding requirements are simulated. Links are built from the current incremental stage of the model and the corresponding provided B component. This simulation allows railway experts to validate the formal models and compare their understandings.

Section 2 discusses the state-of-the-art and describes shortly the case study of ERTMS hybrid level 3. Section 3 explains our iterative approach. Section 4 presents the approach based on the first level of our DSL. Section 5 discusses other refinement levels and provides the lessons learned from this work. It also talks about a new proposed solution to reduce proof obligations. Finally, Sect. 6 draws the conclusions of this paper and discusses its perspectives.

2   Problem Positioning

2.1   Related Works

Taking into account the railway signalling industrial context, the CENELEC 50128[2] norm should be considered. This norm strongly recommends the use of formal methods for critical software development. Among various formal languages that are used in the railway scientific literature, most of the contributions come from the B ecosystem [5]. Though Formal Methods (FMs) prove

---

[2] https://standards.globalspec.com/std/2023439/afnor-nf-en-50128.

a railway system, they do not guarantee its conformity to requirements. The misunderstanding of the user requirements by FM experts may lead to wrong systems. Hence, validation is required to inspect whether specifications meet the user requirements. If looking closer to the CENELEC 50129[3] requirements (dealing with the safety of electronic devices used for signalling), a graphical description of the system, structured specification and formal or semi-formal specification are highly recommended for subsystems having the highest Safety Integrity Level (namely SIL3 and SIL4). A structured specification in the context of CENELEC 50129 means *"design hierarchically broken down and fully traceable back to requirements specification"*. The ERTMS hybrid level 3 specification which is analyzed in our use case is related to the signalling task. Considering CENELEC 50129, this paper provides DSLs together with their visual animation of the ERTMS B specification, ensuring that specifications meet the user requirements.

In the literature, there are tools providing graphical animation and visualization of B specifications such as BRAMA [13], AnimB[4] B-Motion Studio [8], B-Motion Web [9], VisB [14], and animation function in [11]. However, typical drawbacks are associated with these tools: they use scripting or programming languages for visualization or for mapping. This added programming layer is error-prone. The diagnosis becomes more difficult because errors may come from the added programming layer used for visualization. Another limitation of the tools mentioned above is dealing with the visualization of B specifications with multiple components like the ones we are using in this paper. In the existing tools, the formal method expert has to define visualizations for each component but in our approach, (s)he is only responsible for producing mappings in B. We believe that our approach is less error-prone than other tools, as in our approach, mappings are produced by formal method experts themselves. Also, our technique allows domain experts to provide scalable domain representations by themselves rather than defined by formal method experts.

2.2   A Formal B Model of ERTMS Hybrid Level 3 Use Case

ERTMS/ETCS is a set of specifications introduced as a standard for a common inter-operable platform for railway management and signalling systems. It is intended to be adopted in all European countries and to replace the signalling system with a common system in order to reduce cost and provide interoperability. ERTMS/ETCS is segregated into three levels depending on the used equipment and the operating modes. The first two levels are operational while level 3 is still in the design and standardization phase.

In the literature, one can find numerous formal models of ERTMS/ETCS. In this paper, we focus on the one by Mammar *et al.* [12] using Event-B. In their ERTMS/ETCS 3 model, a railway track (a line to run a train) is divided into sections known as Trackside Train Detection (abbreviated as TTD). A TTD is

---

[3] https://standards.globalspec.com/std/10280790/dsf-fpren-50129.

[4] AnimB: https://wiki.event-b.org/index.php/AnimB.

further divided into subsections called Virtual Sub-Section (VSS). An ERTMS train can be fitted with TIMS which stands for Train Integrity Monitoring System reporting its position and integrity to the train supervisor. The latter refers to the system controller. ERTMS trains without TIMS can only report their front position while non-ERTMS trains do not report their position at all to the supervisor. A train's VSS occupation is also determined by the TIMS. In the ERTMS/ETCS, Movement Authority (MA) is the permission assigned to the train to move on specific sections or subsections. In this model, the supervisor assigns the MA (containing VSSs) and sends it to the ERTMS train. A train cannot go beyond the specified VSS in order to avoid collisions.

This model allows the trains to be connected or disconnected. A connected train regularly reports its integrity and position to the supervisor. In order to manage disconnected trains, the model provides the concept of timers. Note that, in this use case, all trains move in the same direction on the same track and MA is chosen non-deterministically. We re-use this model in our approach as a classical-B artifact which consists of four components: an abstract machine (M0) and 3 refinements (M1, M2 and M3). This case study represents a linear route of a railway topology with tracks without any branching points (switches).

3   Towards an Iterative Formal Model-Driven Approach

3.1   Meeduse Language Workbench

This work extends and applies the Meeduse[5] language workbench [7]. The tool is dedicated to formally instrument DSLs using the B-Method. It embeds the ProB [10] model-checker to support animation and verification. The Meeduse approach (taken from [6]) is shown in Fig. 1.

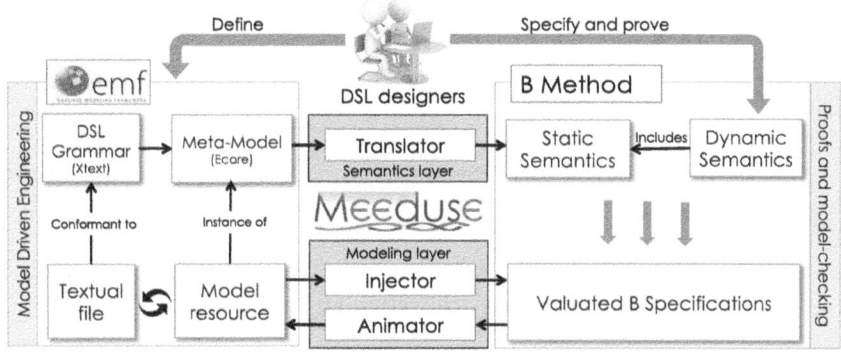

Fig. 1. Meeduse Approach (taken from [6])

The semantics layer translates the meta-model of a DSL into a B specification using the translator component. The resulting B specification defines the static

---
[5] http://vasco.imag.fr/tools/meeduse/.

semantics of the DSL. The Modelling layer includes two components: Injector
and Animator. From a given instance (model resource) of the DSL, Meeduse
first creates a valuated B specification which is an extraction from the translated B model, and then it injects the valuations to populate the various B
data structures. Having this valuated B specification, ProB is applied to animate B operations of any B machine that includes the valuated model. It is
called "Dynamic Semantics" because it confers to the DSL a behavioral character (note that the user specifies the dynamic semantics). At every animation
step, when ProB modifies the internal state of the valuated functional model,
Meeduse translates back this modification to the input model resource, resulting
in an automatic animation of the model resource.

3.2   Using an Existing B Model

In [15], we presented an approach that uses DSLs to visually animate of a given B
specification. The approach introduces a linkage machine that allows the usage of
the B machine as dynamic semantics of the DSL. This approach is extended here
and applied to a realistic case study. We incrementally build our DSL layer and
use refinements and inclusions to make the connection between every increment
of the DSL and the considered B model. The proposed approach is applied to
each abstract/refinement component as depicted in Fig. 2.

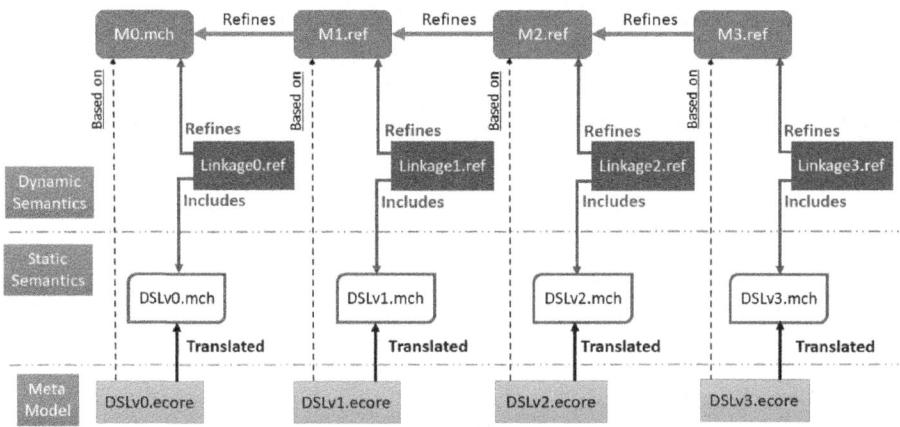

Fig. 2. The Iterative Architecture for the Case Study

First, we develop a DSL meta-model (DSLv0.ecore) based on the initial
abstract component (M0.mch) of the model. Then we provide a linkage machine
(Linkage0.ref) which refines the M0.mch and includes the translated static
semantics of DSLv0.mch (the translation from ecore to B is done using Meeduse). In the next iteration, we update the DSL based on refinement M1.ref
(which is a refinement of the M0.mch machine) and the resulting DSL becomes

DSLv1.ecore. In this iteration, another linkage machine (Linkage1.ref) is introduced which refines the existing refinement M1.ref and includes the translated static semantics DSLv1.mch. The same step is repeated until the final refinement M3. At each iteration, we are able to visually animate the existing component using the corresponding version of the DSL thanks to the execution of the linkage machine in Meeduse. This graphical animation allows the domain expert to check that the verified built specification (existing model) captures the right requirements. The complete B specification of the existing model, the ones generated from the DSL and linkage machines can be found in the Meeduse git repository[6]. The mechanisms of linking B data structures, the initializations and the operations of linkage machines can be also found in our paper [15]. The linkage machines in this case study are generated using our DSL-based Linkage Generator Tool[7], which is built on the definition of generation patterns that are defined by the user and that can be applied (and reused) for various specifications and models.

4   An ERTMS/ETCS Hybrid Level 3 DSL

In a Model-Driven Architecture, a DSL is built from a meta-model. We propose to incrementally create this meta-model based on the existing formal B model, such that each concept in the meta-model corresponds to a concept in the B model. From each version of the meta-model Meeduse generates B static semantics, and the dynamic semantics refers to the corresponding component of the B model. Figure 8 (presented later) shows the whole meta-model where concepts of each refinement are defined using different colors.

4.1   DSL Version 0 (DSLv0)

DSLv0 is built based on abstract machine M0.mch of the existing model. Figure 3 shows the B data structure of M0.mch with its sets, constants and variables. Initialization and operations are not shown here due to space limitation. Machine M0.mch allows free movement of trains on TTDs and collisions are possible at this level. This level contains operations: trainSupervisor, trainEntering, trainMovingInSameTTD, trainMovingFrontNextTTD, trainMovingRearNextTTD, trainExiting, trainConnect, trainDisconnect, and TimerExpiration.

Figure 4 shows the meta-model of DSLv0 where class Railway is the root class. It is composed of class Trackside and class Train. Class Trackside has the attribute TrackStatus which can have the value Free or Occupied from the enumeration Status. Each Trackside has 0 to 2 previous tracksides (association previous) and 0 to 2 next tracksides (association next).

Class Train has two attributes in addition to its identifier: kindOfTrain and Connected. Attribute kindOfTrain refers to three kinds of trains (enumeration

---

[6] https://github.com/meeduse/Samples/tree/main/ETCSLevel3.
[7] https://github.com/meeduse/Samples/tree/main/LinkageGeneratorTool.

```
MACHINE
   M0
SETS
   StateTTD =  {freeT, occupiedT};
   TRAINS ;
   TrainKind = { TimErtms, Ertms, NoErtms}
CONSTANTS
   Ttds, minTTD, maxTTD, trainKind, Trains, Cars
VARIABLES
   stateTTD, trainOccupationTTDFront, trainOccupationTTDRear,
   isConnected, supervisor
INVARIANT
  stateTTD ∈ Ttds → StateTTD
  ∧ trainOccupationTTDFront ∈ TRAINS ⇸ Ttds
  ∧ trainOccupationTTDRear ∈ dom(trainOccupationTTDFront) → Ttds
  ∧ ∀ tr . (tr ∈ dom(trainOccupationTTDFront)
       ⇒ trainOccupationTTDRear(tr) ≤ trainOccupationTTDFront(tr))
  ∧ isConnected ∈ trainKind ⁻¹ [{Ertms, TimErtms}] → BOOL
  ∧ supervisor ∈ BOOL
```

Fig. 3. B data structure of existing abstract machine M0

KindOfTrains): ERTMS, NoERTMS and TIMSERTMS. An TIMSERTMS train is equipped with Train Integrity Management System contrary to simple ERTMS trains. Recall that ERTMS trains equipped with TIMS communicate their location with the supervisor. Attribute Connected is a Boolean that shows the connection of a train with the supervisor. Each train has a head and a tail whose positions are defined with references front and rear to class Trackside.

4.2 Translation of the Meta-model

In order to provide the static semantics as B specifications, Meeduse generates machine DSLv0.mch from DSLv0.ecore, the above ecore meta-model. Figure 5 shows the generated Sets, Properties of the Constants, and the related typing invariants. For more details about this translation we refer the reader to [6,7].

4.3 Linkage Machines

Based on the approach of Fig. 2, linkage B machines are created at every iteration. These machines link the existing B specification (*e.g.* machine M0), which we would like to animate, to the DSL components (*e.g.* machine DSLv0). It refines the existing B model and includes the machine issued from the meta-model of the DSL. The idea is to map concepts from the DSL to concepts from the existing B model. In this sub-section, we give an overview of the mappings used in this first refinement level of our case study. More details about the corresponding approach are provided in [15].

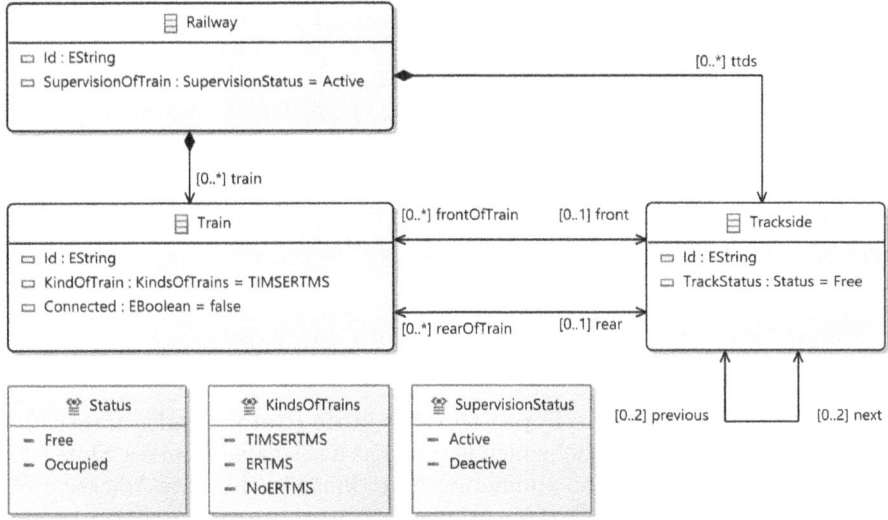

Fig. 4. DSLv0 Meta-Model

Rule 1: EClass to BMachine. In the DSL, class Railway is a root class that contains all other classes. We consider that this class and machine M0 represent the same concept, which is the railway system. Thus, we introduce a constant *Linked_Railway* in the linkage machine which will be helpful later while mapping the underlying concepts.

> CONSTANTS: *Linked_Railway*
> PROPERTIES: *Linked_Railway* ∈ *Railway*

Rule 2: Enumeration to BSet. In machine M0, StateTTD is a set containing values freeT and occupiedT. The similar concept in the DSL is the enumeration Status with the values: Free and Occupied. So, the mapping between an enumeration and a set is done as follows:

> CONSTANTS: *Linked_Status*
> PROPERTIES: *Linked_Status*={*freeT* ↦ *Free, occupiedT* ↦ *Occupied*}

Rule 3: EClass to BSet. In the existing model, constant Ttds is set from constant minTTD to constant maxTTD. The Ttds concept is similar to class Trackside from the DSL. Same like Trains is a finite set of TRAINS and the similar concept from the DSL is class Train. Mapping these sets and classes is done as follows:

> CONSTANTS: *Linked_Trackside, Linked_Trains*
> PROPERTIES:
> > *Linked_Trackside* ∈ *Trackside* ↣ *Ttds* ∧
> > *Linked_Trains* ∈ *Train* ↣ *Trains*

Rule 4: Boolean EAttribute to Boolean BVariable . Attribute isConnected of class Train is a boolean attribute. The similar concept in the existing model is variable Connected. To map these two concepts, we introduce the following invariant:

> INVARIANT: *Connected* = (*Linked_Trains* ; *isConnected*)

Rule 5: EAttribute (EnumType) to BVariable (SetValued). stateTTD is a variable in the existing model which is typed as a set-value from set StateTTD. In the DSL, TrackStatus is an enumeration attribute in the class Trackside. We use composition relation (;) in this mapping as:

> INVARIANT: *TrackStatus* = (*Linked_Trackside* ; *stateTTD* ; *Linked_Status*)

> **Machine**
> > *DSLv0*
> **SETS**
> > *Status* = {*Free, Occupied*};
> > *KindsOfTrains* = {*TIMSERTMS, ERTMS, NoERTMS*};
> > *SupervisionStatus* = {*Active, Deactive*};
> > *RAILWAY; TRACKSIDE; TRAIN*;
> **VARIABLES** [...]
> **CONSTANTS** [...]
> **PROPERTIES**
> > *Railway* ∈ Pow (*RAILWAY*) ∧
> > *Trackside* ∈ Pow (*TRACKSIDE*) ∧
> > *Train* ∈ Pow (*TRAIN*) ∧
> > *KindOfTrain* ∈ *Train* ⇸ *KindsOfTrains* ∧
> > *previous_next* ∈ *Trackside* ↔ *Trackside* ∧
> **INVARIANT**
> > *TrainFront* ∈ *Train* ⇸ *Trackside* ∧
> > *TrainRear* ∈ *Train* ⇸ *Trackside* ∧
> > *SupervisionOfTrain* ∈ *Railway* ⇸ *SupervisionStatus* ∧
> > *TrackStatus* ∈ *Trackside* ⇸ *Status* ∧
> > *Connected* ∈ *Train* ⇸ **BOOL** ∧
> > ∀ *thePrevious*.(*thePrevious* ∈ **ran**(*previous_next*)
> > > ⇒ **card**(*previous_next*$^{-1}$ [{*thePrevious*}]) ≤ 2) ∧
> > ∀ *theNext*.(*theNext* ∈ **dom**(*previous_next*)
> > > ⇒ **card**(*previous_next*[{*theNext*}]) ≤ 2)

Fig. 5. Excerpt of the structural part of machine DSLv0

Rule 6: Attribute (EnumType) to a Boolean Variable. The boolean variable supervisor defines the status of the controller (false if not active and true if it is active). In the DSL, the same concept is defined using an attribute SupervisionOfTrain (enumeration SupervisionStatus) in class Railway. Mapping between enumeration values and the boolean values is established by means of the following invariant:

INVARIANT:
$(supervisor = \textbf{TRUE} \Rightarrow SupervisionOfTrain(Linked\_Railway) = Active)$
$\wedge \ (supervisor = \textbf{FALSE} \Rightarrow SupervisionOfTrain(Linked\_Railway) = Deactive)$

Rule 7: Single Valued EReference to BVariable. Variable trainOccupationTTDFront of the existing model defines the occupation of train's front on a Ttd. In the DSL, this concept is defined with reference TrainFront from class Train to class Trackside. We introduce the following invariant to map both concepts:

INVARIANT:
$\textbf{dom}(TrainFront) = Linked\_Trains^{-1}[\textbf{dom}(trainOccupationTTDFront)]$

4.4   Modeling and Visual Animation

Figure 6 is a graphical model conforming to DSLv0. It features two TIMS trains (Train 1, Train 2) and five tracks (Trackside 1..5). The state of each track is represented in the right-hand side of the figure. In this model all tracks are set to Free; they are not occupied by any of the two trains. Actually, the model of Fig. 6 is drawn by the domain expert to describe an initial state. Once the linkage machine between DSLv0 and M0 is created, this model can be animated in Meeduse. The tool valuates all the B data (variables and constants), initializes the machine and applies ProB for animation. In a classical animation of B specifications, the B method expert has to complete by hand the specifications with valuations. Here, it is the domain expert who created two instances of Train and five instances of Trackside; and then the tool automatically produces the valuated machine together with its initialisation. Figure 7 shows the movement of trains in the graphical representation by triggering the operations of M0.mch mentioned in Sect. 4.1.

The Front of Train 1 occupies TTD 4 while its Rear occupies TTD 1. The Other shows the occupancy of the train on TTDs that are between Front and Rear. For Train 2, the Front is positioned at TTD 2 and Rear at TTD 1. Apart from Trackside 5, all other tracksides are occupied. At this stage, someone can observe the collision between Train 1 and Train 2 as both occupy the TTD 1 and TTD 2. Actually, the safety property regarding the non-collision is only satisfied in the M3 refinement level of the existing model. We chose not to represent other concepts defined in the meta-model such as supervision status, train connect/disconnect, because they are not directly linked to possible collisions.

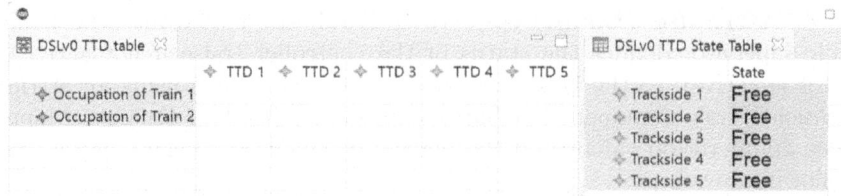

Fig. 6. A model conforming to DSLv0

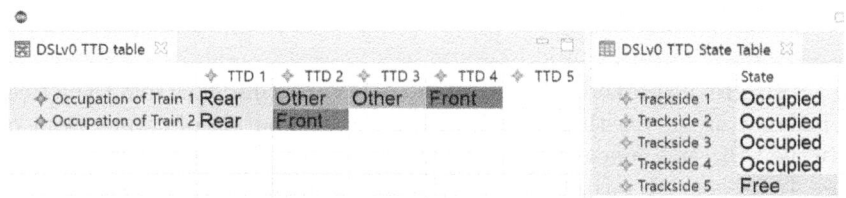

Fig. 7. Animating DSLv0 using M0

5   Application, Findings and Analysis

The complete meta-model of our DSL is shown in Fig. 8. Concepts of every refinement level are presented using different colors. Yellow classes and black associations show DSLv0. Associations in brown color are introduced during DSLv1. Class **VirtualBlock** and associations represented in light blue are from DSLv2. Finally, attributes and associations in purple represent DSLv3.

5.1   Next Iterations

DSLv1. This version is an update of DSLv0 by introducing two new concepts introduced in refinement M1 of the existing specification. Since M1 refines M0, all concepts from M0 are already included in meta-model of DSLv1. Associations frontTrackLocation and rearTrackLocation, illustrated with a brown color in Fig. 8 are added in DSLv1. The frontTrackLocation shows the front location of a train that is communicated to the controller (supervisor) and rearTrackLocation is the communicated location of the train's rear. The inclusion of these concepts updates the graphical representation with Train's location information, as known by the supervisor.

DSLv2. Concepts of DSLv2 are shown in light blue in Fig. 8. Class VirtualBlock has been introduced. It is linked to class Railway using the composition relation. At this level no attributes are included in class VirtualBlock except the Id. A link between class Trackside and VirtualBlock is created where a track side can have many virtual blocks (association virtualblock) and in the opposite each virtual block is associated to at-least one track (association **trackside**). The other introduced associations in this level are those from class Train to class VirtualBlock

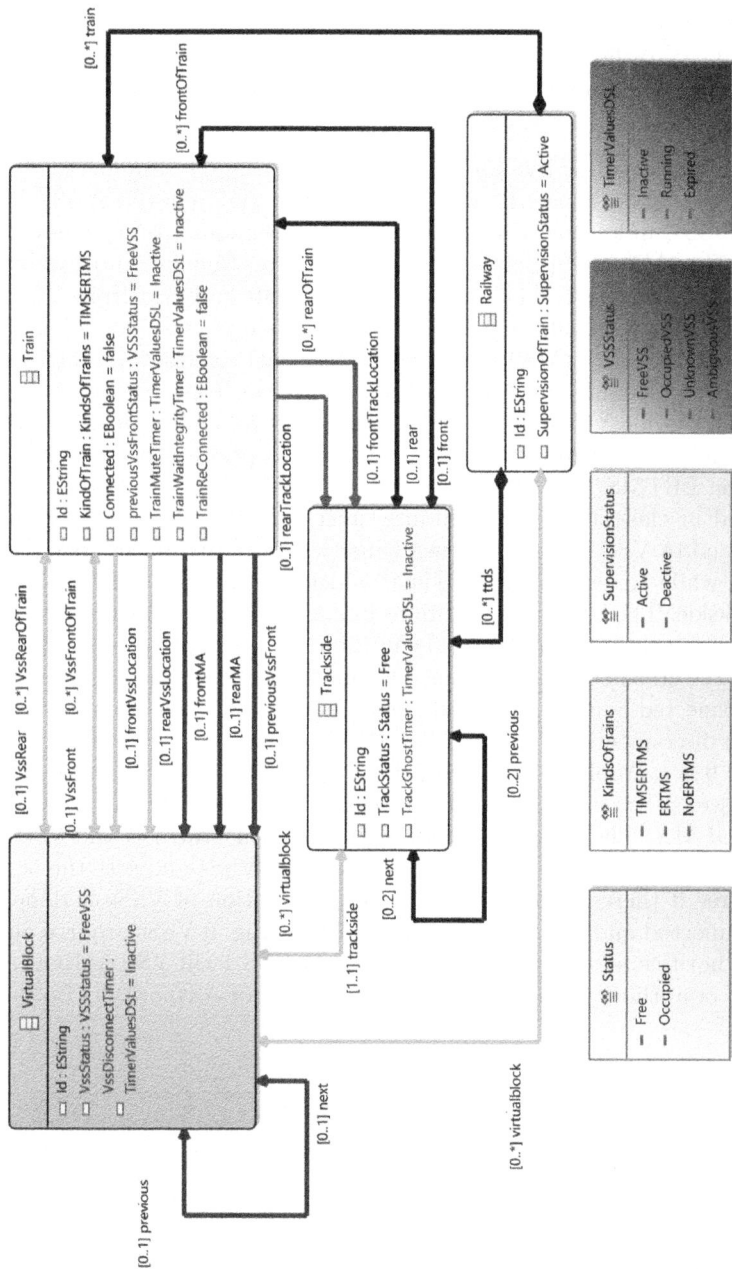

Fig. 8. Whole DSL Meta-Model (Color figure online)

which are VssFront (front of train on virtual block), VssRear (rear of train on virtual block), frontVssLocation (front of train on virtual block communicated to the controller), and rearVssLocation (rear of train on virtual block communicated to the controller). These additional concepts lead to the creation of a new representation containing virtual blocks (VSSs) instead of tracks (TTDs).

DSLv3. Concepts of DSLv3 are shown in violet in Fig. 8. Enumeration VSSStatus gives the state of a virtual block. It can be free (FreeVSS), occupied (OccupiedVSS), unknown (UnknownVSS), or ambiguous (AmbiguousVSS). Enumeration TimerValuesDSL includes values: Inactive, Running and Expired, which are the states of a timer. The associations: previous and next from VirtualBlock to VirtualBlock represents the connections between virtual blocks. The concept of movement authority (MA) is defined by associations: frontMA (MA for train's head) and rearMA (MA for train's rear) from class Train to class VirtualBlock. Association previousVssFront is used to store the value of a previous VSS for a train's front. We introduced the four timer concepts in the DSL as defined in the existing ERTMS/ETCS HL3 specifications. The timers related to the trains are defined in class Train as TrainMuteTimer and TrainWaitIntegrityTimer. The timer related to VSS is VssDisconnectTimer which is introduced inside class VirtualBlock, while timer TrackGhostTimer which is related to TTD is included in class Trackside. The VssStatus attribute in class VirtualBlock shows the status of each virtual block using enumeration VSSStatus. Attribute previousVssFrontStatus stores the status of the previous VSS for the train's front. Finally, attribute TrainReConnected is a Boolean and gets value TRUE when a train is connected back after disconnecting.

Figure 9 is a graphical representation conforming to DSLv3. The figure represents eleven VSSs and two TIMS trains, in addition to train's location, occupation, and MA. The right-hand side represents the states of the VSSs. Before assigning MA, the train supervisor calculates the VSSs and sets the state of all VSSs to free if there is no train. Once the calculation of VSSs is done, a train can be connected and can be assigned an MA. In Fig. 9 VSSs are free and movements authorities are assigned to Train 1 from VSS 1 till VSS 5. Rear of MA is the "start of authority" and Front of MA is "End of Authority" (EoA).

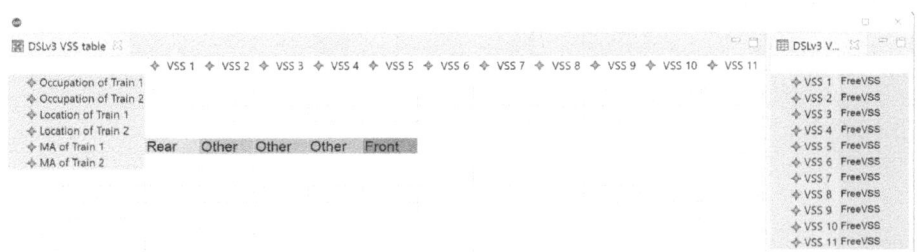

Fig. 9. Assigning MA to Train 1 using DSLv3

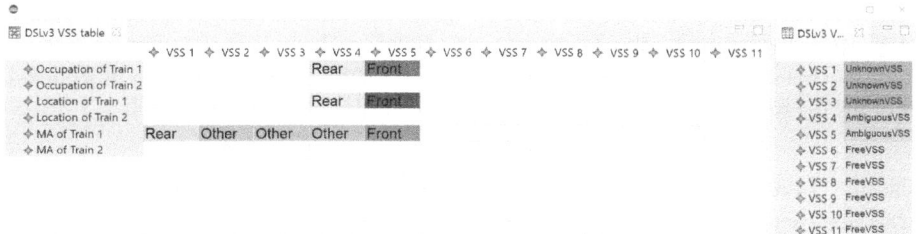

Fig. 10. Train Movement consuming MAs

5.2 Unexpected Behaviors

Figure 10 shows that Train1 has entered and reached EoA. In a normal case, it should be possible to assign new MAs to train 1 to move on the next VSSs, but the animation tells us that this is not possible. The only possible allowed operations are disconnection and connection of Train1. This problem is a deadlock and was identified during the animation of normal scenarios from ERTMS/ETCS. In Fig. 10, a user can observe that some VSSs remain concerned by MAs even after a train has consumed them. It can be seen that VSS 1, 2 and 3 have been released but still they are concerned by the MAs of Train1. This behavior can be considered as a problem from the domain expert point of view. Actually, it reveals some limits (misunderstandings of the requirements) of the existing B specifications.

Another unexpected behavior that we observed is that a VSS never gets the state OccupiedVSS. According to ERTMS/ETCS, when the train's front or rear is over a VSS then the VSS is considered to be occupied. The right side of Fig. 10 shows that VSS 4 and VSS 5 are set to AmbiguousVSS, which should be instead OccupiedVSS as both VSS host the rear and the front of Train1 respectively. In order to test whether this problem is coming from the linkage machine or the existing machine, we analyzed the states of VSSs in the existing machine. We came to the conclusion that the problem is located in the existing machine and this information was communicated to the authors of the existing machine. Note that the authors of the existing model already mentioned in their paper that proof obligations related to VSS state machines were found ambiguous (non-deterministic) and were not discharged.

5.3 Lessons Learned

Lessons Learned from a Formal Methods Expert's Point of View. The existing animation tools (Brama, AnimeB, etc.) do not offer a useful graphical view for the model's animation. Indeed, ProB and AnimB, for instance, only display the values of the variables at each animation step leaving the task of analyzing and explaining them to the user. This is definitely not exploitable for complex systems with several variables like ERTMS 3 system. So, the approach introduced in this paper permits overcoming the drawbacks of the existing tools by providing a useful graphical view of the animation permitting users a better understanding of their systems, and helping them detect errors and bugs. Among others, this approach permits us to detect, for instance, that a VSS never becomes occupied, while this went unnoticed under Rodin and ProB.

Lessons Learned from a Railway Expert's Point of View. Among the three unexpected behaviors, one is to be pointed out at first: VSS never gets the state OccupiedVSS. This is a problem because when a train is on a VSS, this VSS must be occupied. Particularly when the train is connected and sends its position to the control center supervising train movement. In a first analysis, it is surprising that such an evidence is not fulfilled. Analyzing deeper, this is not so surprising. A railway norm is written by railway experts for railway experts. It means that some evidences are not recalled to mind: this is an implicit requirement that everybody is supposed to know.

The paper of Mammar [12], clearly says that in their opinion, the specification is ambiguous. From a methodological point of view, the DSL is run using various scenarios generating behaviors that often surprise a railway expert. Two trains in the same location are correct in the first level because the non-collision mechanisms will be implemented in a later step. The simulation analysis by an expert helps to define the semantic limits of a given DSL but the more interesting part comes when a misunderstanding of the specification is identified. The OccupiedVSS value that is never reached is a good demonstration. Even authors of the code never identified the problem before. It shows that basic specification errors happens and are really difficult to detect without a graphical animation.

5.4   A New Proposed Solution and Proof Obligations

Through an industrial case study (ERTMS 3/ETCS), we showed how a DSL can be used to validate an existing formal B specification. We define links between the concepts defined in our DSLs and those used in the B specification. To verify the various B models, proof obligations (POs) are generated. These POs ensure the correctness of the B specification regarding the linking invariants (see Table 1).

Table 1. Comparison of POs generated using both architectures

Architecture	M0	M1	M2	M3	Linkage0	Linkage1	Linkage2	Linkage3
First Architecture	35	134	153	377	108	168	467	NA
New Architecture	21	83	105	NA	39	120	205	NA

As one can notice, the number of POs is huge; this is due to the fact that the linking invariants of each level include those of the previous one. To overcome this problem, we are working on a new architecture, depicted in Fig. 11), which generates fewer POs and that is even easier to prove.

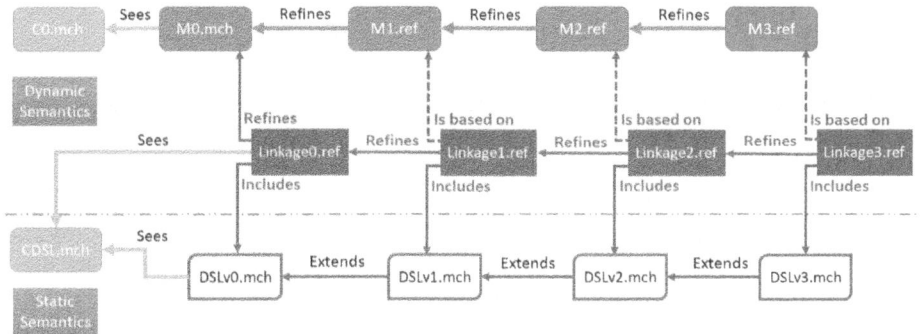

Fig. 11. New Architecture

In this new architecture, first, the concepts of the existing model are kept in a separate context machine called C0.mch and then concepts of the DSL are kept in a context machine called CDSL.mch. In the first iteration, M0.mch sees the C0.mch while DSLv0.mch and Linkage0.mch sees the CDSL.mch. Then the Linkage0.mch refines the M0.mch and includes the DSLv0.mch. In the second iteration, Linkage1.ref (which is based on M1.ref) refines the linkage0.ref and includes the DSLv1.mch (which extends the DSLv0.mch). The same process that is done in the second iteration is applied until the last iteration. The architecture is easier to prove but the animation is complicated using DSLs. In the first architecture shown in Fig. 2, there is one DSL and we are updating the same DSL at each iteration. In the new architecture, at each level, we have a different DSL extending the previous one. The Eclipse Modeling Framework (EMF) [4] does not support such an architecture when designing meta-models of DSLs. But we are able to animate the dynamic semantics of this architecture in Meeduse using a single DSL (using an updated version of DSL at each level). Note that this new architecture is preliminary and not part of this case study.

Currently, the use of this approach on a specification written in a formal language different from the B language incurs additional translation/proof efforts since we have to map the specification into the B language and prove the specification again. For the case study presented in this paper, the B specification provided in [12] has been first rewritten with respect to the B language syntax, and then all the operations have been re-proved under AtelierB by A. Mammar (original author of the existing B specification). Table 1 shows the comparison of number of (POs) generated from both architectures. The table clearly shows that POs generated from the components using the first architecture is significantly higher than the new architecture. AtelierB automatically proved most of the POs. We are not able to generate POs for a few components and in the table, their POs are shown as NA. To reduce this cost, a perspective of this work is to study how the proofs can be performed under Rodin, the Event-B development platform, can be applied to discharge the corresponding proofs under AtelierB.

6   Conclusion

This paper presents an extension of Meeduse, and its application, for validating step-by-step an existing B specification that is developed independently of the Meeduse team. This specification corresponds to the ERTMS/ETCS level 3 case study of the ABZ'2018 conference [2]. The proposed extension brought to the tool the capability to deal with B refinements while animating a DSL. Furthermore, during the application we incrementally built the DSL being guided by the B data that are used in each refinement level. Some specific scenarios were used to show how a visual animation can document border cases of a given refinement level. The railway expert, who is often not an expert in B, is fully aware of the normative framework and may assess and document the hierarchical decomposition with regard to functional safety requirements. An executable DSL allows the railway expert to visualize and discuss requirements such as (non-)collision constraints and train integrity, with the formal methods expert.

While the extension of Meeduse covers B refinements, it does not deal with DSL refinements. This perspective is left to future works. In Sect. 5.4 we discussed a novel architecture and showed that it may reduce the number of POs. However, to make this architecture effective we need to establish a refinement relationship between meta-models of every increment, which requires the extension of the Eclipse Modeling Framework.

References

1. The ERTMS/ETCS signalling system. http://www.railwaysignalling.eu/wp-content/uploads/2016/09/ERTMS_ETCS_signalling_system_revF.pdf. Accessed 16 Mar 2022
2. Butler, M.J., Raschke, A., Hoang, T.S., Reichl, K.: ABZ 2018. LNCS, vol. 10817. Springer, Cham (2018). https://doi.org/10.1007/978-3-319-91271-4
3. Collart-Dutilleul, S., Pereira, D.I.A., Bon, P.: Designing operating rules for ERTMS transnational lines. In: Collart-Dutilleul, S. (ed.) Operating Rules and Interoperability in Trans-National High-Speed Rail, pp. 133–161. Springer, Cham (2022). https://doi.org/10.1007/978-3-030-72003-2_6
4. Steinberg, D., Budinsky, F., Marcelo, P., Merks, E.: EMF: Eclipse Modeling Frameworke, 2nd edn. Addison-Wesley, Boston (2009)
5. Ferrari, A., et al.: Survey on formal methods and tools in railways: the ASTRail approach. In: Collart-Dutilleul, S., Lecomte, T., Romanovsky, A. (eds.) RSSRail 2019. LNCS, vol. 11495, pp. 226–241. Springer, Cham (2019). https://doi.org/10.1007/978-3-030-18744-6_15
6. Idani, A.: Meeduse: a tool to build and run proved DSLs. In: Dongol, B., Troubitsyna, E. (eds.) IFM 2020. LNCS, vol. 12546, pp. 349–367. Springer, Cham (2020). https://doi.org/10.1007/978-3-030-63461-2_19
7. Idani, A., Ledru, Y., Vega, G.: Alliance of model-driven engineering with a proof-based formal approach. Innov. Syst. Softw. Eng. **16**(3), 289–307 (2020). https://doi.org/10.1007/s11334-020-00366-3
8. Ladenberger, L., Bendisposto, J., Leuschel, M.: Visualising event-B models with B-motion studio. In: Alpuente, M., Cook, B., Joubert, C. (eds.) FMICS 2009. LNCS,

vol. 5825, pp. 202–204. Springer, Heidelberg (2009). https://doi.org/10.1007/978-3-642-04570-7_17

9. Ladenberger, L., Leuschel, M.: BMotionWeb: a tool for rapid creation of formal prototypes. In: De Nicola, R., Kühn, E. (eds.) SEFM 2016. LNCS, vol. 9763, pp. 403–417. Springer, Cham (2016). https://doi.org/10.1007/978-3-319-41591-8_27

10. Leuschel, M., Butler, M.: ProB: a model checker for B. In: Araki, K., Gnesi, S., Mandrioli, D. (eds.) FME 2003. LNCS, vol. 2805, pp. 855–874. Springer, Heidelberg (2003). https://doi.org/10.1007/978-3-540-45236-2_46

11. Leuschel, M., Samia, M., Bendisposto, J.: Easy graphical animation and formula visualisation for teaching B (2008)

12. Mammar, A., Frappier, M., Fotso, S.J.T., Laleau, R.: A formal refinement-based analysis of the hybrid ERTMS/ETCS level 3 standard. Int. J. Softw. Tools Technol. Transf. **22**(3), 333–347 (2020). https://doi.org/10.1007/s10009-019-00543-1, https://hal.archives-ouvertes.fr/hal-02975774

13. Servat, T.: BRAMA: a new graphic animation tool for B models. In: Julliand, J., Kouchnarenko, O. (eds.) B 2007. LNCS, vol. 4355, pp. 274–276. Springer, Heidelberg (2006). https://doi.org/10.1007/11955757_28

14. Werth, M., Leuschel, M.: VisB: a lightweight tool to visualize formal models with SVG graphics. In: Raschke, A., Méry, D., Houdek, F. (eds.) ABZ 2020. LNCS, vol. 12071, pp. 260–265. Springer, Cham (2020). https://doi.org/10.1007/978-3-030-48077-6_21

15. Yar, A., Idani, A., Ledru, Y., Collart-Dutilleul, S.: Visual animation of B specifications using executable DSLs. In: Proceedings of the 25th International Conference on Model Driven Engineering Languages and Systems: Companion Proceedings, pp. 617-626. MODELS 2022, Association for Computing Machinery, New York, NY, USA (2022). https://doi.org/10.1145/3550356.3561585

An Efficient Distributed Dispatching Vehicles Protocol for Intersection Traffic Control

Fang Qi[1], Rui Wang[1(✉)], Yong Guan[1], and Xiaoyu Song[2]

[1] Information Engineering College, Capital Normal University, Beijing, China
rwang04@163.com
[2] ECE Department, Portland State University, Portland, USA

Abstract. Passing the intersection is a crucial traffic scenario in urban traffic network. Rapid development of vehicle technologies makes vehicles cross the intersection more efficiently. In this paper, we propose a protocol where vehicles can pass the intersection safely and efficiently at a 6-lane dual carriageway. To ensure the safety of the protocol, we formally model the protocol and verify some necessary properties using the tool Uppaal. To evaluate the performance of our protocol, we conduct extensive experiments. After analyzing experimental data, the average rate of time reduction in a circle is over 92%. The results demonstrate that our protocol is both efficient and effective.

Keywords: urban traffic network · intersection · dispatch vehicles · formal methods · Uppaal

1 Introduction

Over the past decades, the population growth and the increasing number of vehicles made urban traffic congestion inevitable, which is estimated 60% increase by 2030 [6]. Congestion may cause collisions, especially at significant parts of traffic network like intersections. Unfortunately, 40% of all road injury accidents occur at intersections [4]. In addition, traffic congestion caused the deterioration of the ecological situation in cities, reduced the well-being of citizens, worsened the quality of life and the level of satisfaction of needs. Thus, it is crucial for vehicles to pass an intersection safely and efficiently.

To overcome these issues, many researchers and experts have made large studies. Since 1973, the UK Transport and Road Research Laboratory has been researching a vehicle responsive method of signal control called SCOOT (Split, Cycle and Offset Optimization Technical), which is now widely used [2]. Currently, using traffic lights at intersections is still a traditional strategy in our

This paper is supported by NSFC program: Research on Design and Verification Method of Intelligent Vehicle Cockpit Control Driven by Formal Model(62272322).

city. Traffic light control methods are divided into two parts: fixed time traffic control and adaptive traffic control.

The fixed time traffic control uses predetermined green and cycle time that has a fixed duration, and changes of traffic volumes during the day are not taken into account. Some fixed time systems use different preset time intervals for the morning peak hour, evening peak hour, and other busy periods. For this reason, such an approach cannot deal with an unexpected change in traffic demand [5].

The adaptive traffic control means it will monitor the traffic online and optimizes the signal plans to ensure fast responses to changes in traffic demand. Currently, many researchers proposed different approaches to optimize the timing of traffic lights. For instance, paper [3] proposed an improve multi-agent data-driven distributed adaptive coordination control algorithm (I-MA-DACC), which outputs the distributed adaptive green time at each intersection for the purpose of dynamic queue balancing. In paper [7], they combine AI and image processing with the adaptive traffic control system to scale back the traffic congestion. Paper [8] proposed a novel MFAC (Model Free Adaptive Control) traffic signal control scheme. And the prpopsed method can reduce the impact of vehicles on the upstream intersection. However, people find that traffic lights will not work efficiently when traffic volumes are high.

In this paper, we propose a protocol called Intersection Traffic Control Protocol (ITCP) where vehicles can leave with a higher efficiency. Vehicles on the same lane can cross the intersection when the lead car on this lane gets permissions. Based on such a mechanism, traffic control at the intersection will be more efficient.

To ensure the safety of ITCP, we specifically adopt a timed model checker UPPAAL [1] to model and verify the protocol. Details of Uppaal will be introduced in Sect. 2.

The major contributions are summarized as follows:

1. We present a more efficient intersection traffic control protocol without any central controller.
2. The safety and liveness of the protocol are verified in UPPAAL.

The remainder of this paper is structured as follows. In Sect 2, we introduce methodology about model checking and Uppaal. In Sect 3, we give the introduction of protocol, including descriptions and dispatching algorithms. Section 4 presents formal modeling and verification. We discuss performance evaluation between our protocol and traditional traffic lights in Sect 5, and conclude with a summary and future work in Sect. 6.

2   Methodology

2.1   Timed Automata(TA)

The Syntax of TA: A timed automaton is a tuple TA = (Loc, Act, C, $\hookrightarrow$, Loc_0, Inv, AP, L) where

- Loc is a finite set of locations;
- $Loc_0 \in$ Loc is a set of initial locations;
- Act is a finite set of actions;
- C is a finite set of clocks;
- $\hookrightarrow \subseteq$ Loc $\times$ C $\times$ Act $\times$ 2^C $\times$ Loc is a transition relation;
- Inv: Loc $\hookrightarrow$ C is an invariant-assignment function;
- AP is a finite set of atomic propositions, and
- L: Loc $\hookrightarrow$ 2^{AP} is a labeling function for the locations.

The semantics of TA: Any timed automaton can be interpreted as a transition system. Due to the continuous time domain, those underlying transition systems have infinitely many states, and are infinitely branching. Timed automata can thus be considered as finite descriptions of infinite transition systems.

2.2 Model Checking Tool UPPAAL

Uppaal is an integrated tool environment for modeling, validation and verification of real-time systems. It is well suited for verifying such systems that can be modeled as nondeterministic processes. Uppaal uses a client-server architecture, and is divided into three parts: the editor, the simulator and the verifier. To verify properties, Uppaal uses a simplified version of TCTL called the Query Language. Three kinds of properties can be verified in Uppaal: safety, liveness and reachability.

3 Intersection Traffic Control Protocol

3.1 The Description

There are diverse road types in urban traffic network, and in this paper the 6-lane dual carriageway is chosen as our research object. As depicted in Fig. 1(a), there are 3 lanes in each direction: turning right or going straight lane, going straight lane and turning left lane, respectively. We denote the lane numbers from 0 to 11 as $lane_0$ to $lane_{11}$.

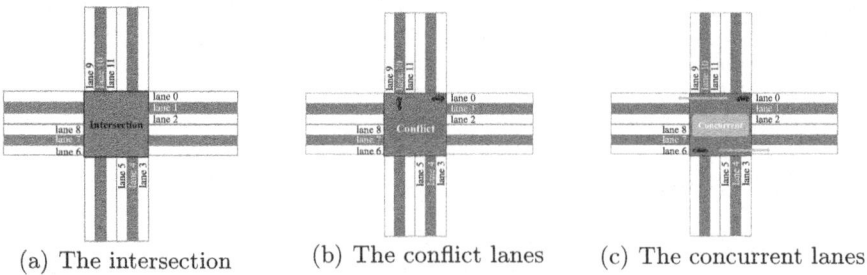

(a) The intersection   (b) The conflict lanes   (c) The concurrent lanes

Fig. 1. Descriptions of different lanes

For $lane_n$ ($n \in N$ and $n \in [0,11]$):

$$n\%3 = \begin{cases} 0, \ lane_n \ \text{means turning right or going straight lane} \\ 1, \ lane_n \ \text{means going straight lane} \\ 2, \ lane_n \ \text{means turning left lane} \end{cases} \quad (1)$$

For each $lane_n$, all vehicles on this $lane_n$ can leave together when the lead vehicle gets permission from other lanes which have already left the intersection. The rectangle in the middle of Fig. 1(a) is the central section of intersection, which is similar to critical resources in computer processes. That is to say, if $lane_i$ and $lane_j$ ($i \neq j$) are conflict, they cannot cross the intersection simultaneously. For example, in Fig. 1(b), vehicles on $lane_0$ and $lane_{10}$ are conflict, which means they should not be in the intersection at the same time, or they will make collisions. However, vehicles on $lane_0$ and $lane_6$ can pass the intersection without any conflict in Fig. 1(c). The relationship between lanes is defined below.

Definition. Concurrent Lanes and Conflict Lanes.
The relationship between lanes is shown in Table 1.

Table 1. Relationship Between Lanes

Current Lane(ID)	Conflict Lanes(ID)	Concurrent Lanes(ID)	Current Lane(ID)	Conflict Lanes(ID)	Concurrent Lanes(ID)
0	3,4,8,9,10,11	1,2,5,6,7	6	2,3,4,5,9,10	0,1,7,8,11
1	3,4,8,9,10,11	0,2,5,6,7	7	2,3,4,5,9,10	0,1,6,8,11
2	3,4,5,6,7,11	0,1,8,9,10	8	0,1,5,9,10,11	2,3,4,6,7
3	0,1,2,6,7,11	4,5,8,9,10	9	0,1,5,6,7,8	2,3,4,10,11
4	0,1,2,6,7,11	3,5,8,9,10	10	0,1,5,6,7,8	2,3,4,9,11
5	2,6,7,8,9,10	0,1,3,4,11	11	0,1,2,3,4,8	5,6,7,9,10

3.2   The Algorithm

To make vehicles cross the intersection efficiently, one vehicle can send messages to other vehicles, such as arrival time, lane number and current states, etc. Hence, other vehicles which are waiting signals to leave will judge if they can leave now. Enabling each vehicle to pass the intersection efficiently without any conflict is the core of our algorithm.

Our dispatching vehicles at intersection algorithm is presented in Algorithm 1. Each vehicle on the lane has five statuses: idle, appr, stopped, passing and passed. When there is no vehicles on the $lane_n$, the status is marked as "idle"; the status will change into "appr" if some vehicles approach on the $lane_n$.

Algorithm 1 Dispatching Vehicles at Intersection

```
begin initialization                              return1;
    InfoVehicles[i] ← null, ∀ i ∈              else if y == (x + a)%12 where a =
{1...,max_lane};                               1, 2, 5, 6, 7 then
end                                                return0;
begin when vehicles enter the intersection     end if
    lane ← getLaneNum();                       else if x%3 == 1 then
    arrivalTime ← getCurrentTime();                if y == (x + a)%12 where a =
if there is no vehicles on the lane, where     2, 3, 7, 8, 9, 10 then
status == idle then                                return1;
    lead ← true;                                   else if y == (x + a)%12 where a =
    arrivalTime_lead ← arrivalTime;            1, 4, 5, 6, 11 then
    if arrivalTime_lead of two different           return0;
    lanes are same then                            end if
        lead ← true, whose number of lane is   else if x%3 == 2 then
        higher;                                    if y == (x + a)%12 where a =
    end if                                     1, 2, 3, 4, 5, 9 then
else                                               return1;
    lead ← false;                                  else if y == (x + a)%12 where a =
end if                                         6, 7, 8, 10, 11 then
status ← appr;                                     return0;
                                                   end if
begin at each cycle                            end if
if PassingCondition() then                     return -1;
    status ← passing;                          end function
    move and cross the intersection;
end if                                         function PassingCondition()
end                                            IsConflict(x);
begin                                          if IsPass == i where i = 1, 2, 3 then
status ← passed;                                   return true;
end                                            else
                                                   return false;
function IsConflict(x)                          end if
if x%3 == 0 then                               end function
    if y == (x + a)%12 where a =
3, 4, 8, 9, 10, 11 then
```

After exchanging some necessary messages, lead car which gets permissions and its following cars can enter the intersection and set their statuses as "passing". Meanwhile, on the basis of two functions in Algorithm 1, other lanes which have one vehicle or more vehicles will compute whether they can leave following the current lane's vehicles. If the function returns true, the corresponding lanes will enter the intersection and set their statuses as "passing". Instead, they set their statuses as "stopped". When all the vehicles on one lane leave the intersection, their statuses are changed as "passed". Up to now, a complete behavior of crossing the intersection has been finished, which is called a circle (Fig. 2).

Now we will mainly explain two functions in Algorithm 1. Whether vehicles on some lane can leave depends on the value of function is true or false. When the value of the function is true, it includes three conditions. In the first case, if statuses of all vehicles on the $lane_i$ which is conflict with current lane are "stopped" and $lane_i$'s $arrivalTime_{lead}$ are greater than current lane's $arrivalTime_{lead}$, only the vehicles on current lane can cross the intersection. The second case and the third case are both for concurrent lanes. In these two cases, $lane_i$ and $lane_j$ (i≠j) are both concurrent with the current lane, their vehicles' statuses are

Table 2. Vehicle Information for Dispatching

Vehicle Information	Note
lane	lane number from 0 to 11
arrivalTime	Arrival time for its own vehicles
lead	True or false
IsPass	Number of lanes leaving the intersection together
status	idle, appr, stopped, passing, or passed

"stopped" and $lane_j$'s $arrivalTime_{lead}$ are greater than $lane_i$'s $arrivalTime_{lead}$. If $lane_i$ is also concurrent with $lane_j$, the vehicles on the three lanes can leave as a whole. Rather, when $lane_i$ is conflict with $lane_j$, the vehicles of the current lane and $lane_i$ are both across the intersection. If the function returns true, the vehicles will change their statuses into "passing" and "passed" when they cross the intersection and pass the intersection, respectively. Next a brief explanation of function IsConflict(x) will be given. The function IsConflict(x) is mainly used to compute if $lane_i$ is conflict with the current lane. If the value of the function is zero, the $lane_i$ is concurrent with the current lane. Instead, it is conflict with the current lane.

4   Formal Modeling and Verification

4.1   Modeling

To demonstrate that the protocol enables vehicles to cross the intersection safely and efficiently, we adopt the formalized method to model the protocol and verify relevant properties using the tool Uppaal. In this work, models of the protocol are built on time and state dimension. There are four automata in our model: Lane, ControllerLane, List_Init, and List_Operation. In addition, the automaton Lane are instantiated as twelve different lanes from $Lane_0$ to $Lane_{11}$. Next the concrete details will be presented below.

Lane Automaton. This automaton in Fig. 2(a) mainly introduces some vehicles' actions within a circle of passing the intersection. In the model, there are five kinds of states: Idle, Appr, Stopped, Passing and Passed. Idle is an initial state which means the $lane_i$ is non-participating. A transition from Idle to Appr happens after sending a synchronization message. For the same reason, other three transitions will be triggered. Uppaal uses a continuous time model, so the clock is a specific feature in Uppaal. In this model, clock x mainly depicts the time where vehicles cross the intersection. Furthermore, we find that arrival time of vehicles on conflict lanes may be same which makes a symmetry. To break this symmetry, lane id can be used. Vehicles whose lane id is higher has higher priority to pass the intersection.

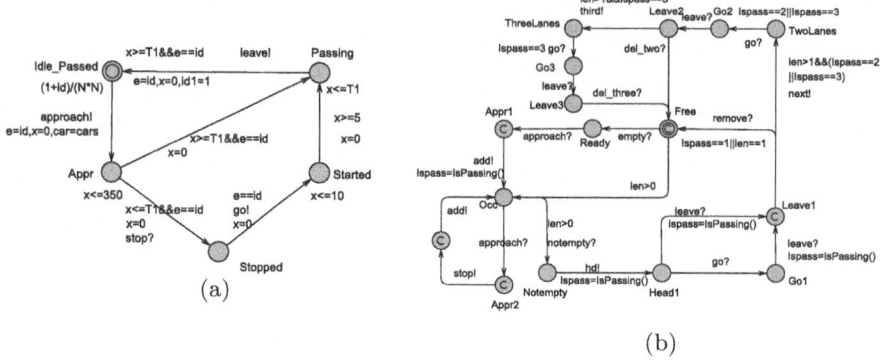

Fig. 2. (a) Lane Automaton (b) ControllerLane Automaton

ControllerLane Automaton. This automaton is designed to mainly control the whole system. Location Appr1 is different from other normal locations. It is an urgent location where transition happens without any delays. Furthermore, the upper section models two functions in Algorithm 1 (Fig. 3).

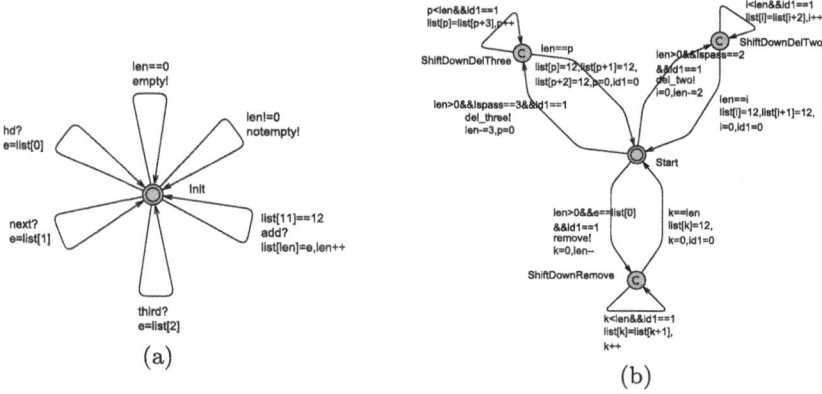

Fig. 3. (a) List_Init Automaton (b) List_DelOperation Automaton

List_Init Automaton. Some basic list operations are shown in the automaton, such as determining whether the list is empty or not, adding nodes, etc. By sending corresponding synchronization messages, list operations can be finished.

List_DelOperation Automaton. There are three cases of deleting nodes which correspond to the function PassingCondition() in Algorithm 1.

4.2 Verification

Based on the modeling, some desired properties are verified in the Uppaal.

1. **Safety Property:**
   P1: A[] not deadlock
   System will not end up in a deadlock state.
   P2:   E[]   not $(Lane_0.Passing \&\&(Lane_3.Passing||Lane_4.Passing||Lane_8.$ $Passing||Lane_9.Passing||Lane_{10}.Passing||Lane_{11}.Passing))$ Consider $Lane_0$
   as an example. $Lane_0$ will not enter the intersection with its conflict lanes at
   the same time. If that happens, vehicles on conflict lanes will collide starkly.
   Other lanes also satisfy this kinds of safety and these properties are denoted
   as P3-P13.
2. **Liveness Property**:
   P14: $A<>Lane_0.Appr$ imply $Lane_0.Idle\_Passed$
   If a vehicle is trying to pass the intersection, then the vehicle must finally
   pass the intersection in finite time. The property liveness defined P15-P25 is
   satisfied for $Lane_0$ through $Lane_{11}$.
3. **Reachability Property**:
   P26: $E<>(Lane_0.Passing||Lane_1.Passing||Lane_2.Passing||Lane_3.Passing||$
   $Lane_4.Passing||Lane_5.Passing||Lane_6.Passing||Lane_7.Passing||Lane_8.Passing||$
   $Lane_9.Passing||Lane_{10}.Passing||Lane_{11}.Passing)$ All lanes will have vehicles
   eventually at a given time.

In this paper, four properties are verified and they are all satisfied. Results
are presented in Table 3.

Table 3. Results of verification

Property	Result	Property	Result	Property	Result	Property	Result
P1	satisfied	P8	satisfied	P15	satisfied	P22	satisfied
P2	satisfied	P9	satisfied	P16	satisfied	P23	satisfied
P3	satisfied	P10	satisfied	P17	satisfied	P24	satisfied
P4	satisfied	P11	satisfied	P18	satisfied	P25	satisfied
P5	satisfied	P12	satisfied	P19	satisfied	P26	satisfied
P6	satisfied	P13	satisfied	P20	satisfied	\	\
P7	satisfied	P14	satisfied	P21	satisfied	\	\

5 Performance Evaluation

In this section, we present experimental results that demonstrate the performance of our protocol for 6-lane dual carriageway intersection traffic control to
validate efficiency and effectiveness. To illustrate the efficiency, we compare our
protocol to traffic lights based on different perspectives.

5.1   Experiments

For experiments, we assume there are 200 vehicles at east-west direction or south-north direction. In addition, combining the actual situations in urban traffic network, we provide three kinds of data: 50 s, 60 s and 70 s. In each case, some sub-cases are given and details are presented below. We take 50 s as an example. In a circle, 50 s are divided into two parts: 40 s and 10 s, 35 s and 15 s, 30 s and 20 s. In each group, smaller seconds are given turning left lanes, and bigger seconds are given going straight lanes and turning right lanes.

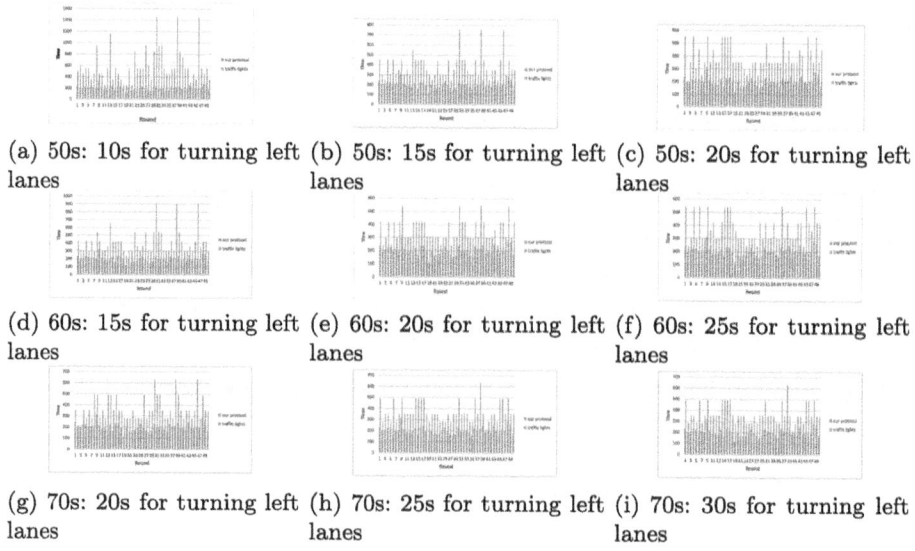

(a) 50s: 10s for turning left lanes (b) 50s: 15s for turning left lanes (c) 50s: 20s for turning left lanes

(d) 60s: 15s for turning left lanes (e) 60s: 20s for turning left lanes (f) 60s: 25s for turning left lanes

(g) 70s: 20s for turning left lanes (h) 70s: 25s for turning left lanes (i) 70s: 30s for turning left lanes

Fig. 4. Results of 50 random sequences under different groups

Based on the above-mentioned data, we first consider 50 random sequences of lanes where there are 12 lanes in each sequence. It is necessary to note that the same lane may occur repeatedly in a sequence. For example, (3,4,9,4,4,1,0,8,9,1,7,4) is a random sequence, in which $lane_4$ occurred four times and $lane_9$ occurred twice. For each sequence, we calculate the time crossing the intersection and final results are shown in Fig. 4. There is a big difference between our protocol and traffic lights, which proves that our protocol is more efficient and effective under the same sequence. The results of other data are also shown.

After analyzing the data in Fig. 4, we find that a extremely large difference appears in round 30, 38 and 46 of picture Fig. 5(a). There is a common feature that turning left lane occurred many times in these three rounds. So we suppose that if there are many turning left lanes in a sequence, the efficiency of our protocol will be more striking. To verify this guess, another comparative experiment is presented. In this experiment, we provide 8 groups of data. In the first 4 groups, there are only vehicles on east-west direction and $lane_2$ and $lane_8$ occurred many times. The concrete results are given in Fig. 5. According to the Fig. 5, it is verified that the more turning left lanes occur in a sequence, the more prominent effectiveness of ITCP is. In the fourth group, we give a extreme case (8,8,2,2,8,2,8) where all lanes are turning left lane. In fact, apart from more turning left lanes, the number of vehicles on turning left lanes is also a key factor. Because during a fixed time, the more vehicles will consume more circles. In Fig. 5, the conclusions mentioned above are proved.

Table 4. Time Performance Analysis

Performance	Rate of Reduction Time		
	50 s	60 s	70 s
random sequences	204.31%	108.42%	86.30%
	44.67%	78.84%	86.45%
	38.76%	89.45%	91.56%
special sequences	1726.90%	349.41%	260.74%
	293.18%	225.07%	215.51%
	221.62%	67.94%	179.95%

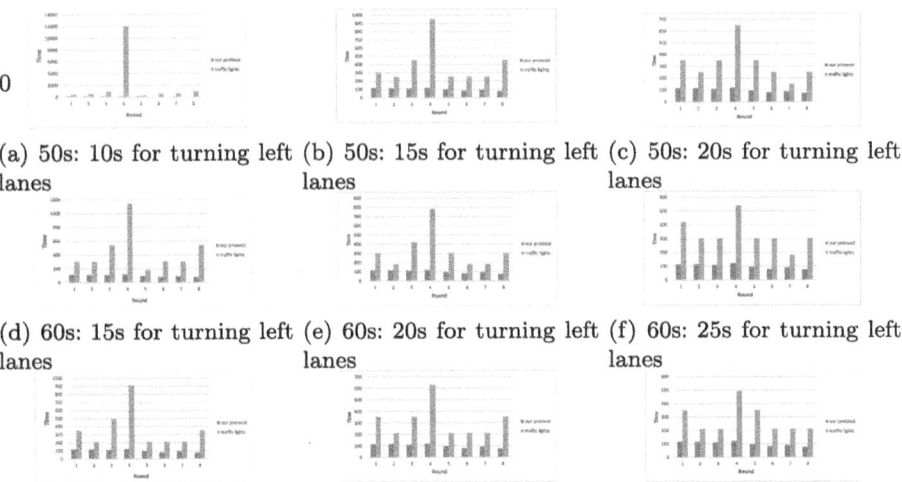

(a) 50s: 10s for turning left lanes (b) 50s: 15s for turning left lanes (c) 50s: 20s for turning left lanes

(d) 60s: 15s for turning left lanes (e) 60s: 20s for turning left lanes (f) 60s: 25s for turning left lanes

(g) 70s: 20s for turning left lanes (h) 70s: 25s for turning left lanes (i) 70s: 30s for turning left lanes

Fig. 5. Results of 8 specific sequences under different groups

5.2 Timing Analysis

In Table 4, we analyze the time performance of experiments based on data. T_{TL} represents average time consumption under traffic lights, and T_{Pro} denotes average time consumption adopting our protocol. The rate of reduction time can be calculated roughly by the formula:

$$RRT = \frac{1}{n} \sum_{i=1}^{n} \frac{T_{TL}(i) - T_{Pro}(i)}{T_{Pro}(i)} \; (n = 8 \, or \, n = 50) \tag{2}$$

6 Conclusion and Future Work

In this paper, we present a protocol where vehicles can cross the intersection safely and efficiently. To further make our dispatching algorithm efficient, we utilize the arrival time of the lead vehicle on each lane, thereby, the following vehicles on the lane can pass the intersection without getting permissions from other vehicles on conflict lanes. After that, we adopt formal methods to model and verify. And verified properties are satisfied. In the future work, we will use autonomous vehicles as our research subject. With the development of high technology, autonomous vehicles will be popular in our life and become main transportation.

References

1. Behrmann, G., David, A., Larsen, K.G.: A tutorial on UPPAAL. In: Bernardo, M., Corradini, F. (eds.) SFM-RT 2004. LNCS, vol. 3185, pp. 200–236. Springer, Heidelberg (2004). https://doi.org/10.1007/978-3-540-30080-9_7
2. Hunt, P.B., Robertson, D.I., Bretherton, R.D.: The scoot on-line traffic signal optimisation technique (glasgow). Traffic Eng. Control 23(4), 190–192 (1982)
3. Ji, H., Yin, H., Ren, Y., Wang, L., Liu, S.: An improved multi-agent based data driven distributed adaptive cooperative control in traffic network signal timing. In: 2022 IEEE 11th Data Driven Control and Learning Systems Conference (DDCLS), pp. 770–775 (2022). https://doi.org/10.1109/DDCLS55054.2022.9858410
4. Noh, S.: Decision-making framework for autonomous driving at road intersections: safeguarding against collision, overly conservative behavior, and violation vehicles. IEEE Trans. Ind. Electron. 66(4), 3275–3286 (2019). https://doi.org/10.1109/TIE.2018.2840530
5. Pavleski, D., Koltovska, D., Ivanjko, E.: Evaluation of adaptive and fixed time traffic signal strategies: case study of skopje (2018)
6. Rafter, C.B., Anvari, B., Box, S.: Traffic responsive intersection control algorithm using GPS data. In: 2017 IEEE 20th International Conference on Intelligent Transportation Systems (ITSC), pp. 1–6 (2017).https://doi.org/10.1109/ITSC.2017.8317795
7. Sirphy, S., Thanga Revathi, S.: Adaptive traffic control system using yolo. In: 2023 International Conference on Computer Communication and Informatics (ICCCI), pp. 1–5 (2023). https://doi.org/10.1109/ICCCI56745.2023.10128619
8. Yin, H., Ren, Y., Wang, L., Ji, H., Liu, S.: Model free adaptive traffic signal control for four-phase intersections. In: 2022 IEEE 11th Data Driven Control and Learning Systems Conference (DDCLS), pp. 752–756 (2022). https://doi.org/10.1109/DDCLS55054.2022.9858479

Security

Confidentiality Management in Complex Systems Design

Michel Bourdellès[1]([envelope]) [iD], Jamal El-Hachem[2] [iD], and Salah Sadou[2] [iD]

[1] Univ. Bretagne Sud, Vannes, France
michel.bourdelles@univ-ubs.fr
[2] IRISA – UMR 6074, Univ. Bretagne Sud, Vannes, France
jamal.el-hachem@irisa.fr, salah.sadou@univ-ubs.fr

Abstract. The use of modelling tools for the design of industrial systems is increasingly replacing the documentation production resulting from a classic system design process. One of the problems to be solved in order to fully succeed in this transition is information confidentiality management.It is also necessary to propose a system design process that allows an evolution through the whole design life cycle, and that guarantees a coherent design for a given level of confidentiality. To the best of our knowledge, neither current Model Based System Engineering (MBSE) tools, nor academic research propose solutions to this specific issue. In this paper, we propose a model based system design process to manage elements belonging to different levels of confidentiality. The process guarantees a plurality of confidentiality levels in line with Bell-Lapadula security policies for the secure management of system design and specification, as required for the protection of national and coalition defense information. It includes the separation into enclaves and adaptations guaranteeing the preservation of confidentiality and integrity to authorized users. We describe how this process can be leveraged for current modelling tools by exploiting existing multi-user functionality.

Keywords: Security of complex systems · System modelling · Product life cycle

1   Introduction

System specification and design documents for sensitive projects often consist of two separate parts: confidential and non-confidential. Among confidential documents, there are security annexes which state the rules to correctly use the whole documents. Thus, identifying data classification levels becomes the first step an organization should consider when developing a data sensitivity program. Then, the organization should apply an access control on data with respect to the level of sensitivity. This corresponds to the "Information Flow Enforcement" access control in the SP800-53 report from the National Institute Standards and Technologies (NIST) [1].

G. Bai et al. (Eds.): ICECCS 2024, LNCS 14784, pp. 303–322, 2025.
https://doi.org/10.1007/978-3-031-66456-4_17

Indeed, to be protected against system attacks, the design of these systems should include a prior security analysis. More precisely, a first security analysis consists in identifying critical assets, performing a risk assessment and taking decisions regarding the security controls to be applied. The latter information being sensitive and impacting the system design, it is important to propose solutions allowing, in a system modelling, to take this information into account and to restrict its access to only authorized persons.

Prospective analysis [2] foresee a greater use of modelling approaches for system design. This is known as Model-Based System Engineering (MBSE). These designs respect normed industrial processes such as ISO/IEC/IEEE 12207 and capture information from initially separate documents (Technical Requirements Specification, System/Subsystem Specification, System/Subsystem Design Description, Security Annexes, Product Breakdown Structure, Interface Control Document). Yet the confidentiality is ensured by a unique level of separation in sensitive documents. However, the information from security analysis documents impacts the overall design and is an integral part of its modelling. In case of sensitive systems, modeling involves adding mechanisms to manage different levels of confidentiality and authorization within the models. It is therefore necessary to extend all the design constraints based on an MBSE approach to introduce a management process considering the confidentiality and the access authorization rules.

Many recent works (UAF [3], SysMLSec [4], MBSEsec [5], Capella Cyber Security viewpoint [6], MBCA [7], SMSA [8], SoSSEC [9]) make it possible to integrate security elements into system modelling designs to secure requirements. However, those approaches, as well as their corresponding modelling tools (Cameo System Modeller, Modelio, Capella/Arcadia, and other SysML modelling tools) work with a "flat" view on the information, which means without managing the confidentiality protection need of specification parts design nor managing access in line with an authorization policy.

To the best of our knowledge, related works not or only partially cover these needs, in particular works based on Model Driven Development [10,11], data obfuscation and filtering [12], enclave partitioning [13], works on database confidentiality management [14–16], or industrial processes [17] do not, or only partially address the previously mentioned challenges.

Thus, in this paper we propose an approach, based on the security enclaves concept, to manage different levels of confidentiality in an MBSE process. This approach is built on top of several years of expertise in design in sensitive systems within a large company in the defense area.

The remainder of the paper is organized as follows: in the next section we provide an example illustrating the research problem of mixing elements of several confidentiality levels in MBSE modelling and an analysis of confidentiality management choices done in the Galileo modelling system design. In Sect. 3, we list a set of criteria derived from industrial experience in a sensitive domain, to which our approach comply. Section 4 details our approach. We show how our approach complies to the criteria in this Sect. 5. We present the application of the proposed approach on the illustrative example in Sect. 6, as well as its integration in current MBSE tools. Before concluding in Sect. 9, a comparison with related work is given in Sect. 7, and a discussion on the integration in MBSE tools in Sect. 8.

2   Illustrative Example and Industrial Solutions Proposal

In this section, we illustrate the problem using a simple example. We then emphasize the limits of the solutions proposed on a complex industrial project, before presenting our approach in the rest of the paper.

2.1   Illustrative Example

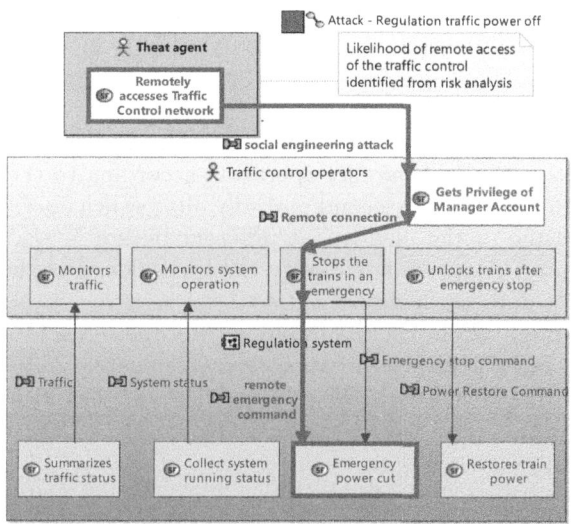

Fig. 1. System modelling design including a remote attack vulnerability identification.

In order to illustrate the problem and the need to distinguish the elements of an MBSE modelling according to the level of confidentiality and to allow a differentiated access in integrity, we present in Fig. 1 a view of the design of a Level-Crossing Control System carried out with the Capella tool[1]. We are amending this design and adding a remote attack vulnerability identified on one of these design elements. This information is issued from a risk analysis and is considered as a sensitive security information.

Consequently, such information should only be accessible by a limited number of authorized users. Further, it should not be visible to the system engineers in

---

[1] https://www.eclipse.org/capella/getstarted.html (access September 2023).

charge of designing the operational part of the Level-Crossing Control System. In addition, to prevent this information from leaking, it should not be stored in the same memory space as the rest of the system design when it is defined. Moreover, designers in charge of modelling the risk analysis elements must not be able to modify the elements of the operational part.

Thereupon, our approach deals with the latter requirements among others. It aims to ensure the non-disclosure of confidential information from global system design models to unauthorized persons.

2.2 Proposed Solutions in the Galileo Complex System

Morlet and all [17] present a feedback from the use of MBSE tools in the system design of the Galileo satellite navigation system. Among all the issues stated is this need for confidentiality management and protection. The two solutions proposed are to first completely validate a design for a given confidentiality level in order to carry out the system design at the higher confidentiality level. The solution proposed in a second step consists of working in parallel on all levels based on the unclassified reference model. A modification to a security level impacting this model is reported to it.

These two solutions have strong drawbacks. The first is the safest but does not allow agile, iterative and parallel operation according to the confidentiality levels of the system design. The second partially allows such operation, but allows working on the same model element at different design levels, and postpones modifications from a given confidentiality level to a lower confidentiality level, with a high risk of data disclosure sensitive.

The solution proposed in this paper aims to propose a solution as secure as the first solution, but offering iterative production capabilities in a collaborative environment specific to the needs of product development in a real industrial context.

More specifically contributions of this paper are as follow: (1) A design process for MBSE modelling with confidentiality inter-enclave management, (2) an implementation of this process based on read/write access to modelling elements carrying confidentiality and integrity information, (3) an application of the proposed approach on a case study, (4) a process allowing the integration of the solution in current MBSE tools, and a concrete integration in the Capella tool as a real proof of concept.

3 Assessment Criteria

As previously mentioned, according to our industrial experience in a sensitive area, regardless of the proposed approach to multi-level confidentiality management, the criteria below must be met. These criteria respond to the changing needs, according to the evolution of the risks of a major European industrial player in the defence and cybersecurity sector in designing its systems. As a former designer in that company, one of this paper's authors has managed these criteria for several years. The first three relate to security requirements, derived from security controls as defined in the NIST SP800-53 framework. The fourth criterion results from the control and handling of accountable information as indicated in the NATO note [18], the latter links a confidential data creation to the level of protection of the physical resource to be processed and stored. The fifth criterion concerns iterative and agile process flow application and the last one assesses the generalisation of the proposal to any kind of models.

- **Confidentiality, Consistency, Integrity assessment (AC1):** The solution should assess the access to authorized people with a strict access in reading and writing. The design at one given confidentiality level should be consistent. By consistency we mean in the one hand the complete design of the system related to requirements up to a confidentiality level, and in the other hand without inconsistencies between the models regardless their level of confidentiality of modelling design.
- **No leak assessment (AC2):** A model element being designed at a given confidentiality level can't be accessed from the lower confidentiality levels of system modelling design.
- **Storage confidentiality assessment (AC3):** A model element being set at a given confidentiality level can't be stored in the same memory space where lower confidentiality levels system modelling designs are stored.
- **No manual labelling assessment (AC4):** The user should work with modelling design tool without being worried on confidentiality labelling. The confidentiality level set of a model element is at the liability of the design process environment.
- **Iterative and adaptive assessment of process flow compliance (AC5):** The solution should be adaptable to any life cycle, with a separation of the specification, design, development, test and validation stages with respect to the confidentiality levels of the system.
- **Genericity of the solution assessment (AC6):** The solution should be generic according to the metamodels and models to which it is applied, and to their number of confidentiality levels.

These criteria are the result of extensive expertise in the construction of sensitive systems that have to withstand real attacks.

4   A Design Process to Ensure Requirements Confidentiality During System Design Modelling

The approach we are proposing consists of a process aimed at satisfying these criteria. This approach ensures the management of a system design combining information from several levels of confidentiality at the design modelling stage.

4.1   Inter-enclave Behavior Justification

The capture of attack scenarios such as the one presented in Sect. 2 comes from a risk analysis. It is clearly stated in the ISO 31000:2018 [19] standard relating to risk management that the creation, storage and processing of documented information resulting from this analysis should take into account the sensitive nature of the information. It is therefore necessary to adapt the access to this information and its processing only to users identified as authorized to access it. This need of protection is even more significant when it comes to systems in the field of defense and whose information (e.g. encryption key, sensitive algorithms, information on the physical characteristics of the secret information transmission) comes under a level of confidentiality (secret, top secret, coalition defense secret) imposing clearance and strict rules regarding the access to information.

The application of this risk management therefore imposes a separation in terms of access and storage of the elements described in the form of requirements. These requirements are the entry documents of the system analysis, and propagate a need for differentiated management in access and storage of the result of the design of these requirements.

It is therefore essential, in the modelling approach used for the systems design considering a risk analysis, to offer the management capacity in separate enclaves for elements from different sensitivity. Although the elements are in different enclaves they still form the same system design.

The first task will take place at the upstream of the system design to properly classify the information on the input requirements provided to the various involved actors in the system design. Indeed, in the example of Sect. 2, it is likely that the information on the potential system vulnerabilities should be hidden from the system engineers in charge of the operational design part. On the other hand, they will surely have information on the identification of the sensitive assets to be protected, as well as their life cycle (creation, storage, transmission and processing). This information is used to apply generic security controls as described in the NIST 800-53 standard [1].

4.2 The Proposed Design Process

This classification of requirements related to their level of confidentiality supposes that the system design is carried out in parallel on separate parts handled by different people. In the long term these parts compose a global design verifying all the input requirements of the system design. We will therefore have an operating mode such as described in Fig. 2, with the following steps:

(1) At the end of the risk analysis, we include an identification of the elements to be injected into the modelling for the system design and their level of confidentiality. These elements are listed to be traced in their injection into a design by MBSE. (2) The system analysis excluding security analysis is carried out in the enclave of lower level of confidentiality, with capture of the design by MBSE. (3) This model with level n of confidentiality is transferred to the n+1 confidence level enclave. (4) Security analysis information (e.g. an attack scenario or the sensitivity level of a system asset) is integrated into the model. Elements of confidence level n can't be modified. (5) In an iterative behavior, the confidentiality level n enclave model evolves. (6) This model is transmitted to the level n+1 enclave in the same way as in step 3. It is possible to identify by

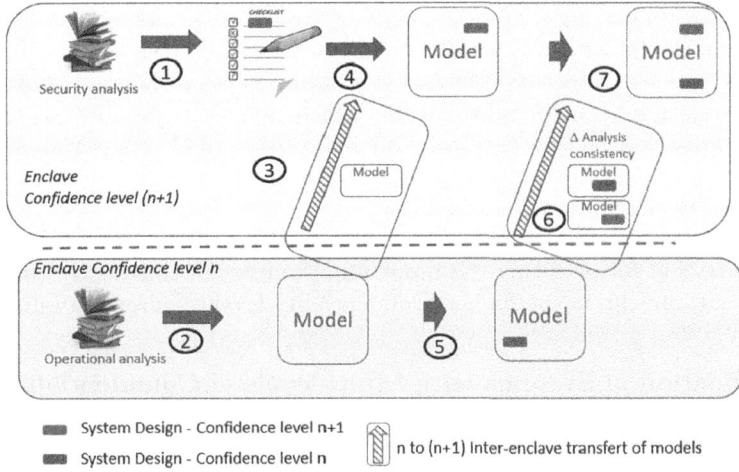

Fig. 2. Process including confidentiality management

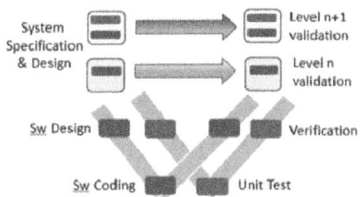

Fig. 3. Validation with multi-levels of confidentiality

comparison the delta between these two models and the impact of the injection of level information confidentiality (n+1). **(7)** In the same way as in step 4, this model can be completed with confidentiality level information (n+1).

In this management, it is not mandatory to give people authorized to write in the confidence level enclave (n+1), to have this right in the confidence level n enclave.

This iterative process applies the no read-up and no write down Bell-Lapadula rules [21]. It forbids storage in an enclave of information of higher confidentiality level. It also provides a strong write integrity rule of limiting write capability in the exact enclave of confidence level of the model element.

4.3   Inter-enclave Designs Inconsistency Mitigation

Note that between 2 iterations of design model feedback (step 6), from confidentiality level n, it is necessary to integrate again and adapt the design information of confidentiality level n+1. This implies a potential redesign work which can be quite expensive. This can also lead to choices that do not allow satisfactory consideration of security requirements.

This situation can however be strongly attenuated by a coarse grain structure of the design model with the prior definition of the interfaces making it possible to work on fixed architectural elements. In addition, the system engineers have part of the design security information, which makes it possible to direct the design towards choices that facilitate the integration of elements resulting from higher level confidentiality requirements. However, the need to modify the lower privacy level design cannot be ruled out.

A restrictive industrial process involving security and product managers must then be provided for adapting the input requirements of this design to the next iteration, without disclosing higher confidentiality level requirement information.

4.4   Validation of Systems with Multi-levels of Confidentiality

The purpose of validation is to demonstrate that a product or product component fulfills its intended use when placed in its intended environment. In the previous example of Level-Crossing Control System, we might wanted to mitigate the remote threat identification by the addition of a specific endpoint detection and response (EDR) proxy, which will be described in the physical refinement view of the system view presented in Fig. 1.

In enclave of confidence level n, the requirements related to this design will be validated. In enclave of confidence level n+1, the validation of specific requirements, as the test of mitigation procedures assessment related to remote attack in the example, should be done.

Table 1. Confidential information in system modelling designs

	Confidential information	Example
1.	Metamodel element	Part of the metamodel describing the capture of instances of the risk analysis part, or certain parts such as the list of encryption algorithms offered
2.	Instances of metamodel element	Instance of this metamodel describing an attack scenario
3.	Attribute	The annotation following the risk analysis as an asset to be secured of a model element
4.	Information broadcast by metamodel transformation	Application in the Capella metamodel of confidential information report actions between operational, system, logical and physical views as proposed in the ARCADIA methodology [20]
5.	Element inconsistent with deletion of confidential elements	A proxy that is only traversed by higher confidential information. Maintaining this element can be considered as a disclosure of confidential information

As shown in Fig. 3, this may be considered as a particular integration test suite mixing elements of level of confidentiality at most n+1.

4.5   Impact on Modelling

Subsection 4.1 justifies the global process based on enclave partitioning. We express here its refinement at modelling level to ensure confidentiality, consistency and integrity. The solution is applied on cases of potential confidential elements in MBSE, reported in Table 1.

We state here our proposal of constraints for multi-enclave management of MBSE designs. Metamodels and models used in a given enclave carry information confidentiality by read-only or read/write access. Any metamodel and model element from lower level enclave is **read-only**. All element created within an enclave has **read and write access** in that enclave. Access Enclave Elements in reading and writing and potential relations towards the elements of models and metamodels from enclaves of the level of less privacy **are stored in the current enclave.**

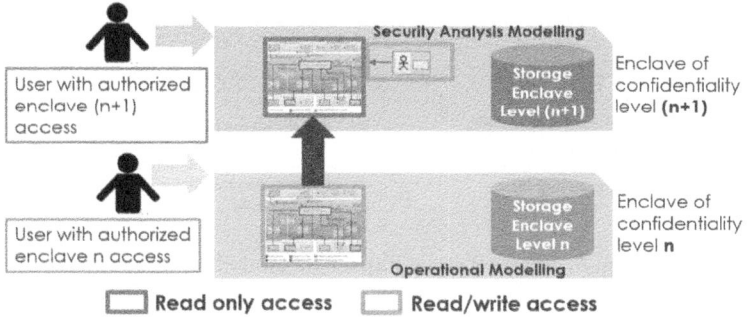

Fig. 4. User data access of multi layered modeling designs.

Figure 4 presents the application of the proposed solution on the example from Sect. 2. Access to the enclaves will be done with a secure access policy of the Mandatory Access Control (MAC) type, and restrictions conforming to the role of the users.

4.6   Modifying the Model Between Iterations

In this section we provide the elements allowing a transfer of design model information to be carried out when receiving a new model of the enclave with a lower sensitivity level. The underlying model for these models is an XMI file which has a tree structure and cannot be modified in higher sensitivity level enclaves. The design complements therefore appear in the XMI file at the extremity of the trees. Frequently, capturing relationships between model elements results in

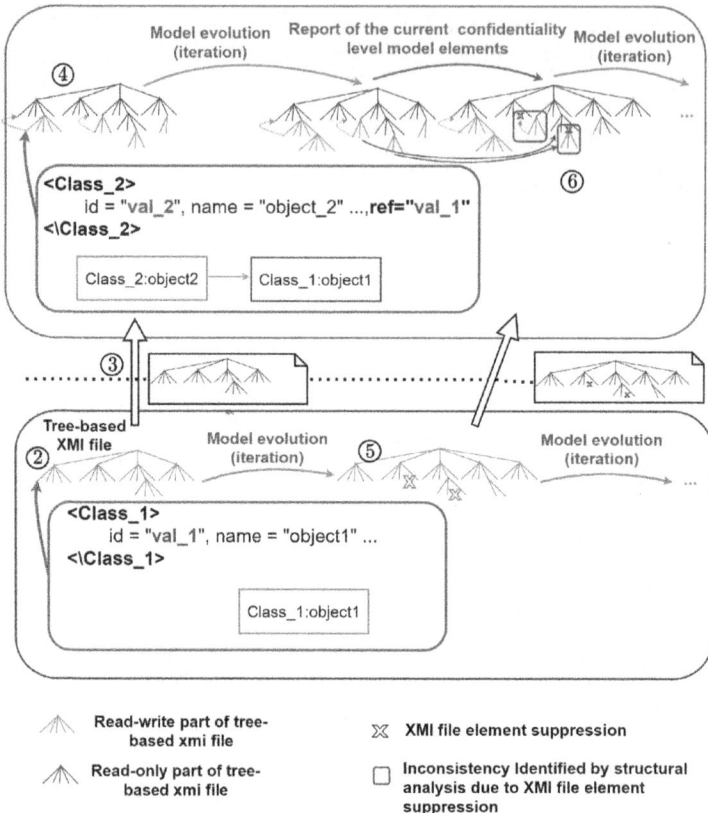

Fig. 5. Figure 2 steps impact at modelling level

memorizing identifiers of other elements. Each element is stored in an XMI tag and has an attribute carrying a unique identifier distinguishing it from other elements. Linking two elements means either containing in this element description attributes memorizing the identifiers of other elements, or including the beacon of one element inner another element, as defined in the tree-based structure of these files.

Figure 5 presents the impacts of the models in their representation in the form of XMI files in their use in the process described in Fig. 2. The steps ② and ⑤ report the evolutions of the model in the lowest confidentiality enclave. Each node of the tree structure corresponds to an XML beacon.

A copy of the elements between two models received will consist of ensuring the presence of the leaf nodes on which the design elements have been added, as well as the referenced beacons.

At one iteration of the model, it is transmitted to the higher level ③ which completes it ④ and enriched with superior confidentiality elements without modifying the model received.

In the next iteration, the model evolved with the addition and removal of
⑤ elements. If the deleted elements are not elements in which elements to be
integrated from the higher level enclave, or referenced elements, are nested, then
the extension is carried out without difficulty with a guarantee of respect for
the grammar of the model. Otherwise, it is easy to identify situations of loss of
consistency ⑥. Either the higher level privacy model is modified to allow the
extension, or a process under the responsibility of security managers is applied
to modify the lower level model.

5   Evaluation of the Proposed Solution

To evaluate the proposed solution, we analyse its compliance to the verification
criteria defined in Sect. 3. Table 2 presents the results on rules corresponding to
criteria refinements.

For each rule, explanatory elements are provided and prove this compliance.
The proposed design process, integrating modeling system design tools, covers
all the needs expressed.

The proposed process processes and stores data at a confidentiality level in a
dedicated enclave accessible only to people authorized to access. This data can
only be transmitted to enclaves with higher confidentiality levels. Confidentiality
protection is well assured.

Maintaining consistency is also well integrated and illustrated in Fig. 5, which
allows validation of the system for a given level of confidentiality.

Integrity is also guaranteed because the modification of an element for a given
confidentiality level is only possible in the enclave allocated to this confidentiality
level. Users of the template at a higher privacy level have read-only access to
this item.

The description of the impacts on the product development process is presented in Fig. 2 and refined in Fig. 5 to describe the impacts in terms of manipulation of system design models. Figure 3 shows the variation in the other stages
of the life cycle, allowing confidentiality protection at each of them. The iterative process applies to any life cycle, notably allowing its application in agile
processes [22]. This process also offers a sharing of design work according to
the confidentiality levels of the product to be produced. This sharing makes it
possible to work in parallel on these different enclaves.

We have not made any presuppositions about the metamodels and models
on which the process applies. The proposed solution does not require annotating
the elements of specific security information models and metamodels. Our proposed approach is therefore generic and applies to any metamodel for which the
identification of the level of confidentiality has not been carried out beforehand,
this can be carried out in the upstream phase for the production of a product
by those responsible. of security. For the same metamodel used, this choice of
division into confidentiality levels may vary depending on the project.

Table 2. Assessment criteria verification

	Rule assessment	Verification
AC1.R1.	Confidentiality	Each element can be annotated with the level of that of the enclave in which it was added, but it is not mandatory with access read-only privacy level items inferior. Implicitly writable elements of the enclave are of the enclave level of confidentiality, the others are of a lower level
AC1.R2.	Consistency	Lower enclave-level consistency and higher level model elements addition consistency guarantee the consistency of the model at any level of confidentiality
AC1.R3.	Integrity	Enclave access control and read-only protection of level-n model elements reported at level (n+1) guarantee the integrity criterion assessment
AC2.R1.	No leak to lower level	It is not possible to add elements with a level of confidentiality higher than that of the enclave
AC2.R2	No miss from lower level	Steps 3 and 6 in Fig. 2 reflect the provision of the level n models in the level enclaves confidentiality (n+1)
AC3.R1	Storage confidentiality	In Fig. 4, we see that at each enclave can only be stored elements up to the compliance level of the enclave
AC4.R1	No labelling error	By construction, any added element is of level of privacy of the enclave
AC5.R1	Iterative process flow	The process presented in Fig. 2 provides such iterative steps
AC5.R2	Multi-levels modification impacts	A comparison between 2 iterations of n-level enclave models helps to identify differences
AC5.R3	Dynamic metamodel modifications impacts	As the level enclave metamodel (n+1) includes the level n enclave metamodel, it can be modified dynamically independently of the treatments lower enclave level. On the other hand, it will be necessary to postpone these metamodel changes to the level enclaves of superior confidentiality
AC5.R4	Validation at a given confidentiality level	The confidentiality level of each system element is defined, and the system is consistent for each confidentiality level. We deduce the validation of the system at a given confidentiality level
AC6.R1	Genericity	The solution is applicable to multi layers designs independent to the models
AC6.R2	Multi-layer management	The solution is applicable to multi layers designs, with no prerequisites to the number of layers

In the following section we show how to represent in a hierarchical form a model described on a single level and mixing information from different confidentiality levels. We then discuss in Sect. 8 how the additional capabilities necessary for this hierarchical representation are already present in modeling tools offering multi-user functionality.

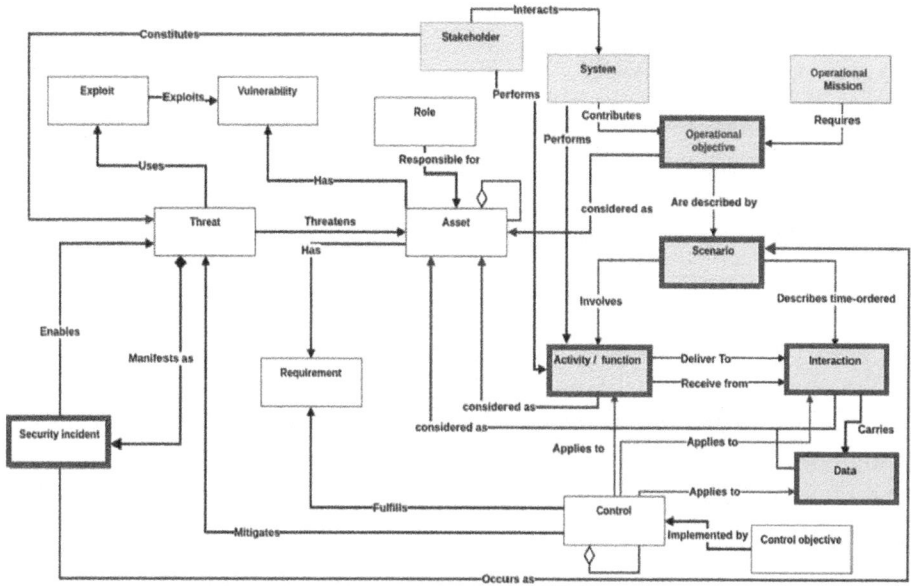

Fig. 6. "Flat" view of the MBCA metamodel from [7]

6 Application on the MBCA Metamodel

We rely on the metamodel elements described in a very recent existing work on the integration of cybersecurity risk assessment into requirement engineering [7] to illustrate the application of the approach. These elements are reported in Fig. 5. In blue we present classes of the metamodel describing the operational features, and in white the classes added to specify features related to security management. We extracted from the metamodel description the part dedicated to capture scenario description information. These scenarios can be either operational scenarios or scenarios describing feared events. These elements are highlighted in purple in the initial MBCA metamodel.

Conforming to the proposed process, we will therefore consider that the part of the metamodel describing the operational scenarios of a confidentiality level n, and the part of the metamodel describing the scenarios of feared events with a

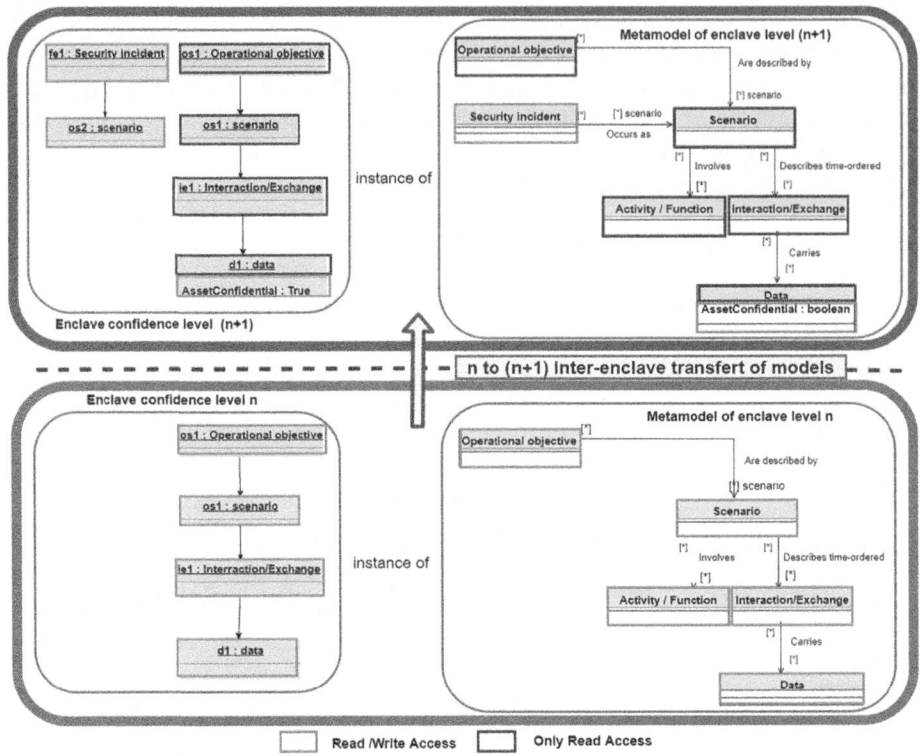

Fig. 7. Hierarchic view of the purple metamodel part of Fig. 6, and of a system design instance

level of confidentiality (n+1). We deduce in Fig. 6 the corresponding metamodel parts and an instance example of it for the enclaves of privacy level n and (n+1) for the purple part of the initial metamodel presented in Fig. 5.

In accordance with the principle set out in Subsect. 4.5, the elements of confidentiality level n are read/write accessible in the privacy level n enclave and are accessible read-only in privacy level enclave (n+1). The confidentiality level elements (n+1) are accessible in read and write in the level enclave of confidentiality (n+1), and are not present in the enclave of confidentiality level n.

7   Related Work

Recent work deals with the capture of security information [3–9] in models. However the multi-level confidentiality of the model is not addressed in these works. The specific problem of managing the confidentiality of information in multi-level confidentiality models is therefore new and, to the best of our knowledge, there are no proposals that specifically address the problem explained in Sect. 3.

Therefore, we analyze different approaches for managing confidentiality or existing masking and see to what extent it is possible to adapt them. We are guided in this by constraints such as those defined by state bodies, for example French Inter-ministerial General Instruction No. 1300 [23] on the protection of national defense secrets.

In a different modelling descriptions as the security models already mentioned, an overview of Model-Driven Developments [10] list several proposals with particularly rich security information stored and processed. But they do not apprehend access right and storage constraints issues of specific modelling elements. It is also the case of Hu et all model [11], which proposes a UML security model but with no separation in enclaves.

As indicated in [12], obfuscation is known as being a good data confidentiality protection. In our case, obfuscation does not obfuscate elements of the metamodel, which may be necessary. Moreover, obfuscation keeps the structure of the model, which can already be considered as a disclosure of confidential information.

Filtering tools would allow confidential elements to be hidden. However, it will be necessary to ensure that the filtering information is not present in the underlying formats of the filtered model displayed. It will also be necessary to ensure a distinct memory storage of the models of different level of confidentiality for their modification in writing. Finally, it is necessary to ensure that the data of a given level of confidentiality resulting from a security analysis is considered, which may be different from the instances of a predefined set of classes of the metamodel.

Johnson and Stevens [13] propose a model confidentiality management between two companies with a common part and specific confidential parts managed either by a two-way transformation relationship, or with specific access to certain parts of the model. There is not in their work a conformity of confidentiality order relationship present in the management of enclaves respecting the rules defined by Bell and Lapadula.

We present in the following security management on databases with similarities with the current work. Denning, Lunt, and all [14] annotated database table entries with items of privacy levels, and views restricted to those levels by anonymization. Their work of defining a multi-level relational data model (MDB) focuses on the impact of these anonymizations in the inter-table inference rules and maintaining consistency with anonymized inputs. These works differ because they are not based on a description of metamodels and consistency rules that benefit MBSE-type solutions, and do not address multi-enclave storage. The modification of access to data according to the level of confidentiality authorized for access has been studied mainly on databases and characterized under the term polyinstantiation. Among these works, [15] analyzes the management of the same database with different levels of confidentiality. In their proposal, the access management with respect of integrity respects the rules of Bell-Lapadula, but with the level of confidentiality carried by the database and not by the storage enclave. These works differ also by information duplication.

Brodsky and all [16] propose to process queries with respect to confidentiality to a database via a Disclosure Inference Engine (DIE). In this proposal the level of confidentiality of the information is explicit by annotation, there is no mention of the modification of this data, nor of the storage in enclaves of different level of confidentiality.

8 Discussion

8.1 Application by Current Modelling Tools

We have presented in Sect. 4 an approach allowing to process models with elements of several levels of confidentiality, ensuring confidentiality of access and storage. This solution applies to any model managing information with a plurality of confidentiality levels. This was barely addressed by the previous work described in Sect. 7.

The application of this solution requires on the one hand the possibility to determine the read-only or read/write access of elements of models and meta-models, and on the other hand the ability to compare two models. These functionalities exist in several modelling tools for systems design. In particular to allow multi-user use which obliges to make accessible only in read-only mode (locker) the elements being modified by a stakeholder. Thus well known tools such as EMFStore [24], ModelCSV [25], Modelio Constellation [26], or even Team for Capella [27] offer such multi-user capabilities.

Let us take the example of Capella. Capella is an open-source MBSE modelling tool and a graphical editor provided with an appropriate engineering method called "Architecture Analysis Design Integrated Approach" (ARCA-DIA). Capella and ARCADIA are designed by THALES, a European leader in cybersecurity.

Capella offers a multi-user capability called Team4Capella. In this capacity, users must authenticate before accessing the model. An item that is modified by one user is not writable by other users until that modification is complete. The Capella tool also allows you to add additional design capabilities that are included in the form of plugins called viewpoints. These viewpoints are extensions of the basic meta-model in order to enrich the model. Among these viewpoints a security viewpoint, called DARC, notably makes it possible to qualify the assets in terms of level and type of vulnerability, as well as the addition of attack scenarios.

To ensure access to model elements only to authorized persons with guaranteed respect for read/write or read-only rights, and in accordance with the proposed approach we define two memory enclaves. In one of these enclaves, in charge of the system design of the operational part of the application, we do not integrate the DARC viewpoint.

At a given deadline, this model is transferred to the other enclave, via its xmi description file. In this enclave the DARC viewpoint is added to the basic Capella metamodel.

This approach is illustrated in Fig. 7, on the crowd surveillance drone case study provided for download with the DARC view point. To allow the reproducibility and replicability of the proposed solution, Team4Capella and DARC are available for download.[2]

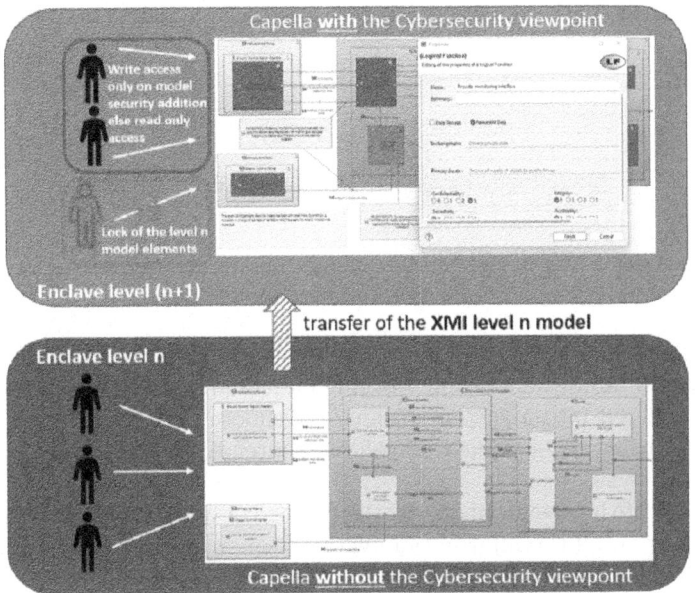

Fig. 8. Integration proposal in Capella

8.2 Security Analysis Based on the Proposed Confidentiality Management

In this paper, we presented a tooled process allowing the design of systems with information confidentiality protection. This work constitutes a first step in the more global tooling of the product development process to ensure the security of systems.

We describe in Fig. 9 this process as part of a more global integration of information elements, processing, production of artifacts in order to improve the product development process to produce equipment from more secure complex systems.

---

[2] https://www.eclipse.org/capella/download.html (access september 2023).

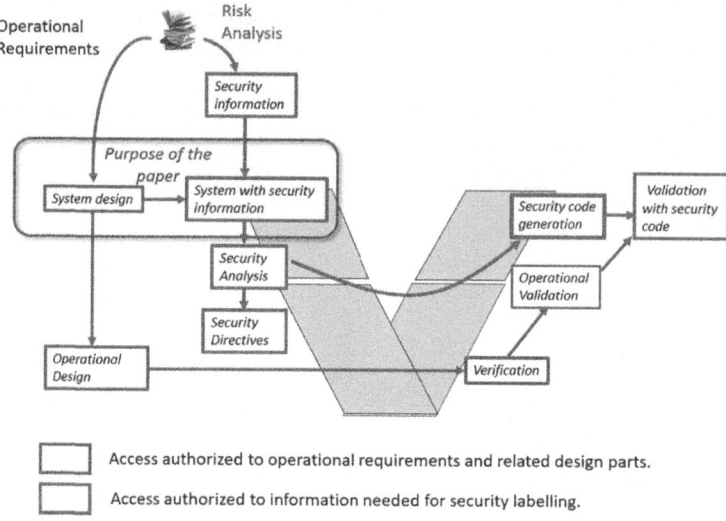

Fig. 9. Security Information exploitation

With this in mind, we are now working on a characterization of the information to be injected into a system design. We choose this information so that it is available during the specification phase, therefore by exploiting information from a risk analysis.

At the end of the design phase, in the enclave of confidentiality level allowing us to benefit from all the information, we work on a security analysis in order to allow on the one hand the production of security directives and on the other hand the production of security code for its use in the validation phase of operational and security requirements.

9   Conclusion

In this paper, we propose a solution to ensure requirements confidentiality in system modelling designs including information with different levels of confidentiality. The proposed solution includes a set of criteria to assess a system multi-level confidentiality, a design process to ensure requirements confidentiality and mitigate inter-enclave design inconsistency.

At first glance, the multi-enclave management of models combining elements of different levels of confidentiality with respect to integrity could be very complex to implement. Our solution shows that not only this is not the case, but that the MBSE tools by their collaborative requirement engineering capabilities often already implement all the functionalities allowing the setting implementation of this solution.

It's true that the solution we propose allows the designer to define the different levels of confidentiality, but it doesn't help her/him to define them consistently with the risk analysis. As future work, we plan to formalise the dependencies between the assets defined during the risk analysis and the design elements. This will give us the opportunity to carry out a consistency analysis between the recommendations of the risk analysis and the definition of the enclaves.

References

1. Security and privacy controls for information systems and organizations. National Institute of Standards and Technology, US, Standard (2020)
2. Voirin, J.-L., Constant, O., Lépicier, E., Maraux, F.: Dream the future: systems engineering in 2030. In: INCOSE International Symposium, vol. 30, no. 1, pp. 771–782 (2020). https://incose.onlinelibrary.wiley.com/doi/abs/10.1002/j.2334-5837. 2020.00754.x
3. UAF. Unified Architecture Framework (UAF) Domain Metamodel. OMG (2022). https://www.omg.org/spec/UAF/1.2
4. Roudier, Y., Apvrille, L.: Sysml-sec: a model driven approach for designing safe and secure systems. In: 2015 3rd International Conference on Model-Driven Engineering and Software Development (MODELSWARD), pp. 655–664 (2015)
5. Mažeika, D., Butleris, R.: MBSEsec: model-based systems engineering method for creating secure systems. Appl. Sci. **10**, 2574 (2020)
6. Navas, J., Voirin, J.-L., Paul, S., Bonnet, S.: Towards the integration of cybersecurity risk assessment into model-based requirements engineering. In: INCOSE International Symposium, Orlando, USA, September 2019, vol. 29, no. 1, pp. 850–865. Wiley (2019). https://doi.org/10.1002/j.2334-5837.2019.00639.x
7. Naouar, D., Hachem, J.E., Voirin, J., Foisil, J., Kermarrec, Y.: Towards the integration of cybersecurity risk assessment into model-based requirements engineering. In: 29th IEEE International Requirements Engineering Conference, RE 2021, Notre Dame, IN, USA, 20–24 September 2021, pp. 334–344. IEEE (2021). https://doi. org/10.1109/RE51729.2021.00037
8. Derdour, M., Alti, A., Gasmi, M., Roose, P.: Security architecture metamodel for model driven security. J. Innov. Digital Ecosyst. **2**(1), 55–70 (2015). https://www. sciencedirect.com/science/article/pii/S2352664515000206
9. El Hachem, J., Pang, Z.Y., Chiprianov, V., Babar, A., Aniorte, P.: Model driven software security architecture of systems-of-systems. In: 2016 23rd Asia-Pacific Software Engineering Conference (APSEC), pp. 89–96 (2016)
10. Fernández-Medina, E., Jurjens, J., Trujillo, J., Jajodia, S.: Model-driven development for secure information systems. Inf. Softw. Technol. **51**(5), 809–814 (2009). sPECIAL ISSUE: Model-Driven Development for Secure Information Systems. https://www.sciencedirect.com/science/article/pii/S0950584908000761
11. Hu, X., Zhuang, Y., Cao, Z., Ye, T., Li, M.: Modeling and validation for embedded software confidentiality and integrity. In: 12th International Conference on Intelligent Systems and Knowledge Engineering, pp. 1–6 (2017)
12. Xu, H., Lyu, M.R.: Assessing the security properties of software obfuscation. IEEE Secur. Priv. **14**(5), 80–83 (2016)
13. Johnson, M., Stevens, P.: EnglishConfidentiality in the process of (model-driven) software development. In: Proceedings of 2nd International Conference on the Art, Science, and Engineering of Programming(Companion), pp. 1–8. ACM (2018). https://2018.programming-conference.org

14. Denning, D.E., Lunt, T.F., Schell, R.R., Heckman, M.R., Shockley, W.R.: A multilevel relational data model. In: 1987 IEEE Symposium on Security and Privacy, pp. 220–220 (1987)
15. Sallam, A.I., Elrabie, S.M., Faragallah, O.S.: Comparative study of polyinstantiation models in mls database. In: International Computer Engineering Conference (ICENCO) 2010, pp. 158–165 (2010)
16. Brodsky, A., Farkas, C., Jajodia, S.: Secure databases: constraints, inference channels, and monitoring disclosures. IEEE Trans. Knowl. Data Eng. **12**(6), 900–919 (2000)
17. Morlet, C., et al.: Embarking to end-to-end system modelling for galileo second generation (2022)
18. NATO. Security within the North Atlantic treaty organization (NATO), C-M(2002)49-REV1 (2020). https://www.nbf.hu/docs/C-M(2002)49-REV1.pdf
19. Risk management - guidelines. International Organization for Standardization, Geneva, CH, Standard (2018)
20. Voirin, J.-L.: Model-based system and architecture engineering with the arcadia method (2017)
21. Bell, D., LaPadula, L.: A mathematical model, Technical report esd-tr-278, vol. 2. The Mitre Corporation, Bedford (1973)
22. Beck, K.L., et al.: Manifesto for agile software development (2013). https://api.semanticscholar.org/CorpusID:109006295
23. IGI-1300. Instruction générale interministérielle sur la protection du secret de la défense nationale. Secrétariat général de la défense et de la sécurité nationale (2020). http://www.sgdsn.gouv.fr/uploads/2016/10/igi-1300-20210809.pdf
24. Koegel, M., Helming, J.: Emfstore: a model repository for emf models. In: Proceedings of the 32nd ACM/IEEE International Conference on Software Engineering, ser. ICSE 2010, vol. 2, pp. 307–308 (2010). https://doi.org/10.1145/1810295.1810364
25. Kramler, G., Kappel, G., Reiter, T., Kapsammer, E., Retschitzegger, W., Schwinger, W.: Towards a semantic infrastructure supporting model-based tool integration. In: Proceedings of the 2006 International Workshop on Global Integrated Model Management, ser. GaMMa 2006, pp. 43–46 (2006). https://doi.org/10.1145/1138304.1138314
26. Garcia-Dominguez, A., Barmpis, K., Kolovos, D.S., da Silva, M.A.A., Abherve, A., Bagnato, A.: Integration of a graph-based model indexer in commercial modelling tools. In: Proceedings of the ACM/IEEE 19th International Conference on Model Driven Engineering Languages and Systems, ser. MODELS 2016, pp. 340–350. Association for Computing Machinery, New York (2016). https://doi.org/10.1145/2976767.2976809
27. Boudjennah, C., Combemale, B., Exertier, D., Lacrampe, S., Peraldi-Frati, M.A.: CLARITY: open-sourcing the model-based systems engineering solution capella. In: Proceedings of the International Workshop on Open Source Software for Model Driven Engineering Ottawa, Canada (2015)

Analyzing Excessive Permission Requests in Google Workspace Add-Ons

Liuhuo Wan[1]([✉]), Chuan Yan[1], Mark Huasong Meng[2], Kailong Wang[3], and Haoyu Wang[3]

[1] The University of Queensland, Brisbane, Australia
liuhuo.wan@uq.edu.au
[2] National University of Singapore, Singapore, Singapore
[3] Huazhong University of Science and Technology, Wuhan, China
wangkl@hust.edu.cn

Abstract. In the digital era, business collaboration platforms have become pivotal in facilitating seamless remote work and virtual team interactions. These platforms, typified by Google Workspace, offer an integrated suite of tools (such as Google Docs, Slides, and Calendar) that significantly enhance business operations. They often extend their functionality through the integration of third-party applications, known as "add-ons". Google Workspace exemplifies this trend, blending traditional business solutions with advanced, add-on-driven capabilities. While this greatly augments productivity and collaboration for online personal or team work, concerns about the excessive use of data and permissions have been raised by both users and legislators, as add-ons can utilize the granted permissions to access and manipulate files managed by business collaboration platforms.

In this work, we propose an end-to-end approach to automatically detecting excessive permissions among add-ons. It advocates *purpose limitation* that the requested permissions of the add-on should be for its specific functionality and in compliance with the actual needs in fulfilling the functionality. Our approach utilizes a hybrid analysis to detect excessive permissions, including analysis of the add-on's runtime behavior and source code, and state-of-the-art language processing techniques for textual artifact interpretation. This approach can serve the users, developers and store operators as an efficient and practical detection mechanism for excessive permissions. We conduct a large-scale diagnostic evaluation on 3,756 add-ons, revealing that almost half of existing add-ons contain issues of excessive permissions. We further investigate the root cause of excessive permissions and provide insights to stakeholders. Our work should raise the awareness of add-on users, service providers, and platform operators, and encourage them to implement solutions that restrict the excessive permissions in practice.

G. Bai et al. (Eds.): ICECCS 2024, LNCS 14784, pp. 323–345, 2025.
https://doi.org/10.1007/978-3-031-66456-4_18

1 Introduction

Business collaboration platforms represent a complex web-based application paradigm, epitomizing the concept of a "super platform". Platforms like Google Workspace [8] and Zoom [17] exemplify this model by hosting a diverse array of third-party applications, referred to as "add-ons". These add-ons integrate deeply with the platform's core services, offering users a seamless and efficient experience. Google Workspace, in particular, is a prime example of this ecosystem, providing a suite of applications that cater to various business needs. Notable add-ons within Google Workspace include Auto-LaTeX Equations [2] for Google Docs, facilitating the insertion of mathematical equations, and No Phishing for Gmail, enhancing email security [13]. The escalating demand for remote work, online education, and virtual collaboration has significantly propelled the adoption and evolution of these platforms, establishing them as mature ecosystems in the contemporary digital landscape.

The add-on runs with a strong dependence on the host platform. It is typically executed as an extension without its own process address space, requesting resources through the APIs provided by the host. Since cross-domain data and control flow are involved, the OAuth protocol [14] has been commonly adopted for performing multi-party authorization among the add-on, the host and the user. More specifically, developers simply declare the requested permissions through the OAuth scope fields, in a similar manner to the Android's manifest file. As illustrated by the example shown in Fig. 1, the add-on declares three permissions, including the ability to view email messages, send emails and view user's spreadsheets, inside `oauthScope` in the `appsscript.json` file. Despite the convenience for authorization, the coarse-grained permission management inevitably introduces potentially over-privileged or excessive permissions [23,43]. As a result, dishonest developers could inadvertently request more permissions than those required by the functionality of their add-ons, rendering ill-purposed exploitation possible. This clearly violates the principle of least privilege, and may also cross the boundary for user data protection drawn by the strictly implemented and enforced privacy regulations/laws around the world [3,7].

As efforts towards a better understanding of the underlying permission management risks [44] in business collaboration platforms, several empirical studies have focused on examining the inconsistencies and weaknesses in their permission authorization, such as privacy violation [22,25,27,36], unprompted permission authorization [47,48], and phishing [21]. They attribute the user data leakage to loose permission scopes [21,48], without touching on the essential causes of excessive permission request issues, such as coarse-grained bundled permissions.

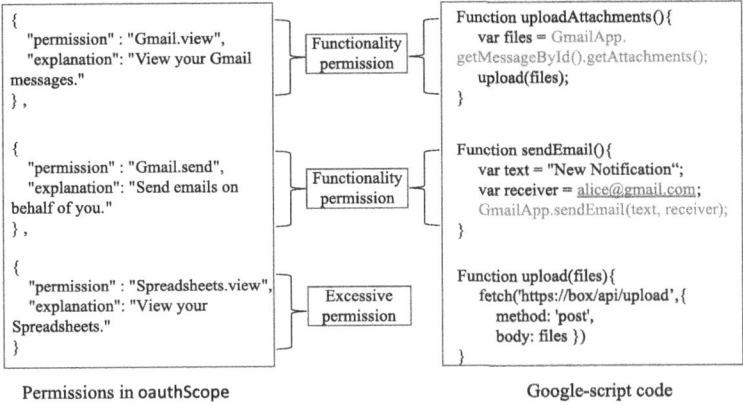

Fig. 1. An example of permission requests from an add-on

Our Work. This work introduces a novel, scalable end-to-end approach to automatically detect issues of excessive permissions among Google Workspace add-ons. The objective is to provide an efficient and practical solution for evaluating excessive permissions for various stakeholders, including users, add-on developers, and the platform operators. With our approach, users can better understand excessive permissions before installing the add-on, and platform operators can scan for potential excessive permissions to enhance the permission and privacy compliance of their system. For add-on developers, this approach can mitigate the threats of excessive permissions before releasing their products.

We implement our approach as PEDEC, a *permission excess detector* for Google Workspace add-ons. PEDEC consists of two major components, the *functionality extractor* and the *excessive permission detector*. First, due to the partial availability of source code for these add-ons (**Challenge #1**), the functionality extractor extracts all functionality-related materials, including the interacted DOM, through web testing. We acknowledge that web testing has limitations and may not cover all paths. To supplement, we include functionality-related materials such as descriptions, tutorials, or demonstration videos. For the partially available add-ons with source code (228), we develop a lightweight static analysis tool to extract their functionality-related host APIs. Second, the excessive permission detector checks the consistency of functionality permissions and requested permissions to detect excessive permissions. The functionality permissions, sourced primarily from the materials extracted by the functionality extractor, primarily comprise natural language descriptions with varying format and quality among add-ons (**Challenge #2**). To address this, we employ the latest natural language inference (NLI) model to infer the functionality-related (or unrelated) permissions.

To understand the current status of excessive permissions among add-ons, our proposed approach is utilized to comprehensively screen 3,756 add-ons in the Google Workspace Marketplace [12]. The results show that a significant number (2,091) of add-ons have excessive permission issues.

Contributions. To the best of our knowledge, we are the first to diagnose the excessive permission issues of add-ons from Google Workspace at scale. In this work, we particularly focus on Google Workspace due to its popularity, which constitutes a substantial and representative market share of around 70% [8] in business collaboration platforms. In summary, this work mainly contributes to the following aspects.

- **Understanding the excessive permission issues for add-ons hosted on Google Workspace.** We first formally formulate and study the permission excessive issue, which has been a key cause for private data mismanagement in add-ons.
- **A systematic assessment approach.** We propose an automatic tool named PEDEC to detect excessive permissions among add-ons. We implement a prototype of PEDEC that checks permission compliance based on a hybrid analysis. The ground-truth evaluation on 100 add-ons shows that PEDEC achieves a 100% TNR (true negative rate) and 92% TPR (true positive rate) for detecting the excessive permission issue.
- **Revealing prevalence of the excessive permission issue in real-world Google Workspace add-ons.** Our large-scale study on Google Workspace reveals that permission management of add-ons is problematic (Section 4). Our investigation reveals that bundled permission declaration is the major cause of such an issue. Our findings should raise an alert to the users, and encourage add-on developers and platform operators to redesign their interfaces.

2   A Running Example and Definitions

2.1   A Running Example

We delineate the overarching workflow of the Google Workspace add-on as follows. The user interacts with the add-on client on their computer or mobile device. Their request is sent to the workspace server as an *event*, which is then handled by a corresponding *event handler* provided by the add-on developer. The event handler serves the request, and during this process, it invokes APIs provided by the workspace to access the user's data managed by the workspace. The obtained data may be sent back to the add-on server for further processing in some computation-intensive tasks. After processing the request, the workspace server sends a response back to the add-on client and updates the user's client page.

For example, when the user clicks the "save attachments" button (shown in the sidebar of Fig. 2), the event is transmitted to the Google server, triggering the event handler of the Box add-on, this event handler calls the `GmailApp.getMessageById().getAttachments()` to retrieve attachments of the current Gmail message and uploads them to Box cloud storage.

2.2 Problem Definition

PEDEC focuses on addressing the issue of permission excess, specifically determining whether the add-on unnecessarily requests functionality-irrelevant permissions. To facilitate the understanding of our work, we formally define the target problem.

Definition 1 (Functionality Permissions). *Permissions must be granted to the add-on for it to execute its functionality correctly, denoted as $\mathbb{P}_{fun}$. The functionality of the add-on can be reflected by its source code, runtime behaviour, and textual description, and we use $\mathbb{P}_s$, $\mathbb{P}_r$, and $\mathbb{P}_d$ to indicate the permission sets that are associated with each of them, respectively. Therefore, we have $\mathbb{P}_{fun} = \mathbb{P}_s \cup \mathbb{P}_r \cup \mathbb{P}_d$.*

Definition 2 (Requested Permissions). *The set of permissions that the add-on requests through `oauthScope` field is called requested permissions, denoted as $\mathbb{P}_{req}$. They are specified by developers (the left side of Fig. 1) and vetted by platform operators.*

Definition 3 (Excessive Permissions). *We name $\hat{\mathbb{P}} = \mathbb{P}_{req} - \mathbb{P}_{fun}$ as excessive permissions. Intuitively, $\hat{\mathbb{P}}$ is defined as the set of permissions that can be removed from $\mathbb{P}_{req}$ without affecting the normal functionality of the add-on. These permissions are deemed excessive because they are unnecessary for the add-on to execute its intended functionality, and their inclusion raises security and privacy concerns for the users and their data.*

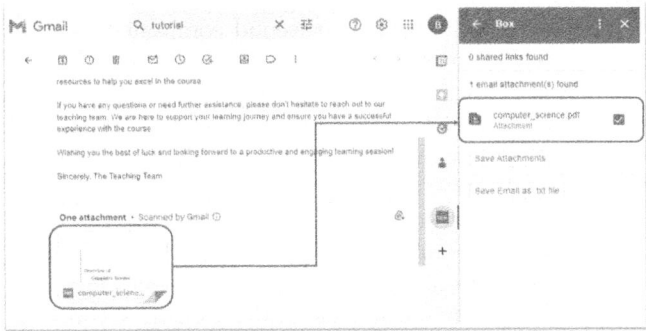

Fig. 2. Example of the Box add-on

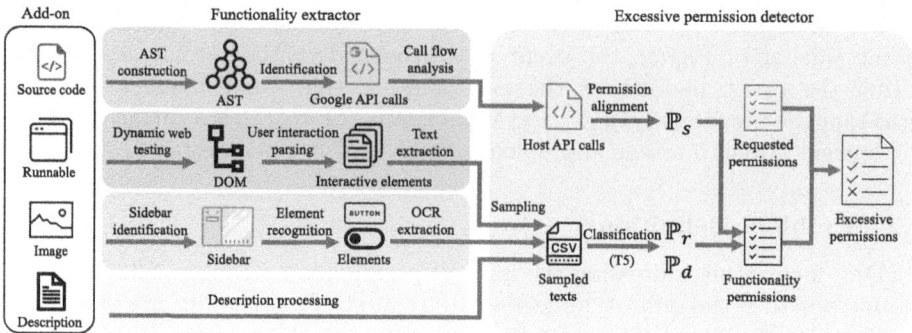

Fig. 3. Workflow of PEDec

3 Our Approach

3.1 Approach Overview

We propose a fully automatic analysis approach, leveraging techniques of hybrid functionality analysis and state-of-the-art natural language inference and Text-To-Text-Transfer-Transformer (T5) models. The overall architecture of our approach is depicted in Fig. 3, which composes the following components.

Functionality Extractor. This component aims to extract functionality-related materials. The Document Object Model (DOM) tree of a dynamically generated HTML file provides an accurate hierarchical representation of the available operations on the add-on interface, allowing us to infer permissions related to the add-on's functionality (**addressing Challenge #1**). For the partially available add-ons with source code, similar to Android, specific actions or functions performed through API calls can reveal fine-grained permissions.

Excessive Permission Detector. The second component is an excessive permission detector that compares the functionality permissions based on extracted materials (**Definition 1**) with the declared permissions (**Definition 2**) of add-ons. However, the detection of excessive permissions remains a challenge. Previous efforts [20,23,26,42,47] have approached this issue using applications' counterparts [41,42,49] - applications that share similar functionality - as a reference. However, this approach is not suitable for Google Workspace, given the limited number of add-ons (3,756) compared with other problem settings (millions of mobile and browser applications). Therefore, we propose a text-based technique to extract permission-related information (**addressing Challenge #2**) from application pages, images, and description (to be detailed in Sect. 3.3).

3.2   Functionality Extractor

Dynamic Analysis for Element Discovery. PEDEC extracts functionality-related materials, such as interacted DOM trees, through dynamic web testing for the majority of add-ons in the absence of source code. While web testing has limitations in covering paths requiring semantic-sensitive input, we supplement functionality-related materials with descriptions and tutorial videos provided by add-on developers. These materials serve as guidelines for new users, outlining the supported functionality of add-ons.

Dynamic Web Testing. We perform dynamic web testing on add-ons to trigger a wide range of behaviours and HTML elements, deriving the functionality permissions required by them. For example, buttons related to creating or deleting user documents imply permissions related to document management. To this end, we develop a tester based on Selenium [15] to automatically simulate user interactions. In greater detail, the tester follows a step-by-step approach. Initially, it installs the add-on, initiates its host application (e.g., Gmail), and subsequently activates the add-on to conduct a thorough scan of all user-interactive elements on the interface, such as buttons and input .fields. Following that, it sequentially interacts with these elements, mimicking the user's perspective in a usage scenario. To thoroughly explore these pages, the tester employs the depth-first search (DFS) algorithm. All dynamically generated HTML files are recorded at the same time.

To expedite the dynamic analysis process, we restrict the maximum HTML page search depth to ten, based on our observation from a small-scale empirical study that 100% of sampled add-ons (500 randomly sampled) fall within this limit. Previous studies also reveal that add-on developers aim to optimize user experience by constricting the complexity of their products [39,45,48].

Enhancing Interaction During Testing. The effectiveness of dynamic interaction is intrinsically limited in scenarios where logical dependencies among events are crucial. For example, certain add-ons require users to set the names of columns in the spreadsheet before launching specific functions. Basic and straightforward dynamic interactions may fail to capture such subtle relationships [32], resulting in problems like the inability to navigate to the next page. Additionally, some add-ons have complex functions that demand domain knowledge (i.e., insert math equations into a Google Doc) and cannot be triggered by simple interaction rules. While manual efforts might address these issues, they are unsuitable for large-scale analysis. As a mitigation strategy, we introduce a novel performance enhancement technique that incorporates add-ons' tutorials and demonstration videos/images.

We start by retrieving all videos and images from the homepage of each add-on. Next, we employ the openCV [10] library to divide the video into image sequences, sampling one frame per second, and filter out identical or similar images. Subsequently, we utilize the YOLO object recognition method to locate the sidebar of add-on and its interactive elements from the images. Finally, we apply Optical Character Recognition (OCR) to identify the text description for the target interactive elements.

Locating Add-on Sidebar. Figure 2 illustrates how the add-on appears as a sidebar in Gmail, enhancing the functionality of the host application. To accurately recognize the add-on in the derived images, we utilize the state-of-the-art object detection algorithm YOLOv8 [28]. To train the model, we manually label 150 images collected from the Google Workspace Marketplace and split them into training (80%), validation (15%), and test (5%) datasets. Our model achieves a 100% accuracy rate with a total processing and prediction time of less than 7ms. We have publicly shared our dataset[1] for further research.

Locating Interactive Elements. To build our model dataset, we utilize HTML pages successfully triggered during **Dynamic Web Testing** and extract the attributes of interactive elements using Selenium. This dataset includes fundamental details about interactive elements in HTML pages, such as type (e.g., button, input field), explanatory text, and coordinates. Additionally, we convert HTML pages into screenshots using selenium, associating element attributes as the ground truth of each screenshot.

We gather a total of 3,553 different HTML pages containing 18,195 interactive elements during **Dynamic Web Testing**. Using the same experimental settings as in the sidebar location, we achieve a high accuracy of 92%.

Code Analysis for API Call Identification. The add-on is developed in Google Script, based on JavaScript, and incorporates Host API specifications for interacting with Google services. While it shares syntax similarities with JavaScript, it introduces host-related features that pose critical challenges to traditional static analyzers. Our goal is to identify **ONLY** the Host API calls that the add-on requires for its functionality, e.g., `GmailApp.getActive Thread().getEmails()` shown in Fig. 4. Traditional analysis (e.g., control flow and data flow analysis) of JavaScript code becomes complicated and ineffective for add-ons. Therefore, we use syntax-level parsing to track the call chain in the add-on's script code.

The workflow of the code analysis is summarized in Algorithm 1. First, we leverage esprima [6] to convert all Google Script code of the add-on into Abstract Syntax Trees (ASTs), which offer a clear view of function and class hierarchy. Second, we identify entry-point classes[2] of Host APIs [9] from the ASTs and record all relevant variables (as shown in line 5–12 in Algorithm 1). We track both intra-function (lines 9–12) and inter-function data flow (lines 7–8) to determine all related Google API calls, which serve as references for its $\mathbb{P}_s$.

---

[1] https://app.roboflow.com/addon/sidebar.
[2] Seven are available, each representing a Workspace app, e.g., DocumentApp and DriveApp.

```
1 attachment = GmailApp.getMessageById().getAttachments();
2 attachment.upload('Google Drive Folder');
3 attachment.getMetadata().getSender().getEmail()
4
5 func current_active_thread(){
6     return GmailApp.getActiveThread()
7 }
8
9 emails = current_active_thread().getEmails()
10 emails.delete()
```

Fig. 4. An example of script invoking Google APIs

Intra-function Call Flow. We record variables related to the entry-point classes as local variables within the function (line 10 in Algorithm 1), and then traverse all statements and analyze the CallExpression or AssignExpression to determine whether any value propagation (e.g., line 1, 9 in Fig. 4) or function call occurs (e.g., line 2, 3, 10 in Fig. 4). For complex statements (e.g., line 3, 6 in Fig. 4), we apply depth-first search (DFS) of the AST to identify the call chain.

Inter-function Call Flow. We record the variables that appear in the return statement of a function as global variables along with the function name (line 7–8 in Algorithm 1, line 5–7 in Fig. 4). Next, we traverse all statements in other functions, including the main function, to check whether the recorded function name (line 9 in Fig. 4) is called. This enables us to identify the flow of call chain between functions in the Google Script code and track the usage of global variables, allowing us to capture dependencies among functions and identify the Google APIs used across functions.

3.3   Excessive Permission Detector

Following the definition of excessive permissions in Sect. 2.2, we need to explicitly determine the requested and functionality permissions respectively.

Requested Permissions. The requested permissions are listed on the homepage of each add-on, following Google's permission restrictions [5] as per the standard and can be accurately parsed.

Functionality Permissions. The functionality permissions include permissions identified from source code and runtime behaviours of add-ons. The detected APIs from source code also exhibit no ambiguity by referring to Google developer documentation [9], since Google lists the required permission for each API. Unfortunately, determining permissions for runtime behaviours is challenging due to the heterogeneity of our analysis (i.e., testing and interaction enhancement).

Algorithm 1. Code analysis for API calls

    Input: Source code C
    Output: Array of permission-related $APIs$
1: $AST \leftarrow parse(C)$, $G_{API}Global \leftarrow dict\{\}$, $G_{API}Local \leftarrow dict\{\}$
2: $G_{API}Entry \leftarrow [CalendarApp, DocumentApp, DriveApp, FormApp, GmailApp,$
    $SlidesApp, SpreadsheetApp]$
3: **function** Proc_function()
4:    **for** $stmt$ in AST **do**
5:       $API, G\_related \leftarrow$ Google_Related_API($stmt$)
6:       **if** $G\_related = True$ **then**
7:          **if** $stmt.type = Return$ **then**
8:             $G_{API}Global[function] \leftarrow API$
9:          **if** $stmt.type \in Variable$ **then**
10:            $G_{API}Local[variable] \leftarrow API$
11:          **if** $stmt.type \in Expression$ **then**
12:            $APIs \leftarrow APIs \cup API$
13:      $APIs \leftarrow APIs \cup G_{API}Local$
14:      $G_{API}Local \leftarrow \{\}$
15: **function** Google_Related_API($stmt$)
16:    Get API_calls of $stmt$
17:    $name \leftarrow API\_calls$, $G\_related \leftarrow False$
18:    **if** $name \in (G_{API}Entry \vee G_{API}Global \vee G_{API}Local)$ **then**
19:       $G\_related \leftarrow True$
20:    **return** $name$, $G\_related$

As a result, it is not practical to simply detect excessive permissions through well-defined techniques such as keywords [32,43] or rule-based [18] matching (which may lead to high false positives) or human inspection (which is difficult to scale). To address the challenge in runtime permission detection, we employ a natural language inference method to infer the functionality permissions existing in the extracted materials. We first construct a manually labelled dataset for runtime permissions as the ground truth and train a classifier for accurate permission prediction at runtime.

Data for Determining Runtime Permissions. The key challenge in this pertains to the dataset construction, as only a part of the dynamic analysis results provides permission-relevant context. For example, many DOM pages are permission-irrelevant, such as add-ons settings, advertisements, and membership purchases. The unevenly distributed permissions further exacerbate the issue, with some permissions being more commonly seen than others. In this work, we leverage the self-descriptive nature of triggered HTML pages and images, which contain contextual and semantic significance to runtime permissions, as discussed by Miniukovich [39]. Based on the technique detailed in Sect. 3.2, we convert images and HTML pages derived in dynamic analysis into natural language description texts. We also include functionality descriptions along with them for complementary explanations as some dynamically triggered pages might not be self-explanatory. To further preserve the special contextual meaning of DOM elements such as buttons, inputs, and radios, we apply an auxiliary processing method using simple rules [30]. For example, a button named "send an email" would be converted to "This button helps you send an email", while an input with the description "email address" would be converted to "You need to provide an email address to this application".

We rely on the Natural Language Inference (NLI) [34] to determine if the derived texts entail specific permissions. This involves deciding whether a natural language hypothesis can be inferred (i.e., true/entailment) or not (i.e., false/-contradiction or undetermined/neutral) from a given premise. For example, consider a premise that states, "This application helps you to edit cell values in your spreadsheet and send an email notification to your collaborators". A hypothesis claiming, "This application can send emails on your behalf" would be labelled as entailment. As an illustration, we provide some of the used permission hypotheses in Table 1. These hypotheses are derived from Google's official developer documentation without any modification [5]. We utilize the pre-trained NLI model used by a recent study [26]. This model is trained on MultiNLI [12], a dataset comprising 433K sentence pairs from various domains, making it suitable for our functionality-diverse add-ons. We query the NLI model with each pair of add-on's functionality description (i.e., premise) and its requested permission (i.e., hypothesis), then record the returned response from this model (i.e., entailment, neural, or contradiction). After filtering out neutral labels, this results in 7,722 pairs (premise and hypothesis) labelled as entailment or contradiction.

Table 1. Permission hypotheses

Scope	Hypotheses
drive.readonly	This add-on can see all of your Google drive files
	This add-on can download all of your Google drive files
drive.file	This add-on can see only the specific Google drive files used with itself
	This add-on an edit only the specific Google drive files used with itself
	This add-on can create only the specific Google drive files used with itself
	This add-on can delete only the specific Google drive files used with itself
drive	This add-on can see all of your Google drive files
	This add-on can edit all of your Google drive files
	This add-on can create all of your Google drive files
	This add-on can delete all of your Google drive files
spreadsheets.readonly	This add-on can see all your spreadsheets
spreadsheets	This add-on can see all of your spreadsheets
	This add-on can edit all of your spreadsheets
	This add-on can create all of your spreadsheets
	This add-on can delete all of your spreadsheets
userinfo.email	This add-on can see your primary Google account email address
userinfo.profile	This add-on can see your personal info that is publicly available
gmail.settings.sharing	This add-on can manage your sensitive mail settings, including who can manage your mail
gmail.send	This add-on can send email on your behalf
gmail.insert	This add-on can add emails to your Gmail mailbox

Runtime Permission Detection. Due to the uneven distribution of the hypotheses (representing permissions) in the constructed dataset, some permissions receive more entailment or contradiction labels than others. To prevent bias towards any specific permission, we randomly sampled 40 instances (half entailment and half contradiction) for each permission. This results in 800 pairs for manual labelling. To avoid ambiguity in the prediction results, we only label permission as entailment (indicating non-excessive) or contradiction (indicating excessive), avoiding assigning neutral labels.

We proceed with the manual verification of the description for each involved add-on, followed by installation and interaction with it to label the functionality permissions and excessive permissions. The corpus is independently labelled by two authors, both with a relevant computer science background. A tiny proportion of 0.37% (3 out of 800) of hypotheses received different labels from them. In cases of disagreement, they would discuss together to make the final decision. The dataset is split into three parts: 80% for training, 15% for validation, and 5% for testing. We use a large language model named Text-To-Text Transfer Transformer (abbreviated as T5), known for its good performance in such inference tasks [26], and compare its performance with other mainstream models, including DistilBERT [11] and GPT-based models [4].

- **Fine-tuned T5 model**: We utilize the pre-trained T5-base [16] model and fine-tune it on our labelled corpus. Our model achieves an accuracy of 86.9%.
- **DistilBERT**: We evaluate the DistilBERT [11] model trained on the GLUE dataset and fine-tuned on our corpus. While BERT [29] has been a well-known large language model from 2018 to 2022 (before the rise of ChatGPT), our testing shows that it achieves an accuracy of 82.8%, which is lower than our T5 model.
- **ChatGPT**: We also evaluate the performance of ChatGPT, one of the most popular GPT models. However, as ChatGPT is a black-box model, we apply prompt engineering [33] to guide it. To avoid bias, we use the same prompt as T5 and choose GPT-3.5 [4] as the base model, which is widely used. Despite this, it achieves an accuracy of only 65.6%, possibly due to the domain-specific nature of our task, which may not be well understood by GPT-3.5, as it primarily focuses on question-answering tasks.

4   Evaluation

To evaluate PEDEc, we have collected a large-scale dataset of publicly available Google Workspace add-ons (Sect. 4.1). Following this, we scrutinized the dataset for excessive permission issues, focusing on three key research questions (RQs).

- **RQ1.** What is the performance of PEDEc in detecting excessive permissions for add-ons (refer to Sect. 4.2)?
- **RQ2.** What is the proportion of add-ons that requested excessive permissions, as reported by PEDEc (Sect. 4.3)?
- **RQ3.** What are the root causes for excessive permissions (Sect. 4.4)?

RQ1 assesses the reliability and effectiveness of PEDEC, while **RQ2** provides insights into the compliance of add-ons. Finally, **RQ3** is essential for developers of Google Workspace and add-ons to address the issue of excessive permission requests.

4.1 Dataset Overview

We employ a web crawler to gather data from the Google Workspace Marketplace. Due to the absence of a comprehensive list of add-ons, we leverage its search functionality, utilizing the 10,000 most common English words provided by Google [1] as keywords. Following a de-duplication process, we acquire a total of 3,756 add-ons. Among the acquired add-ons, 123 add-ons are excluded as they are tailored for non-English users. Additional 228 add-ons are identified with publicly accessible source code post-installation in Google developer mode, consequently selected for our code analysis. The majority of the add-ons (1,414) are developed for Google Spreadsheets, with a smaller number designed for Google Forms, as detailed in Table 2. Notably, two-thirds of the add-ons (2,515) are available for free, while 239 add-ons offer free trials, and 198 add-ons feature premium services.

Table 2. Dataset overview

Feature	Category	Number of add-ons
Host	Google drive	681
	Gmail	430
	Google calendar	167
	Google doc	566
	Google sheets	1414
	Google slides	274
	Google forms	171
	Google classroom	12
	No Host	408
Charge	Free	3285
	Paid	34
	Free trial	239
	Free charge paid features	198

Table 3. Performance over sets of `oauthScope` permissions

Permission set	TP	TN	FP	FN	TPR	TNR	Accuracy
See	16	80	3	1	94%	96%	96%
Edit	15	79	3	3	83%	96%	94%
Create	37	58	2	3	93%	97%	95%
Delete	42	56	1	1	98%	98%	98%
Manage	21	71	1	7	75%	99%	92%
Total	**131**	**344**	**10**	**15**	**90%**	**97%**	**95%**

4.2 RQ1: PEDec Accuracy

Ground Truth Annotation. Given the absence of prior work and benchmarks, we manually curate a benchmark dataset. We randomly select 100 add-ons from our collected dataset and engage three annotators with computer science backgrounds-comprising one final-year undergraduate, one master's student, and one PhD student. To standardize the assessment of excessive permissions, we formulate explicit labelling instructions[3] drawing inspiration from established definitions in the literature and common practices [20,26]. The instructions include definitions for excessive permissions and functionality permissions, aiming to address individual perceptions. To clarify nuanced cases, such as permissions on all files or currently interacted files, we provide illustrative examples. Throughout the annotation process, annotators adhere to the guidelines, install and interact with add-ons using our testing account, and identify instances of excessive permissions. In disagreements, annotators and we collectively resolve differences, seeking input from a security expert with eight years of experience (2015-) as needed. The Kappa score for annotator agreement is 0.844, signifying nearly perfect agreement (above 0.81) according to established criteria [38].

PEDec Performance. We initiate our analysis with the 100 annotated add-ons. In Table 3, we present the true positive (TP), false positive (FP), true negative (TN), and false negative (FN) rates of PEDec. We define the excessive permission as positive and the non-excessive permission as negative. Thus, TP denotes that the excessive permission is detected by both PEDec and human annotators, while FP indicates that PEDec marks the permission as excessive, whereas human annotators determine it as non-excessive.

PEDec achieves a promising performance, we evaluate PEDec's performance on each positive and negative case using the top-5[4] requested permission sets. As shown in Table 3, PEDec achieves an average of 90% TPR and 97% TNR. Notably, for the highly sensitive permission like *delete*, PEDec can achieve 98% TPR and 98% TNR. The performance of PEDec on the add-on level is presented in Table 4. It showcases the ability to accurately identify add-ons with (or without) excessive permissions, achieving a high accuracy of 95%.

---

[3] Available at https://github.com/CoTestAccount/third-party-application.

[4] The remaining one is "send', comprising only one record.

Table 4. Overall performance of PEDEC on Ground Truth

TP	TN	FP	FN	TPR	TNR	Accuracy
61	34	0	5	92%	100%	95%

We conduct a manual inspection to investigate the low TPR associated with the "manage" permission, as illustrated in Table 3. Discussions with annotators reveal that the complexity lies in the compound nature of "manage", encompassing a series of permissions that vary across different host applications. Further scrutiny of the Google Workspace developer page [5] unveils that "manage" includes permissions like "modify" and "share with other users spreadsheets that this application has been installed in". While we did sample "manage" hypotheses in Sect. 3.3 and assigned the correct label as training data, our model struggles to capture such fine-grained scopes without additional information. As depicted in Table 3, "manage" exhibits a relatively low TPR compared to others. A potential solution involves splitting the compound permission into a series of permissions, as demonstrated in Table 5, necessitating additional manual effort for corpus relabeling.

API Call Identification Performance. We additionally evaluate the accuracy of our lightweight code analysis tool. We randomly sample 20% (50 out of 228) of add-ons and perform a manual inspection of the code analysis performance. Leveraging Google's detailed mapping from each Host API to the required permissions in the Google Developer Documentation [9], we employ the fine-grained Host API Coverage (detector/ground-truth) as a metric to gauge PEDEC's effectiveness in capturing all functionality-related Host API.

PEDEC fails to detect API calls in four add-ons. After investigating through source code screening, we discover that two add-ons employ code obfuscation, rendering PEDEC designed for readable source code ineffective. Additionally, two add-ons utilize outdated APIs no longer supported by Google. For the remaining 46 add-ons, the API coverage stands at 100%.

4.3   RQ2: Landscape of Add-Ons

We then apply PEDEC to analyze all collected add-ons. It identifies that 56% (2,091 out of 3,756) of them have excessive permissions. Table 7 breaks down the service with excessive permissions across different installation ranges and particular permissions. The most common (40%, 1,484 out of 3,756) excessive permission is *delete*, where the add-on does not provide delete files or folder functionality but requests this permission. We observe that the issue of excessive permissions is quite prevalent at 17% for editing files, 16% for managing (including sharing) files, and 20% for seeing files. Despite its prevalence, excessive permissions remain a significant issue in many widely-used add-ons, particularly those with over one million installations. Table 7 shows that add-ons designed for Google

Table 5. Manage under different scenarios

Manage permission	Compound permissions
Manage spreadsheets that this add-on has been installed in	Modify spreadsheets that this add-on has been installed in Share with other users spreadsheets that this add-on has been installed in
Manage your forms in Google drive	Create new forms Modify existing forms Submit form responses Process form responses Share forms with others
Manage forms that this add-on has been installed in	Modify forms that this add-on has been installed in Share with other users forms that this add-on has been installed in Process responses to forms that this add-on has been installed in

Table 6. Co-occurrence analysis of excessive permission set

Co-occurrence	See	Edit	Create	Delete	Manage
See	772	587	635	660	130
Edit	–	631	611	622	113
Create	–	–	1257	1220	172
Delete	–	–	–	1484	186
Manage	–	–	–	–	603

Co-occurrence	Sheets	Slides	Docs	Form	Gmail
Sheets	1204	101	229	87	16
Slides	–	246	101	29	4
Docs	–	–	532	52	5
Form	–	–	–	158	4
Gmail	–	–	–	-	45

Table 7. Proportion of excessive permission requested by add-ons. "E" represents "Excessive" (i.e., the number of add-ons that contain excessive permissions), "N" represents "Non-Excessive" (i.e., the number of add-ons that contain no excessive permissions) ,and "T" represents "Total" (i.e., the total number of add-ons)

App	Add-ons' installation distribution									Add-ons with excessive permissions on the oauthScope set									
	0-1000			1000-1 Million			1 Million+			See		Edit		Create		Delete		Manage	
	E	N	E/T	E	N	E/T	E	N	E/T	E	E/T	E	E/T	E	E/T	E	E/T	E	E/T
Gmail	39	127	0.23	65	178	0.27	8	13	0.38	71	0.17	25	0.06	35	0.08	55	0.13	17	0.04
Google Docs	111	35	0.76	259	90	0.74	56	15	0.79	150	0.27	128	0.23	275	0.49	290	0.51	157	0.28
Google Sheets	417	142	0.75	598	173	0.78	68	16	0.81	353	0.25	306	0.22	713	0.51	835	0.59	342	0.24
Google Slides	41	14	0.75	149	27	0.85	36	7	0.84	102	0.37	78	0.28	156	0.57	163	0.59	82	0.30
Google Forms	11	2	0.85	84	30	0.74	35	9	0.80	44	0.26	44	0.26	70	0.41	73	0.43	99	0.58
Google Calendar	27	43	0.39	40	50	0.44	1	6	0.14	29	0.17	11	0.07	12	0.07	41	0.24	7	0.04
Google Drive	38	50	0.43	153	432	0.26	26	68	0.28	150	0.20	91	0.12	127	0.17	136	0.18	36	0.05
No Host	47	88	0.35	137	12	0.92	12	17	0.41	92	0.23	52	0.13	59	0.14	114	0.28	11	0.03

Docs, Sheets, Slides, and Forms have a higher proportion with excessive permissions, considering their rich functionality. We conclude the root cause in **RQ3** as an exploration and alert for developers of Google Workspace and add-ons.

We conduct an analysis of the distribution of co-occurrence of detected excessive permissions, as depicted in Table 6. To streamline the presentation, we exclusively showcase the upper triangular matrix due to its symmetrical nature. It is noteworthy that when the permission "edit" is present, "delete" is identified as an excessive permission in nearly 98.6% of cases (622 out of 631). Although "delete" is the most frequently detected excessive permission, it often co-occurs with "edit" (41.9%, 622 out of 1484), "manage" (12.5%), and "see" (44.5%). Additionally, a noteworthy observation is that a substantial portion of add-ons designed for Google Sheets are flagged for excessive permissions, frequently requesting such permissions for Docs (19.0%, 229 out of 1204), Slides (8.4%), and Gmail (1.3%).

4.4   RQ3: Root Causes (RCs) and Case Study

After discovering that approximately half of the add-ons request excessive permissions, we delve deeper into their Root Causes (RCs) to gain a comprehensive understanding of excessive permissions. We randomly select and manually inspect 50 (out of 2,091) services with excessive permissions. Based on the types of RCs, we categorize them into the following three groups. The first two types stem from permission management issues on the platform operator's side (i.e., Google in our study), while the last type is associated with poor implementation practices from the add-on developers. Each type of RC is elucidated through detailed example.

RC1: Permission Bundle (30 Add-Ons). When permissions are grouped into bundles, there is an increased risk of excessive permissions. Specifically, the add-on has to request the entire group of permissions even if only part of them is functionality-required. This all-or-nothing implementation of permissions contributes to security threats in Google Workspace [21]. This Root Cause (RC) is the primary factor behind excessive permissions in popular add-ons like Box (more than one million installations, 1M+ for short hereafter), DocuSign eSignature (2M+), and Auto-LaTex Equations (10M+). Based on their types, the permission bundle can be further classified into two categories: *bundled operations* and *bundled objects*. MathType (10M+) functions as a math equations and chemical formulas editor for Google Docs. The correct permissions for its functionality are "see" and "modify". However, the issue arises as "see" and "modify", and "share" permissions are bundled into one group. This bundling requires MathType to request the excessive "share" permission, even though it is not strictly necessary for its functionality.

RC2: Coarse-grained Permission Management (20 Add-Ons). Google's current permission management system employs a general and broad permission

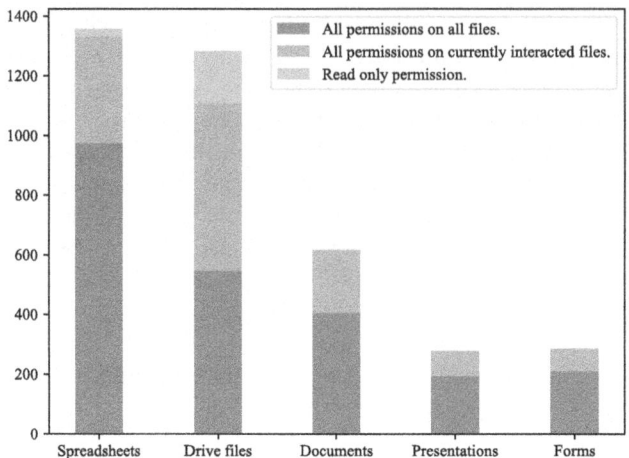

Fig. 5. Distribution of permission scope

coverage, inadvertently exposing more data than intended by the developers. Zoom for Gmail (12M+) enables users to respond to meeting schedules. Only the email title is needed for setting the meeting topic, and the remaining content (i.e., body content, attachment) is non-relevant to the functionality. However, Google provides no fine-grained permission to exclusively view the email title. By requesting the "see" permission, Zoom can access all the content of emails.

RC3: Extra Permissions Requested by Add-ons (12 Add-Ons). There is a tendency for the add-ons to frequently request excessive permissions under a loosely managed permission scheme [47]. As depicted in Fig. 5, around two-thirds of add-ons request "All permissions on all files" instead of the more restricted permission option of "All permissions on currently interacted files". Bjorn's NAV-thumbs (607) offers features for navigation in Google Slides. While it requires permission to "see and edit all your Google Slides presentations" for its functionality, it also requests excessive permissions permissions such as "see, edit, create and delete all your Google Docs documents", raising privacy and security concerns. It is alarming that such requested permissions passed the vetting process.

5   Limitations and Discussion

5.1   Limitations

Quality of Tutorial Images and Videos. Developers are required to supply demonstration images, but the use of low-quality images may adversely affect

PEDEC performance. Our evaluation indicates that developers of popular services with a large user base tend to offer high-quality images and comprehensive demonstration videos. In contrast, developers with a smaller user base may not provide clear tutorials.

NLP Performance. PEDEC relies on the functionality materials and employs the Natural Language Inference (NLI) and T5 methods to extract excessive permissions. However, there is still a failure rate of 13% due to inherent limitation of Natural Language Processing (NLP) techniques.

5.2 Discussion

PEDEC heavily relies on dynamic analysis of runtime behavior and NLP techniques to infer excessive permissions. Each time the add-ons receive a new update, the tool must be rerun to check for excessive permissions. In the future, we plan to explore strategies that only consider the new update to re-detect excessive permissions. Additionally, since Google frequently updates their host API documentation, PEDEC should be able to capture such updates in a timely manner.

6 Related Work

Static Code Analysis. To assess the security of software permissions and safeguard user data, many researchers explore static code analysis [19] and API call graphs. Taint analysis is widely employed in add-ons for mobile apps, identifying numerous permission escalation [21,45]. Given our emphasis on Host API call detection, existing call-graph or data-flow analysis designed for JavaScript, as demonstrated in previous research [32], are unsuitable for our tasks. In our work, we employ a lightweight API call analysis to recognize all functionality-related Host API calls.

Dynamic Web Testing and Enhancement. To ensure the quality and usability of web applications, many researchers explore dynamic testing in both black-box and white-box contexts [31–33,35,37]. Given that human interaction serves as the primary communication between web applications and users [39], the most common automated testing methods [40] are based on user profiles (e.g., name, gender and license number) or scenarios (e.g., shopping websites). Gao et al. [24] investigate the ecosystem of account registration bots. Their experiments suggest that the most prevalent human verification methods, including SMS, CAPTCHA, and IP restrictions, can be successfully bypassed by anti-human verification services. However, owing to the complexity and customization of modern web applications, dynamic automated GUI testing tools [33] that simulate human operations (e.g., login, click or scroll) cannot guarantee the triggering of all web paths. We acknowledge this inherent limitation and utilize tutorial images/videos as supplements.

Excessive Permission Detection in Android. Many studies have analyzed excessive permissions in Android apps [41,42,46,49]. Whyper [41] and Autocog [42] leverage the app's description and counterpart to infer permissions that are irrelevant to functionality. Pscout [19] and Taintmini [45] construct call graphs and track sensitive data flow to detect excessive permissions. These approaches require access to source codes or a large app base for accuracy and reliability in clustering app functionality. However, addressing the domain-specific challenges existing in Google Workspace is necessary due to the limited size and closed nature of the data.

7 Conclusion

The add-on enhances customization options but also introduces new security concerns related to excessive permissions. They may request access to more permissions than necessary, potentially putting user's data at risk. It is important to carefully review the permissions requested by the add-on and consider whether they are necessary for its functionality. To our best knowledge, no prior work has provided such automated checks for excessive permissions of the add-on that may lead to security issues, and our work addresses this gap. In this paper, we design a tool called PEDEC to automatically check whether the add-on requests excessive permissions. Ground truth evaluation shows that PEDEC performs well, achieving 100% TNR and 92% TPR in checking excessive permissions.

Acknowledgement. We would like to thank anonymous reviewers for improving this manuscript. This research has been partially supported by the University of Queensland and the Huazhong University of Science and Technology.

References

1. 10,000 most common English words Repo (2023). https://github.com/first20hours/google-10000-english
2. Auto-Latex Equation (2023). https://workspace.google.com/marketplace/app/autolatex_equations/850293439076
3. California Consumer Privacy Act (2023). https://oag.ca.gov/privacy/ccpa
4. ChatGPT fine tune reference (2023). https://platform.openai.com/docs/api-reference/fine-tunes
5. Developer Reference: OAuth 2.0 scopes (2023). https://developers.google.com/identity/protocols/oauth2/scopes
6. Esprima (2023). https://esprima.org
7. General Data Protection Regulation (2023). https://gdpr-info.eu/
8. Google Workspace Market Share (2023). https://6sense.com/tech/office-suites/google-workspace-market-share
9. Google Workspace Reference Overview (2023). https://developers.google.com/apps-script/reference/

10. Homepage of OpenCV (2023). https://opencv.org/
11. Hugging Face DistilBERT (2023). https://huggingface.co/docs/transformers/model_doc/distilbert
12. HuggingFace: roberta-large-mnli model (2023). https://huggingface.co/roberta-large-mnli
13. No Phishing! (2023). https://workspace.google.com/marketplace/app/no_phishing/149706744667
14. OAuth 2.0 Rich Authorization Requests (2023). https://datatracker.ietf.org/doc/html/draft-ietf-oauth-rar-12
15. Selenium Dev (2023). https://www.selenium.dev/
16. T5 base (2023). https://huggingface.co/t5-base
17. Zoom Apps (2023). https://marketplace.zoom.us/
18. Andow, B., et al.: Policylint: investigating internal privacy policy contradictions on google play. In: USENIX Security Symposium, pp. 585–602 (2019)
19. Au, K.W.Y., Zhou, Y.F., Huang, Z., Lie, D.: Pscout: analyzing the android permission specification. In: Proceedings of the 2012 ACM Conference on Computer and Communications Security, pp. 217–228)(2012)
20. Balash, D.G., Wu, X., Grant, M., Reyes, I., Aviv, A.J.: Security and privacy perceptions of {Third-Party} application access for google accounts. In: 31st USENIX Security Symposium (USENIX Security 22), pp. 3397–3414 (2022)
21. Chen, Y., Gao, Y., Ceccio, N., Chatterjee, R., Fawaz, K., Fernandes, E.: Experimental security analysis of the app model in business collaboration platforms. In: 31st USENIX Security Symposium (USENIX Security 2022), pp. 2011–2028 (2022)
22. Edu, J., Mulligan, C., Pierazzi, F., Polakis, J., Suarez-Tangil, G., Such, J.: Exploring the security and privacy risks of chatbots in messaging services. In: Proceedings of the 22nd ACM Internet Measurement Conference, pp. 581–588 (2022)
23. Fang, Z., Han, W., Li, Y.: Permission based Android security: issues and countermeasures. Comput. Secur. **43**, 205–218 (2014)
24. Gao, Y., Xu, G., Li, L., Luo, X., Wang, C., Sui, Y.: Demystifying the underground ecosystem of account registration bots. In: Proceedings of the 30th ACM Joint European Software Engineering Conference and Symposium on the Foundations of Software Engineering, pp. 897–909 (2022)
25. Harkous, H., Aberer, K.: "if you can't beat them, join them" a usability approach to interdependent privacy in cloud apps. In: Proceedings of the Seventh ACM on Conference on Data and Application Security and Privacy, pp. 127–138 (2017)
26. Harkous, H., Peddinti, S.T., Khandelwal, R., Srivastava, A., Taft, N.: A deep learning system for navigating privacy feedback at scale. In: 2022 IEEE Symposium on Security and Privacy (SP), pp. 2469–2486. IEEE (2022)
27. Jannett, L., Mladenov, V., Mainka, C., Schwenk, J.: Distinct: identity theft using in-browser communications in dual-window single sign-on. In: Proceedings of the 2022 ACM SIGSAC Conference on Computer and Communications Security, pp. 1553–1567 (2022)
28. Jocher, G., Chaurasia, A., Qiu, J.: YOLO by Ultralytics (2023)
29. Devlin, J., Kenton, M.W.C., Toutanova, L.K.: Bert: pre-training of deep bidirectional transformers for language understanding. In: Proceedings of NAACL-HLT, vol. 1, p. 2 (2019)
30. Khandelwal, R., Linden, T., Harkous, H., Fawaz, K.: {PriSEC}: a privacy settings enforcement controller. In: 30th USENIX Security Symposium (USENIX Security 2021), pp. 465–482 (2021)
31. Ling, Y., Hao, Y., Wang, Y., Wang, K., Bai, G., Dong, J.S.: Essential or excessive? mindaext: measuring data minimization practices among browser extensions (2024)

32. Ling, Y., Wang, K., Bai, G., Wang, H., Dong, J.S.: Are they toeing the line? diagnosing privacy compliance violations among browser extensions. In: Proceedings of the 37th IEEE/ACM International Conference on Automated Software Engineering (ASE) (2022)

33. Liu, Z., et al.: Fill in the blank: context-aware automated text input generation for mobile gui testing. In: 2023 IEEE/ACM 45th International Conference on Software Engineering (ICSE), pp. 1355–1367. IEEE (2023)

34. MacCartney, B.: Natural language inference. Stanford University (2009)

35. Mahadewa, K., et al.: Scrutinizing implementations of smart home integrations. IEEE Trans. Softw. Eng. **47**(12), 2667–2683 (2019)

36. Mahadewa, K., et al.: Identifying privacy weaknesses from multi-party trigger-action integration platforms. In: Proceedings of the 30th ACM SIGSOFT International Symposium on Software Testing and Analysis, pp. 2–15 (2021)

37. Mahadewa, K.T., Wang, K., Bai, G., Shi, L., Dong, J.S., Liang, Z.: Homescan: scrutinizing implementations of smart home integrations. In: 2018 23rd International Conference on Engineering of Complex Computer Systems (ICECCS), pp. 21–30. IEEE (2018)

38. McHugh, M.L.: Interrater reliability: the kappa statistic. Biochemia Medica **22**(3), 276–282 (2012)

39. Miniukovich, A., De Angeli, A., Sulpizio, S., Venuti, P.: Design guidelines for web readability. In: Proceedings of the 2017 Conference on Designing Interactive Systems, pp. 285–296 (2017)

40. Nasraoui, O., Soliman, M., Saka, E., Badia, A., Germain, R.: A web usage mining framework for mining evolving user profiles in dynamic web sites. IEEE Trans. Knowl. Data Eng. **20**(2), 202–215 (2007)

41. Pandita, R., Xiao, X., Yang, W., Enck, W., Xie, T.: Whyper: towards automating risk assessment of mobile applications. In: USENIX Security Symposium, Washington, DC, vol. 2013 (2013)

42. Qu, Z., Rastogi, V., Zhang, X., Chen, Y., Zhu, T., Chen, Z.: Autocog: measuring the description-to-permission fidelity in android applications. In: Proceedings of the 2014 ACM SIGSAC Conference on Computer and Communications Security, pp. 1354–1365 (2014)

43. Wan, L., Wang, K., Mahadewa, K.T., Wang, H., Bai, G.: Don't bite off more than you can chew: investigating excessive permission requests in trigger-action integrations. In: Proceedings of the ACM Web Conference 2024 (2024)

44. Wan, L., Wang, K., Wang, H., Bai, G.: Is it safe to share your files? an empirical security analysis of google workspace. In: Proceedings of the ACM Web Conference 2024 (2024)

45. Wang, C., Ko, R., Zhang, Y., Yang, Y., Lin, Z.: Taintmini: detecting flow of sensitive data in mini-programs with static taint analysis. In: 2023 IEEE/ACM 45th International Conference on Software Engineering (ICSE), pp. 932–944. IEEE (2023)

46. Wang, L., Wang, H., Luo, X., Sui, Y.: Malwhiteout: reducing label errors in android malware detection. In: Proceedings of the 37th IEEE/ACM International Conference on Automated Software Engineering, pp. 1–13 (2022)

47. Zha, M., Wang, J., et al.: Hazard integrated: understanding security risks in app extensions to team chat systems. In: Network and Distributed Systems Security (NDSS) Symposium (2022)

48. Zhang, J., Yang, L., Han, Y., Xiang, Z., Hei, X.: A small leak will sink many ships: vulnerabilities related to mini-programs permissions. In: 2023 IEEE 47th Annual Computers, Software, and Applications Conference (COMPSAC), pp. 595–606. IEEE (2023)
49. Zhou, L., et al.: {POLICYCOMP}: counterpart comparison of privacy policies uncovers overbroad personal data collection practices. In: 32nd USENIX Security Symposium (USENIX Security 2023), pp. 1073–1090 (2023)

Formal Verification Techniques for Post-quantum Cryptography: A Systematic Review

Yuexi Xu[1], Zhenyuan Li[1], Naipeng Dong[1($\boxtimes$)], Veronika Kuchta[2], Zhe Hou[3], and Dongxi Liu[4]

[1] The University of Queensland, Queensland, Australia
n.dong@uq.edu.au
[2] Florida Atlantic University, Boca Raton, USA
[3] Griffith University, Nathan, Australia
[4] Data61, CSIRO, Eveleigh, Australia

Abstract. In the quantum computing era, the imperative role of post-quantum cryptography in securing digital communications has led to the development of computer-aided cryptography verification tools. These tools simplify the verification of post-quantum cryptography primitives and protocols, alleviating the challenges associated with manual proofs. This paper systematically reviews research in four main areas: quantum computing, post-quantum cryptography, cryptanalysis, and verification, establishing a foundation for future research. Emphasising the significance of challenges in post-quantum cryptography, we outline the current state of research on cryptography primitives and protocols. Categorising state-of-the-art computer-aided cryptography verification tools based on assumptions, models, and application levels, our analysis delves into each tool's features, including modelling, adversary models, security properties, validation, and an in-depth analysis of their limitations. This comprehensive analysis offers insights into the nexus of post-quantum cryptography and computer-aided verification. Concluding with recommendations for researchers and practitioners, this paper explores potential future research directions.

Keywords: Formal Verification · Post-quantum Cryptography · Post-quantum Protocol · Quantum Threats · Review · Survey

1 Introduction

Cryptography holds significant importance in safeguarding confidential data and communication, serving as the fundamental building block for secure systems. In the rapidly evolving landscape of modern cryptography, the emergence of quantum computing necessitates a comprehensive understanding of their latest

---

Y. Xu and Z. Li—are co-first authors.

© The Author(s), under exclusive license to Springer Nature Switzerland AG 2025
G. Bai et al. (Eds.): ICECCS 2024, LNCS 14784, pp. 346–366, 2025.
https://doi.org/10.1007/978-3-031-66456-4_19

advancements and security implications, as the proliferation of quantum computing threatens the security foundations of certain classical cryptographic schemes,
calling for an urgent exploration of alternative paradigms.

Simultaneously, there is a pressing need for the evaluation of the robustness
of cryptographic primitives and protocols (applications of cryptographic primitives). Thus, cryptanalysis has gained paramount importance, ensuring their correctness, resilience, and alignment with desired security properties. Researchers
have turned to formal verification techniques, which provide a systematic and
rigorous approach to verifying their reliability and correctness. These techniques
play a crucial role in identifying vulnerabilities, errors, and weaknesses that
might otherwise go unnoticed.

This survey paper embarks on an ambitious journey through the realms of
post-quantum cryptography with breakdowns in four areas: post-quantum cryptography, quantum computing, cryptanalysis, and formal verification, to unravel
the intricate tapestry that underpins these domains. Comprehensive research of
these areas is necessary because the interplay between them has created a complex and rapidly changing landscape. A deep understanding of the key elements
in play is crucial to staying ahead of potential threats and ensuring the robustness of cryptographic applications in this shifting environment. By doing so,
this survey provides an overarching perspective that contributes to the ongoing
efforts to fortify the digital security infrastructure.

Existing surveys (detailed in Sect. 3) offer valuable insights into specific facets
of quantum computing, post-quantum cryptography, cryptanalysis, and verification, but exhibit limitations in their integration—the formal verification of post-
quantum cryptography. In addition, they are limited in coverage and depth.
Notably, these surveys tend to emphasize theoretical explorations, often over-
looking the practical dimensions essential for real-world applications. The challenges and opportunities transitioning the verification approaches from classical
to post-quantum cryptography remain largely unexplored, leaving a critical void
in our understanding.

To address these gaps, this paper undertakes a comprehensive and metic-
ulous analysis, delving into various verification methods within the context
of post-quantum cryptography and protocols. This exploration encompasses a
wide range of verification approaches, including manual, automated, and semi-
automated methods, while also extending its focus to encompass both classical
and quantum paradigms. The survey aims to provide a holistic understanding of
the intricacies and potential remedies that define the cryptographic landscape.
Through this endeavour, the paper aims to offer insights that will not only illu-
minate the current panorama but also guide future research directions in the
pursuit of robust cryptographic systems capable of withstanding the challenges
of an increasingly complex and quantum-powered world.

2   Survey Scope

The emergence of **quantum computing** presents a formidable challenge
to classical cryptography and its applications, as it can break widely-used

cryptography algorithms, such as the RSA signature and the Elliptic Curve Diffie-Hellman (ECDH) based cryptography, by leveraging its superior computational capabilities, specifically for factoring large numbers and solving discrete logarithm problems efficiently. Quantum computing introduces a pressing need for post-quantum cryptography, which aims to develop cryptographic algorithms resilient to quantum attacks, and necessitates a reevaluation of cryptographic systems and their vulnerabilities. Thus, understanding quantum computing is pivotal to grasping the current and future cryptography landscape.

As a countermeasure to the quantum threats, **post-quantum cryptography** (PQC) is proposed that focuses on developing cryptographic schemes and their applications that can withstand quantum attacks. With the rapid improvement of quantum computing, the need for post-quantum cryptographic solutions is urgent [25]. Understanding post-quantum cryptography is also vital to the development and analysis of secure sensitive information and communication systems in the face of emerging quantum threats.

As the security of cryptography is ensured generally by **cryptanalysis** that uncovers vulnerabilities and weaknesses of the cryptography, cryptanalysis of post-quantum cryptography is indispensable for assessing the robustness of the proposed post-quantum cryptographic schemes and the effectiveness of their applications. As such, we will also explore the surveys that evaluate the strengths and weaknesses of both classical and post-quantum cryptographic solutions.

Among various cryptoanalysis, **formal verification** techniques provide a systematic and rigorous approach to verifying the reliability and correctness of cryptographic schemes and applications. It is becoming a norm that a cryptographic scheme must come with a proof that it satisfies some standard security properties to be accepted. These techniques are essential for ensuring that cryptographic systems (including cryptographic schemes and applications) operate as intended and are free from vulnerabilities and errors. Therefore, the study of verification methods, both manual and automated, is crucial to understanding their capabilities, pros and cons in proving correctness and security, contributing to their overall reliability and effectiveness.

This survey explores the above four topics due to their paramount significance and interrelated nature. We examine the interplay between the four topics as they collectively shape the evolving landscape of future digital security. In particular, our focus is on *formal verification techniques for post-quantum cryptography*.

Data Collection. Our investigation encompassed an examination of scholarly works published from 2020 onwards for the above four topics. In the pursuit of a robust comprehension of verification methods, we broadened our scope to encompass literature dating back to 2010, as discerned by the paucity of available information within the field. Employing a methodical keyword-based search strategy, we navigated the scholarly terrain via Google Scholar and databases including ACM, IEEE, ANST and Scopus employing the related keywords, and refined the search results by screening their titles and abstracts.

3   Existing Surveys

There have been surveys on each individual area of quantum computing, post-quantum cryptography, cryptanalysis and formal verification, presented as follows. However, there is no survey focusing on the formal verification of post-quantum cryptography, which is the key motivation of this work.

Quantum Computing. Surveys within the domain of quantum computing offer insights into emerging technologies and fields such as machine learning [5,46], post-quantum cryptography and quantum entanglement [47], and financial field [35] e.g., blockchain [25]. Each paper undertakes a comprehensive review of recent developments and research within its field and provides a discussion of potential areas for future research. Nonetheless, these papers have insufficient coverage of practical implementation challenges, scalability issues, and ethical considerations. Furthermore, these papers tend to prioritize the theoretical physics aspects of quantum computing over other aspects.

Post-quantum Cryptography. Four survey papers are examined on this topic, covering different aspects. The work of Tan et al. [53] focuses on the challenges of implementing post-quantum digital signature algorithms in real-world applications, while the work by Kumari [39] covers the suitability of post-quantum cryptography techniques for securing communication in resource-constrained IoT devices. The work conducted by Hasija and other researchers [34] covers the third-round candidates for post-quantum cryptography selected by NIST and their algorithmic structures, security properties, and implementation details. Finally, the paper of Zeydan et al. [56] covers the recent advances in post-quantum cryptography for network security, including key exchange, signature schemes, and encryption schemes. However, none of the papers discuss the challenges of transitioning from classical to post-quantum cryptography, such as the compatibility of post-quantum algorithms with existing systems or the impact of quantum computers on current cryptographic protocols.

Cryptanalysis. There are several papers on cryptography that cover various topics such as encryption methods, key management, digital signatures, and hash functions. Research conducted by Mnkash [42] discusses traditional and advanced encryption techniques, cryptographic protocols, and key distribution methods. Abinaya and Prabakeran [4] focus on lightweight block ciphers and their suitability for IoT devices, covering their design principles, security features, and potential attacks. Wang's work [54] surveys the use of lattice-based cryptosystems in the standardization processes of cryptographic algorithms, highlighting the security properties, potential advantages, ongoing efforts, and open problems in this area. The last paper is a bibliometric analysis of research papers related to the cryptanalysis of block ciphers in cyber security [9], covering the evolution of block ciphers and cryptanalysis techniques, their application in cyber security, and research trends and patterns. However, these surveys are also theory-focused but light on practical concerns.

Formal Verification. Previous survey papers on cryptographic primitive and protocol verification have primarily focused on the blockchain field [40], while others have concentrated on specific areas such as IoT devices [37, 39], algorithms [24] and networks [48]. Matteo et al. [7] conducted a comprehensive analysis of various automated protocol provers used for classical cryptography. Their analysis evaluated RCF, Applied pi-calculus, CryptoVerif, ASPIER, Horn clauses, First-order logic (FOL), and LySa with respect to both computational and symbolic models. The study elaborated on the operational patterns, languages, and formal semantics of each model. In addition to protocol provers, their survey paper also covered code generation approaches. Some survey papers toughed a bit on post-quantum cryptography; for instance, Kumari et al. [39] covered post-quantum cryptographic techniques but specific to IoT devices.

One can observe that there is limited research on formal verification for post-quantum cryptography. Therefore, this paper endeavours to bridge the gap by conducting an extensive survey of verification approaches in the context of post-quantum cryptographic primitives and protocols. Through this exploration, the paper seeks to shed light on the multifaceted challenges, solutions, and potential avenues for future research of post-quantum cryptographic verification.

Compared with the above formal verification surveys, our work takes a more extensive approach and explores the complete landscape of provers for both primitives and protocols. The scope of our survey is not limited to automated provers but also considers semi-auto and manual provers. Most importantly, we focus on the quantum adversary in our analysis, rather than the polynomial adversary in the classical setting. In contrast to the verification of a specific cryptographic scheme, our paper emphasizes the generic approaches.

4   Post-quantum Cryptography

Cryptographic primitives and cryptographic protocols are two fundamental components in the field of cryptography. Cryptographic protocols, built on top of primitives, provide rules guiding secure communication between entities. In essence, protocols are applications of primitives. We discuss their state-of-the-art separately in the following subsections.

4.1   Post-quantum Cryptographic Primitives

Post-quantum cryptographic primitives are often categorized into five classes based on hard problems: Hash-based, Code-based, Multivariate, Lattice-based and Isogenies [25]. Many have been proposed but later found flawed e.g., in the process of NIST competition, highlighting the importance of cryptanalysis. Hence, we only consider the candidates selected by NIST because they have a high chance of being adopted in the real-world applications.

The National Institute of Standards and Technology (NIST) initiated the PQC Standardization project to identify and promote cryptographic algorithms that can defend the evolving threat in the face of quantum computing advancements. After six years of cryptanalysis, in 2022, NIST made a significant stride in this endeavour by officially unveiling a set of cryptographic algorithms deemed as the "selected candidates" in round three. There are four notable candidates [44]: CRYSTALS-KYBER, CRYSTALS-DILITHIUM, FALCON, and SPHINCS+. They fall into two major categories: public-key encryption and key-establishment, and the digital signature. CRYSTALS-KYBER is the exclusive algorithm within the category of public-key encryption and key-establishment. Notably, while the other candidates are rooted in lattice-based cryptographic principles, SPHINCS+ stands out as a hash-based cryptographic solution.

NIST has already initiated the process of collecting feedback and comments for the upcoming fourth round [43], aiming at more public-key encryption and key-establishment algorithms. The submissions in this round encompass BIKE, Classic McEliece, HQC, and SIKE. Note that SIKE employs isogeny-based cryptographic techniques, while the other candidates are code-based.

4.2   Post-quantum Cryptographic Protocols

In the field of post-quantum cryptographic protocols, current research results are mainly focused on TLS and blockchain, illustrating the limited practical applications of post-quantum cryptography in solving real-world quantum threats.

The Transport Layer Security (TLS) protocol, especially version 1.3, is the foundation for securing data transmission over the internet [6,52]. Amidst the evolution of quantum computing, TLS 1.3 needs to be validated and improved against quantum threats. Post-quantum TLS candidates are proposed by integrating post-quantum cryptographic primitives [6,52]. These works also conducted performance experiments, providing a well-reasoned reference for future research and practical preparations in the quantum era.

Blockchain technology has become a dynamic focus in post-quantum protocol studies, evident in numerous papers and studies [25,30,45]. The growing volume of research and applications underscores the industry's recognition of blockchain's pivotal role in fortifying the foundations against quantum uncertainties. Researchers are enhancing cryptographic primitives to secure blockchain transactions, focusing on various signature schemes. The convergence of blockchain and post-quantum cryptography addresses the urgent security needs of digital currencies, reshaping trust structures for decentralized systems in a quantum future.

5   Formal Verification Techniques: Taxonomy

Formal verification is a promising way to verify the security of post-quantum cryptographic primitive and protocols. As a rigorous approach, formal verification needs to clearly define the *assumptions* of the results. Given the assumption, and a rigorous *model* representing the system behaviour as well as the security claims, formal verification applies various *verification techniques* to get

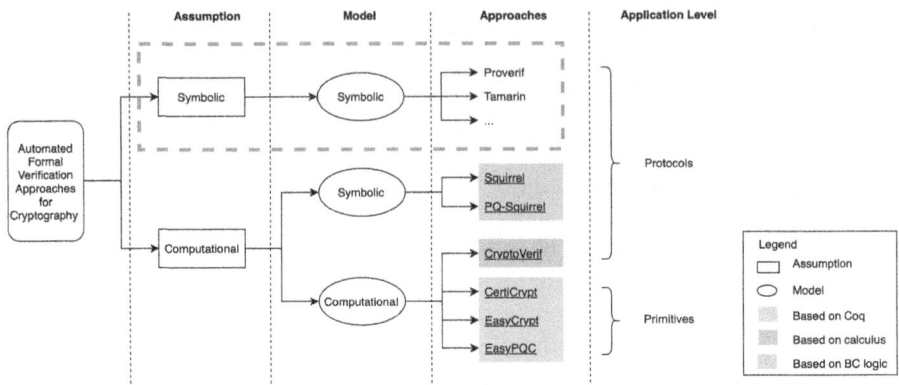

Fig. 1. Taxonomy of Formal Verification Techniques for Post-Quantum Cryptography

the results—whether the model satisfies the security property under the given assumption, with the assistance of computers. We classify the existing formal verification methods for post-quantum cryptography based on the above steps (Fig. 1). The initial classification pertains to the foundational assumptions, differentiating between symbolic and computational approaches. Directly verifying a cryptography-involved system using computers are complex. This is true even in the proving of a cryptographic primitive, e.g., encryption and signature schemes, not mentioning protocols that use these cryptographic primitives. Therefore, abstraction techniques emerge targeting verifying cryptographic protocols. Starting from the Dolev-Yao model [29], a line of research has been developed and matured on automated verification of logic flaws in cryptographic protocols. Notable tools include ProVerif [1] and Tamarin [3]. This research made the following assumptions: 1) The data in a cryptographic protocol are atomic symbols. 2) The cryptographic primitives are assumed to be perfect. 3) The attacker controls the entire network. This approach is limited to protocols only, as verifying cryptographic primitives requires to represent the data in its original form as bitstring (thus, cryptographic primitives are assumed to be perfect). And it limits the verification capability to logic flaws only.

Due to the high level of abstraction, the symbolic approach is far from the view usually adopted by cryptographers. On the contrary, the computational approach treats data as bitstrings and, therefore, is able to represent cryptographic primitives and be able to prove a different type of security properties represented as games (see Sect. 6). Note that we refrain from a detailed examination of tools based on symbolic assumptions and models, such as ProVerif and Tamarin, as these have been comprehensively surveyed and studied [11]. Furthermore, this classification predominantly focuses on classical cryptography, which does not align with the core objectives of our research.

Under the computational assumption, the second classification pertains to the underlying models that describe the behaviour of the cryptography. The first class inherits the symbolic models but is able to prove computational properties.

Notable approaches are PQ-Squirrel [28], based on the work Squirrel [2]. The other class represents the behaviour in computational models.

Subsequently, the computational model category has two categories depending on their verification approaches: CryptoVerif [22] and the EasyPQC [12]. EasyPQC is based on previous works CertiCrypt [18] and EasyCrypt [14].

Finally, from the application perspective, we distinguish the approaches into two categories: one focuses on verifying cryptographic primitives; the other on the protocols, as they adopt different adversary and abstraction levels. Before detailing the verification techniques, we introduce some basic concepts.

6   Game-Based Cryptography Verification

Proving security of cryptographic primitives can be tedious work that tends to be error-prone and difficult to read, due to the non-trivial mathematics that are involved such as number theory, group theory and probability theory. To reduce the proof complexity, a game-based approach is proposed in 2004 [49]. The game is played between an *adversary* and some benign entity called the *challenger*, where the adversary and challenger are probabilistic processes that communicate with each other. Security, in this setting, is defined as some particular events S occurring (e.g., adversary guessing a bit) is bounded by a target probability that is normally very small [49], meaning that the adversary does not gain any advantage in guessing the secret message.

The proof follows a refinement approach of games, sketched as follows: Security of cryptography is based on the hard maths problems that are difficult to solve. If a problem is hard enough, it would take too long for an attacker with even quantum computers to solve it. Therefore, proving the security of a cryptographic algorithm boils down to proving its equivalence to the underlying hard maths problems. Therefore, to prove the security property, the game-based approach creates a sequence of games, denoted as G_0, G_1, ..., G_n. The G_0 is the original game describing the target cryptographic primitive that needs to be proved. In the sequence, a successive game only modifies a small detail or a little step of the previous game. Normally there are three types of transitions between the successive games: 1) transitions based on indistinguishability where the changes are indistinguishable by an efficient distinguisher (the adversary); 2) transitions based on failure event where the two games are identically unless a certain "failure event" occurs and thus probability is bounded by the probability of the "failure event"; and 3) bridging steps which aim to prepare the ground for a transition of one of the above two types.

This refinement of the game-based approach is less error-prone and more easily verifiable, even mechanically verifiable [21]. Following the direction of mechanical verifiability, Bellare and Rogaway deemed games and adversaries as programs and proposed a programming language to represent a game [20]. Halevi took a step further and proposed the idea of creating a computer-aided tool for generating proofs to reduce human errors and increase proof readability [33]. The basic idea is that a game is automatically generated (by the tool) that consists of a main loop where each iteration calls an adversary routine, supplying it with the results of the last iteration and getting back the results of the current iteration. The output is typically the adversary output of the last iteration. To prove the security of the game, the tool generates a sequence of the above games, each time changing some aspects of the current game where the changing is within a list of permissible transformations. The proof proceeds until they are reduced to the empty game, where nothing is left in the code to analyze.

Based on the above idea, David proposed a refinement framework for generating proofs using the proof assistant Coq. Corin and Hartog extended the probabilistic Hoare Logic with functions to represent attackers with arbitrary behaviour and orthogonality that allows them to reason about the game transformation [26]. Courant et al. [27] proposed an automated procedure based on Hoare Logic, dedicated to analyzing asymmetric encryption schemes. Later, these methods were adopted in [31,32] to verify symmetric encryption modes and message authentication codes. Based on these game-based approaches[1], formal verification of post-quantum cryptography is developed, detailed in Sect. 7.

7 Verification of Cryptographic Primitives

In this section, we discuss the verification techniques focusing on proving post-quantum cryptographic primitives—EasyPQC. To do so, we first introduce the two prior works—EasyCrypt and CertiCrypt, on which EasyPQC is based.

7.1 CertiCrypt

Following the game-based approach, CertiCrypt is proposed, aiming to build an automated framework designed to facilitate the verification of (classical) cryptography utilizing the Coq proof assistant. Compared to other works, CertiCrypt has improved support for proof automation and wide applications e.g., being able to handle random oracles, security assumptions such as Diffie-Hellman hardness assumption, and probabilistic polynomial time complexity [18]. It adopts the code-based approach, meaning that the security goals and hardness problems are modelled as probabilistic programs with unspecified adversary code, uses tools issued from program verification and programming language theory to rigorously check cryptographic reasoning.

To verify a cryptographic primitive, the analysts need to rigorously specify the following components: a model representing the cryptography algorithm, an adversary model, and the security property. Given these information, CertiCrypt developed verification techniques to automate the proofs. We detail the above components as follows:

---

[1] Formal verification approaches exist that are not game based, for instance, a type system that tracks whether values are uniform and fresh, or adversarial controlled, is proposed and used for classical cryptographic primitive verification [36,41].

Modeling. To specify the game/code, CertiCrypt proposed an imperative programming language pWHILE with probabilistic assignments, structured datatypes, and procedure calls [18]. It is constructed upon the WHILE-programming language, structured to encompass functionalities of assignments, if-then-else statements, and while loops [50]. In addition, CertiCrypt ensures well-typed expressions and commands and allows user-defined types, operators, as well as general types, including booleans, bitstrings, natural numbers, pairs, lists, and elements in a group. For the game formalization, CertiCrypt additionally offers a definition that maps a procedure in games to commands that are consistent with its parameters, body and return expression.

Adversary Model. CertiCrypt assumes a common adversary whose computational complexity is bounded by a polynomial function. This is achieved by specifying variables and procedures that are accessible to adversaries. The adversary can call oracles, but any other procedures it calls must follow a given set of rules that guarantee that each time the adversary reads and writes a variable, the adversary has the required permission. Additional constraints may be imposed on adversaries and are formalized using lists that record the oracle calls and verify that the calls are legitimate.

Security Property. The main difference between CertiCrypt and other tools is that CertiCrypt provides exact security rather than showing asymptotically negligible advantages for any effective adversary against the security of a cryptographic system [18]. The exact security is to provide a concrete lower and upper bound for the advantage of an adversary execution, where the lower bound is the probability that the adversary successfully breaks the hard problem and the upper bound is the execution time [18].

Verification. CertiCrypt uses a code-based technique that relies on programming theory to justify the process of proof. It offers concrete tools to reason about the equivalence of probabilistic programs involving relational Hoare logic, observational equivalence theory, reasoning based on sequences of games technique, and verified program transformation [18].

CertiCrypt develops a Probabilistic Relational Hoare Logic (pRHL) to reason about the equivalence of programs, by extending the relational Hoare logic (RHL) [18]. Both RHL and pRHL are extensions of the classical Hoare logic; the main difference is that RHL is used for deterministic programs while pRHL is used for probabilistic programs. In detail, in RHL, a program fragment starts in a state satisfying the precondition and will terminate in a state satisfying the postcondition; while in pRHL, the logic additionally deals with assertions about probabilistic properties, for example, the probability of a certain event happening with a certain value. The reasoning of pRHL relies on *judgments* in the form of $c_1 \sim c_2 : \Phi \Rightarrow \Psi$, where c_1 and c_2 are probabilistic programs, and both Φ and Ψ are first-order relational assertions where Φ is the pre-condition and Ψ is the post-condition [14,38].

In this context, games G_1 and G_2 are equivalent w.r.t. pre-condition Ψ and post-condition Φ iff for any initial memories m_1 and m_2 satisfying the pre-condition $m_1\Psi m_2$, if the evaluations of G_1 in m_1 and G_2 in m_2 terminate with final memories m_1' and m_2' respectively, and $m_1'\Phi m_2'$ holds, which is formalised as the following judgment:

$$\models G_1 \sim G_2 : \Psi \Rightarrow \Phi \stackrel{\text{def}}{=} \forall m_1 m_2.\ m_1\Psi m_2 \Rightarrow (G_1)m_1 \sim_\Phi (G_2)m_2.$$

CertiCrypt developed a set of derived rules to reason about whether the above equivalence can be achieved. It formalised a theory of observational equivalence which is an instance of the above judgment where pre and post-conditions are limited to relations based on equality over a subset of variables. All the derived rules can be specialized to the case of observational equivalence. CertiCrypt implements a calculus of variable dependencies and two functions, such that given a command and a set of output (input) variables, it computes a set of input (output) variables such that the two games are observational equivalent. In addition, CertiCrypt provides a set of tactics and algebraic equivalences to automate the bridging steps.

Summary. CertiCrypt is a fully formalized framework dedicated to cryptographic primitives, including OAEP [17], FDH [55], and zero-knowledge protocols [19]. Its limitations are twofold: it can only be used for classical cryptography; creating machine-checked proofs is time-consuming and demands a significant level of proficiency in formal proof.

7.2   EasyCrypt

EasyCrypt is a dedicated tool for cryptographic primitives, aiming at reducing the effort and expertise required for adopting formal verification techniques by providing enhanced automation, addressing the limitation of CertiCrypt that is not user-friendly. Instead of generating high-guarantee proofs, EasyCrypt builds machine-checked proofs from proof sketches and offers a machine-processable representation of the essence of a security proof [16].

EasyCrypt is built based on the same principle as CertiCrypt; thus, they share the same adversary model and support the same security properties. Hence we omit these two components and focus on the modeling and verification.

Modeling. Compared to CetiCrypt, EasyCrypt uses functional language that mainly has two types of declarations: basic declarations and game declarations [14]. EasyCrypt can declare types, constants, and operators as its basic declarations, for example, some basic types: unit, bool, int, real, bitstring, list, and finite map. The game declarations are used to model the games during the security proof involving probabilistic statements such as while, if, function definition with keyword *fun*, adversary declaration with keyword *adversary*, and game definition with keyword *game* [14]. Typically, a game is defined as a module (module) with some procedures (proc) in EasyCrypt. In addition, EasyCrypt inherits some language properties from Coq, for example, it can use tactics.

Verification. Similar to CertiCrypt, EasyCrypt supports interactive construction in the form of games where the theoretical foundation for game transitions is probabilistic Relational Hoare Logic (pRHL) and pRHL judgments. Unlike CertiCrypt, EasyCrypt separates the program verification and information-theoretic—while the pRHL is used for logical relations connection between games, pRHL judgements are applied for information-theoretic reasoning of the events [16].

EasyCrypt is more effective compared to CertiCrypt, in that EasyCrypt adopts first-order logic to represent the verification conditions—sufficient and valid conditions for a pRHL judgement, and implements an automated procedure that computes the verification conditions. Since the logic in EasyCrypt does not involve probability, general-purpose theorem provers, for example, the SMT (Satisfiability Modulo Theories) solver, can easily be adopted for generating the proofs. In contrast, CertiCrypt sometimes needs probability-involving proofs and thus is less efficient than EasyCrypt.

Compared to CertiCrypt, EasyCrypt is more user-friendly. Since it is set up based on CertiCrypt, and the verifiable proof sketches are compiled into the CertiCrypt framework for automatically verification by provers such as SMT solver [16], EasyCrypt inherits the modelling flexibility of CertiCrypt, but its adoption of the general-purpose assistants improves the proofs' readability [16].

Summary. EasyCrypt supports various cryptographic primitives including public-key encryption schemes, block cipher modes of operation, digital signature schemes, and hash function designs [14]. However, most of them are limited to cryptographic primitives. It is not intuitive to implement cryptographic protocols using the EasyCrypt because cryptographic protocols are more complex and consider different adversary. Proving the security of a protocol involves not only proving the security of individual primitives but also considering their interactions and the composition of these primitives. And analyzing these interactions may require a higher level of abstraction and modeling.

7.3   EasyPQC

EasyPQC is an extension of the EasyCrypt that supports post-quantum cryptography proofs [12]. The main challenge is the ability to reason about quantum adversary that can simultaneously query the quantum random oracle with the constraint that queries cannot be retried [12]. EasyPQC addresses the challenge by extending relational logic from pRHL to qpRHL. The theoretical foundation that empowers EasyPQC is the QROM [23].

QROM. A random-oracle model (ROM) plays an important role in the analysis of cryptographic primitives, providing random values in response to queries of an adversary. A ROM is a 'black box' that hides secret information from the adversary, simulating functions that the adversary can call but knows nothing of the internal information. For example, an adversary can query a hash of a message from ROM without knowing the original message. The QROM is a quantum ROM that represents the adversary that is able to call quantum programs/functions. Differing from classical programs, quantum programs allow multiple states (represented as quantum bits) to exist simultaneously. Thus, for a given query, QROM provides the distribution of the final simultaneous multiple states of the quantum program, instead of a single state.

Modeling. The EasyPQC defines quantum procedure calls enabling the adversary query the QROM, indicated by the keyword *quantum*. In addition, EasyPQC allows defining quantum variables by providing the commands of quantum initialization (assigning a classical value to quantum variables), unitary transformation (quantum operations) and quantum measurement (assigning quantum values to classical variables).

Verification. EasyPQC proposes a moderate variant of EasyCrypt by modifying the theoretical game transition techniques from pRHL to post-quantum relational Hoare logic (pqRHL). It proves that pqRHL is sound for reasoning about quantum adversaries; and in consequence, EasyPQC is sound for post-quantum cryptography (PQC) security proofs [12].

The key technique is that the pqRHL enables probabilistic reasoning of quantum adversaries. To this end, pqRHL first extends the relational equality assertions by defining a global equality operation to represent the quantum assertions. The classical assertions cannot be applied because the state of a quantum adversary is non-deterministic and multiple states exist simultaneously [12]. Subsequently, EasyPQC defines two extra rules for reasoning of adversary that incorporates the quantum assertions, and defines rules to support reasoning of the above quantum commands. Other than the above, the rest of the rules are the same as in pRHL; and EasyPQC can adopt the existing proof system of pRHL.

Note that in the proving process, EasyPQC faces another challenge due to the quantum setting. The execution path of the probabilistic/quantum programs is non-deterministic. Consequently, the final output state after a conditional operation (if-then-else) is a mixture of every possible output from each branch. In other words, the program doesn't produce a single deterministic output but rather a mixture of possible outcomes. To ensure that the mixture of outputs also satisfies the preconditions, a set of side conditions named the CM conditions is needed. In the probabilistic setting (classical cryptography), the CM can be checked for each relational proof, however, this is impossible in the quantum setting due to the simultaneously existing states. To address this challenge, EasyPQC develops a theory proving that it is sufficient to additionally checking the satisfaction of CM at the final post-condition. This enables the reuse of the existing proof systems of pRHL by exempting the CM checking in each relational proof and only checking CM for the final post-condition.

Summary. Evidenced by the full domain hash signature example, EasyPQC can be used to prove the security of post-quantum cryptography against quantum adversaries [23]. However, only three cryptographic schemes are verified by lifting the classical setting to post-quantum setting, they are PRF-based MAC, Full Domain Hash and GPV08 identity-based encryption [12]. There is no guarantee that other cryptography can be verified in a post-quantum setting. In addition, there is no guarantee that EasyPQC can achieve the formal verification of cryptographic protocols since all EasyPQC's case studies are limited to cryptographic primitives [28].

8   Verification of Cryptographic Protocols

This section introduces the verification techniques for post-quantum cryptographic protocols—PQ-Squirrel. Since it is based on the BC logic and is an extension of the Squirrel verifier, we introduce the BC logic and Squirrel first.

8.1   BC Logic

Bana-Comon (BC) logic is a formal system designed to analyze and reason about (classical) cryptographic protocols [10]. Differing from the mainstream protocol verification techniques that make symbolic assumptions, such as ProVerif and Tamarin, BC logic is built upon a computational model. Also, BC logic focuses on capturing the interactions and computations of the protocol participants as well as the adversary.

Modeling. BC logic provides a way to express properties and assertions about the behaviour and security of protocols. It extends the standard first-order logic with additional constructs and operators specific to protocol analysis.

Adversary Model. BC logic introduces the notion of a "symbolic attacker" who is capable of performing symbolic computations and reasoning. The attacker is Turing-complete. This enables the analysis of cryptographic protocols under a wide range of computational scenarios and provides a stronger notion of security than previous models.

Verification. BC logic incorporates concepts from computational complexity theory, such as polynomial-time reductions and NP-complete problems, to reason about the computational aspects of protocol security. It provides a framework to formally specify equivalence properties of protocols, such as strong equivalence and trace equivalence. Strong equivalence relates to the indistinguishability of protocol executions from an attacker's perspective, while trace equivalence considers the equivalence of protocol traces under various attacks.

8.2   Squirrel

Squirrel [8] is an interactive and semi-automatic prover whose primary goal is to assist in the formal verification of (classical) cryptographic protocols. Built upon the BC logic, it adopts a symbolic model with a computational security guarantee, which allows for reasoning about the protocol's behaviour in a step-by-step manner.

Modeling. Squirrel is based on the pi-calculus and first-order logic. The symbolic Squirrel model captures the behaviour of the protocol by considering the messages exchanged between participants and their computational capabilities.

Adversary Model. Squirrel takes into account the presence of an adversary who can manipulate the messages to break the security properties of the protocol.

Verification. The prover employs an interactive approach, where the user and the prover engage in a dialogue to verify the protocol. The user provides the prover with high-level properties that the protocol should satisfy, typically expressed in a formal logic or specification language. The prover then employs various developed techniques and algorithms to reason about the protocol's execution and attempts to prove or disprove the given properties.

Summary. Squirrel provides a rich set of built-in cryptographic constructs and primitives that can be used in the protocol specification, and supports verification of various cryptographic primitives and protocols, including encryption, key exchange, digital signatures, secure multi-party computation.

8.3   PQ-Squirrel

Classical proof techniques in Squirrel do not carry over to quantum case, as they are incapable of addressing quantum adversaries, because quantum information cannot be copied and measurements destroy information [28]. PQ-Squirrel extends the Squirrel tool with features that simplify protocol specification and verification under post-quantum security assumptions.

Modeling. It simplifies the process of specifying protocols by automatically generating attacker terms from input and output commands, assuming the existence of a single attacker. This design choice aligns well with post-quantum requirements and prevents users from inadvertently modelling a weaker threat model with multiple disjoint attackers.

Verification. PQ-Squirrel offers two verification modes: classic mode and post-quantum mode. In the post-quantum mode, PQ-Squirrel restricts the use of tactics and axioms to those that have been proven to be post-quantum sound. This is achieved by performing synchronization checks for every indistinguishability appearing in a proof, ensuring that the specified side conditions are met. The main idea is to identify a minimal set of syntactic conditions, resulting in a concise extension comprising only a few hundred lines of additional code. The verification in PQ-Squirrel depends on the PQ-BC logic, which is an extension of the BC logic, as detailed below.

PQ-BC Logic. It was observed that the original proofs in BC can directly apply to the post-quantum setting if there exists an instantiation of the assumption that satisfies the requirement against a quantum attacker. Based on this insight, PQ-BC extends the BC logic by considering the post-quantum soundness of BC rules with respect to cryptographic assumptions.

The cryptographic assumptions supported in PQ-BC include PRF (Pseudorandom Function), IND-CCA (Indistinguishability under Chosen Ciphertext Attack), EUF-CMA (Existential Unforgeability under Chosen Message Attack), ENC-KP (Encryption Key Privacy), INT-CTXT (Integrity of Ciphertext), and

OTP (One-Time Pad). To ensure post-quantum security, these assumptions must be instantiated in a manner that is secure against post-quantum attackers. However, the instantiation of the DDH (Decisional Diffie-Hellman) assumption, which is known to be secure against classical attackers, currently lacks a post-quantum secure instantiation and is therefore excluded from the list of allowed cryptographic assumptions in PQ-BC.

9   Potential Alternatives

In addition to the above dedicated approaches for post-quantum cryptography, we observed an opportunity that proofs in the classical setting may be "lifted" as the proofs in the quantum setting. The researchers in [51] proposed a general framework where quantum security proofs are decomposed into a series of classical security reductions. The study defines the sufficient conditions under which classical reductions can be "lifted" into a quantum environment, including the equivalence of games, the preservation of reduction properties, the structure of linear reductions, and assumptions about particular classes of machines must be met. For instance, the case studies of EasyPQC satisfy these conditions, thus being verified straightforwardly in EasyPQC. Once proved satisfying these sufficient conditions, the various formal verification approaches for classical cryptographic primitives discussed previously can be applied to prove post-quantum cryptography. It also provides opportunities for proving cryptographic protocols. Since CryptoVerif applies to classical protocol proofs, proving that protocols satisfy the above conditions allows them to be "lifted" to the post-quantum environment by re-using the classical CryptoVerif proofs.

9.1   CryptoVerif

Modeling. The games in CryptoVerif are presented in a process calculus [22], which is developed based on the pi calculus. Messages are represented as bitstrings, and cryptographic primitives are functions that transform bitstrings into other bitstrings. It extends the pi-calculus with a probabilistic choice operator to support the modelling of random variables and defines probabilistic semantics. Indistinguishability is the primary method used to establish security features.

Verification. CryptoVerif makes use of a collection of game transformations.

- **Game Initialization**: Cryptoverif typically begins by defining an initial game, often denoted as G_0, which represents the starting point of the protocol analysis. This may involve setting up the protocol's initial state and defining the security properties to be verified.
- **Game Execution** The protocol's progress is modelled as transitions from one game state to another, symbolized by $\rightarrow$. These transitions capture protocol actions and adversarial moves, demonstrating the evolving nature of the analysis: $G_i \xrightarrow{\text{protocol action/adversarial move}} G_{i+1}$.

– **Game Termination Condition** CryptoVerif defined conditions for game termination e.g., restricting sessions.

An essential kind of transformation takes use of the security assumptions made on cryptographic primitives via the use of observational equivalence. Other game transformations are called syntactic transformations, which are used 1) to simplify the game that is acquired after applying an assumption to a cryptographic primitive or 2) to enable the application of an assumption on a cryptographic primitive. These presumptions are defined in the CryptoVerif library. Users are also permitted to change the libraries in order to incorporate cryptographic primitives that are absent from the default library.

These game transformations are structured using a proof approach: when a transformation fails, it proposes alternative transformations that should be applied in order to allow the transformation that is wanted - and this led to the automatic proof generation and automatic game generation of CryptoVerif. A key difference is that CryptoVerif is able to automatically reason about protocols that involve the generation and distribution of random values, such as public key protocols that use randomized public keys or random session keys. This enables CryptoVerif to reason about the security of protocols that rely on randomness in a rigorous and automated way.

9.2   Specific Approach for Lattice-Based Cryptography

In contrast to the above approaches that work for general cryptography, there is a notable work that proposed a symbolic logic to prove the correctness specifically for lattice-based cryptography [15]. Although not directly applicable to post-quantum cryptography, it provides a foundation that could be extended for other lattice-based cryptography for instance the three NIST post-quantum standards. This work is built based on a logic called Computational Indistinguishability Logic (CIL), which is proposed for reasoning about cryptographic primitives in computational models [13]. Game-based proofs are achieved by transforming an oracle system (modelling the initial game) to another bisimilar system until failure. Other works based on CIL exist for proving cryptographic constructions/protocols; but since they are not particularly for post-quantum cryptography, we do not detail them here.

10   Gaps, Limitations and Opportunities

In the process of the survey, we identified the following gaps and limitations of existing works with respect to the post-quantum applications, verification case studies, verification capabilities and tool usability.

Application Gaps. We have noticed that there is a gap between theories and applications. While many post-quantum protocols have been proposed in academia, the practical application and verification of these protocols remain limited. The paucity of concrete examples hinders the ability to demonstrate the security properties required to prove the effectiveness of the protocols.

Case Study Gaps. Among all the examples provided by various tools, there are not enough up-to-date examples in the NIST final list. In addition, existing theoretical frameworks (e.g., lifting theorems) lack automated application methods and there are gaps in the practical application of these theories.

Verification Capability Limitations. The previous sections have illustrated that formal verification of post-quantum cryptography is still in its infancy. There is no theoretical proof of the verification capabilities of existing approaches, nor sufficient case studies to draw a line in the sand. There are limited real-world post-quantum cryptographic applications that are rigorously tested and verified against the desired security properties.

Usability. Stemming from the tools' intricate design and complex configuration, they are difficult for researchers to navigate. In addition, limited customization options and platform support limit the adaptability of these tools to different research needs and technical environments. The different program languages used by the tools can also make them less efficient for researchers unfamiliar with them. For instance, EasyCrypt uses Coq and CryptoVerif uses applied pi-calculus; they both have a considerable learning curve.

Opportunities. Based on our observation and analysis, the following directions are promising to enhance the post-quantum cryptography and its verification.

- As stated in Sect. 3, existing surveys are insufficient in the related topics of quantum computing, post-quantum cryptography and cryptanalysis of post-quantum cryptography. This survey only addresses the gaps in formal verification of post-quantum cryptography. Comprehensive surveys on the remaining topics help accelerate the adoption of more secure solutions withstanding in the quantum are.
- Introducing and enhancing cryptographic tools by providing researchers and practitioners with more robust resources for analysis and implementation, including making them user-friendly by 1) developing fully automated verification algorithms and providing better tutorials; 2) clarifying their capabilities theoretically or in practice by introducing more case studies.
- As stated in the previous section if there exists a systematic approach to guide users through the validation of conditions specified in [51], the need for introducing a new tool becomes less imperative, thereby offering a pathway to simplify the entire workflow.
- Encouraging and supporting the real-world applications of post-quantum cryptography, fostering their integration into practical systems to validate their effectiveness beyond theoretical frameworks.

References

1. Proverif. https://bblanche.gitlabpages.inria.fr/proverif/
2. The squirrel prover. https://github.com/squirrel-prover/squirrel-prover/
3. Tamarin prover. https://tamarin-prover.com/
4. Abinaya, M., Prabakeran, S.: Lightweight block cipher for resource constrained IoT environment-an survey, performance, cryptanalysis and research challenges. In: ICICNIS, pp. 347–365 (2022)
5. Abohashima, Z., Elhosen, M., Houssein, E.H., Mohamed, W.M.: Classification with quantum machine learning: a survey. arXiv preprint arXiv:2006.12270 (2020)
6. Alnahawi, N., Müller, J., Oupický, J., Wiesmaier, A.: SoK: post-quantum TLS Handshake. Cryptology ePrint Archive (2023)
7. Avalle, M., Pironti, A., Sisto, R.: Formal verification of security protocol implementations: a survey. Formal Aspects Comput. **26**, 99–123 (2014)
8. Baelde, D., Delaune, S., Jacomme, C., Koutsos, A., Moreau, S.: An interactive prover for protocol verification in the computational model. In: S&P (2021)
9. Bagane, P.A., Kotrappa, S.: Bibliometric survey for cryptanalysis of block ciphers towards cyber security. Library Philosophy and Practice, pp. 1–18 (2020)
10. Bana, G., Comon-Lundh, H.: A computationally complete symbolic attacker for equivalence properties. In: CCS, pp. 609–620 (2014)
11. Barbosa, M., et al.: SOK: computer-aided cryptography. In: S&P (2021)
12. Barbosa, M., et al.: EasyPQC: verifying post-quantum cryptography. In: CCS, pp. 2564–2586 (2021)
13. Barthe, G., Daubignard, M., Kapron, B., Lakhnech, Y.: Computational indistinguishability logic. In: CCS, pp. 375–386 (2010)
14. Barthe, G., Dupressoir, F., Grégoire, B., Kunz, C., Schmidt, B., Strub, P.-Y.: EasyCrypt: a tutorial. In: Aldini, A., Lopez, J., Martinelli, F. (eds.) FOSAD 2012-2013. LNCS, vol. 8604, pp. 146–166. Springer, Cham (2014). https://doi.org/10.1007/978-3-319-10082-1_6
15. Barthe, G., Fan, X., Gancher, J., Grégoire, B., Jacomme, C., Shi, E.: Symbolic proofs for lattice-based cryptography. In: CCS, pp. 538–555 (2018)
16. Barthe, G., Grégoire, B., Heraud, S., Béguelin, S.Z.: Computer-aided security proofs for the working cryptographer. In: Annual Cryptology Conference (2011)
17. Barthe, G., Grégoire, B., Lakhnech, Y., Zanella Béguelin, S.: Beyond provable security verifiable IND-CCA security of OAEP. In: Cryptographers' Track at the RSA Conference, pp. 180–196 (2011)
18. Barthe, G., Grégoire, B., Zanella Béguelin, S.: Formal certification of code-based cryptographic proofs. In: POPL, pp. 90–101 (2009)
19. Barthe, G., Hedin, D., Béguelin, S.Z., Grégoire, B., Heraud, S.: A machine-checked formalization of sigma-protocols. In: CSF, pp. 246–260 (2010)
20. Bellare, M., Rogaway, P.: Code-based game-playing proofs and the security of triple encryption. Cryptology ePrint Archive, Paper 2004/331 (2004)
21. Bellare, M., Rogaway, P.: The security of triple encryption and a framework for code-based game-playing proofs. In: EUROCRYPT 2006, pp. 409–426 (2006)
22. Blanchet, B.: CryptoVerif: a computationally-sound security protocol verifier. Technical Report (2017)
23. Boneh, D., Dagdelen, Ö., Fischlin, M., Lehmann, A., Schaffner, C., Zhandry, M.: Random oracles in a quantum world. In: ASIACRYPT, pp. 41–69 (2011)
24. Chareton, C., Bardin, S., Lee, D., Valiron, B., Vilmart, R., Xu, Z.: Formal methods for quantum programs: a survey. arXiv preprint arXiv:2109.06493 (2021)

25. Ciulei, A.T., Crețu, M.C., Simion, E.: Preparation for post-quantum era: a survey about blockchain schemes from a post-quantum perspective. Cryptology ePrint Archive (2022)
26. Corin, R., den Hartog, J.: A probabilistic hoare-style logic for game-based cryptographic proofs (extended version). Cryptology ePrint Archive, Paper 2005/467 (2005). https://eprint.iacr.org/2005/467
27. Courant, J., Daubignard, M., Ene, C., Lafourcade, P., Lakhnech, Y.: Towards automated proofs for asymmetric encryption schemes in the random oracle model. In: CCS, pp. 371–380 (2008)
28. Cremers, C., Fontaine, C., Jacomme, C.: A logic and an interactive prover for the computational post-quantum security of protocols. In: S&P, pp. 125–141 (2022)
29. Dolev, D., Yao, A.: On the security of public key protocols. IEEE Trans. Inf. Theory **29**(2), 198–208 (1983)
30. Fernandez-Carames, T.M., Fraga-Lamas, P.: Towards post-quantum blockchain: a review on blockchain cryptography resistant to quantum computing attacks. IEEE Access **8**, 21091–21116 (2020)
31. Gagné, M., Lafourcade, P., Lakhnech, Y.: Automated security proofs for almost-universal hash for MAC verification. In: ESORICS, vol. 8134, pp. 291–308 (2013)
32. Gagné, M., Lafourcade, P., Lakhnech, Y., Safavi-Naini, R.: Automated security proof for symmetric encryption modes. In: Annual Asian Computing Science Conference, vol. 5913, pp. 39–53 (2009)
33. Halevi, S.: A plausible approach to computer-aided cryptographic proofs. IACR Cryptol. ePrint Arch. **2005**, 181 (2005)
34. Hasija, T., Ramkumar, K., Kaur, A., Mittal, S., Singh, B.: A survey on NIST selected third round candidates for post quantum cryptography. In: ICCES (2022)
35. Herman, D., et al.: A survey of quantum computing for finance. arXiv:2201.02773 (2022)
36. Hoang, V.T., Katz, J., Malozemoff, A.J.: Automated analysis and synthesis of authenticated encryption schemes. IACR Cryptol. ePrint Arch, p. 624 (2015). http://eprint.iacr.org/2015/624
37. Hofer-Schmitz, K., Stojanović, B.: Towards formal verification of IoT protocols: a review. Comput. Netw. **174**, 107233 (2020)
38. Kfoury, A.: Hoare logic and variations: probabilistic, relational, probabilistic+ relational (2018). https://www.cs.bu.edu/faculty/kfoury/UNI-Teaching/CS512/AK_Documents/Hoare_Logic/main-post.pdf
39. Kumari, S., Singh, M., Singh, R., Tewari, H.: Post-quantum cryptography techniques for secure communication in resource-constrained internet of things devices: a comprehensive survey. Softw. Pract. Experience **52**(10), 2047–2076 (2022)
40. Liu, J., Liu, Z.: A survey on security verification of blockchain smart contracts. IEEE Access **7**, 77894–77904 (2019)
41. Malozemoff, A.J., Katz, J., Green, M.D.: Automated analysis and synthesis of block-cipher modes of operation. IACR Cryptol. ePrint Arch, p. 774 (2014)
42. Mnkash, S.H.: Survey of different cryptography methods. Resmilitaris **12**(2), 495–516 (2022)
43. NIST: Round 4 submissions - post-quantum cryptography: CSRC. https://csrc.nist.gov/Projects/post-quantum-cryptography/round-4-submissions. Accessed 20 Oct 2023
44. NIST: selected algorithms 2022 - post-quantum cryptography: CSRC. https://csrc.nist.gov/Projects/post-quantum-cryptography/selected-algorithms-2022. Accessed 20 Oct 2023

45. Parida, N.K., Jatoth, C., Reddy, V.D., Hussain, M.M., Faizi, J.: Post-quantum distributed ledger technology: a systematic survey. Sci. Rep. **13**(1), 20729 (2023)
46. Ramezani, S.B., Sommers, A., Manchukonda, H.K., Rahimi, S., Amirlatifi, A.: Machine learning algorithms in quantum computing: a survey. In: IJCNN (2020)
47. Shannon, K., Towe, E., Tonguz, O.K.: On the use of quantum entanglement in secure communications: a survey. arXiv preprint arXiv:2003.07907 (2020)
48. Shim, K.A.: A survey of public-key cryptographic primitives in wireless sensor networks. IEEE Commun. Surv. Tutorials **18**(1), 577–601 (2015)
49. Shoup, V.: Sequences of games: a tool for taming complexity in security proofs. IACR Cryptol. ePrint Arch, p. 332 (2004)
50. Sieber, K.: The foundations of program verification (2013)
51. Song, F.: A note on quantum security for post-quantum cryptography. In: Post-Quantum Cryptography, pp. 246–265 (2014)
52. Sosnowski, M., et al.: The performance of post-quantum TLS 1.3. In: CoNEXT (2023)
53. Tan, T.G., Szalachowski, P., Zhou, J.: Challenges of post-quantum digital signing in real-world applications: a survey. Int. J. Inf. Secur. **21**(4), 937–952 (2022)
54. Wang, A., Xiao, D., Yu, Y.: Lattice-based cryptosystems in standardisation processes: a survey. IET Inf. Secur. **17**(2), 227–243 (2023)
55. Zanella-Béguelin, S., Barthe, G., Grégoire, B., Olmedo, F.: Formally certifying the security of digital signature schemes. In: S&P, pp. 237–250 (2009)
56. Zeydan, E., Turk, Y., Aksoy, B., Ozturk, S.B.: Recent advances in post-quantum cryptography for networks: a survey. In: MobiSecServ, pp. 1–8 (2022)

Program Analysis

AutoWeb: Automatically Inferring Web Framework Semantics via Configuration Mutation

Haining Meng[1,2], Haofeng Li[1], Jie Lu[1], Chenghang Shi[1,2], Liqing Cao[1,2],
Lian Li[1,2,3(✉)], and Lin Gao[4]

[1] SKLP, Institute of Computing Technology, CAS, Beijing, China
{menghaining,lihaofeng,lujie,shichenghang21s,caoliqing19s,
lianli}@ict.ac.cn
[2] University of Chinese Academy of Sciences, Beijing, China
[3] Zhongguancun Laboratory, Beijing, China
[4] TianqiSoft Inc, Beijing, China
gaolin@tianqisoft.cn

Abstract. Web frameworks play an important role in modern web applications, providing a wide range of configurations to streamline the development process. However, the intricate semantics, facilitated by framework configurations, present substantial challenges when conducting static analyses on web applications. To mitigate this issue, existing approaches resort to manually modeling framework semantics for static analysis tools. Unfortunately, these manual works are both time-consuming and error-prone, especially considering the vast array of web frameworks and their frequent updates.

In this paper, we present the first automated method for inferring web framework semantics. Our innovative approach can automatically deduce framework specifications by mutating configurations. We have developed a prototype called AutoWeb and performed extensive experiments on three popular Java web frameworks. The empirical results show that AutoWeb is comparable to these manual approaches in terms of precision, with a false negative rate of 8.2% and no false positives.

Keywords: static analysis · framework modeling · web framework · Java

1 Introduction

Modern web applications are commonly built on top of web frameworks. Those frameworks (e.g., Spring [25] and Spring Boot [24]) offer high-level abstractions for common web tasks, thereby greatly simplifying the development process. For instance, many frameworks provide concepts such as *controllers* in the popular model-view-controller (MVC) design pattern: developers can simply declare handler methods for a particular URL without knowing the intricate request dispatching mechanism, and the framework is responsible for processing incoming requests and dispatch them to corresponding handlers.

G. Bai et al. (Eds.): ICECCS 2024, LNCS 14784, pp. 369–389, 2025.
https://doi.org/10.1007/978-3-031-66456-4_20

To effectively analyze web applications, it is crucial to precisely model the semantics of their underlying frameworks, representing the possible frameworks behaviors at runtime. Otherwise, many existing static analyses become inapplicable. For instance, web applications are driven by frameworks to interact with user requests, and there is no `main` method to start with. Furthermore, existing static analyses often fail to capture the dynamically introduced points-to relations and call relations that arise from dependency injection and dynamic dispatching mechanisms within frameworks, resulting in unsound and imprecise results. Unfortunately, frameworks are notoriously hard to analyze statically. They often employ hard-to-analyze dynamic patterns (e.g. reflection) to interact with application code, and their concrete semantics are customized by the application via configuration files or annotations. It is a daunting task, if not impossible, to automatically analyze frameworks with good precision.

In practice, researchers resort to manually modeling framework semantics to analyze web applications. Framework features were hard-coded in the analysis implementation [5,21,29]. In addition, researchers have proposed several more general solutions to specify framework semantics effectively modeling behaviors under given configurations [1,26]. For instance, IBM's F4F [26] defined the Web Application Framework Language (WAFL) to express framework-related behaviors for specific web applications, where WAFL specifications are generated by hand-crafted generators (one for each framework). JackEE [1] declares framework-related behaviors using Datalog rules, effectively mapping framework configurations to static relations, which can be processed by the Doop [4] analysis engine. Nevertheless, those existing approaches still need manual efforts and can be labor-intensive and error-prone, particularly when frameworks undergo frequent updates.

This paper presents an automatic method for inferring web framework semantics. To the best of our knowledge, this is the first approach of such an attempt. Since it is generally infeasible to directly obtain the concrete semantics by analyzing the complex implementation details of web frameworks, we do not consider various framework-specific concepts such as *filters* and *controllers*. Instead, we focus on the relations that affect the application code, which are framework-introduced but commonly required by static analyses, i.e., entry points, points-to relations, and call relations[1]. Previous work [27,30] has also shown that the above relations are crucial for the static analysis of web applications. Web frameworks provide developers with hundreds of configuration parameters, allowing for the customization of applications that exhibit rich semantics. Our goal is to automatically deduce the semantics under given configurations, i.e., demystifying how entry methods and call/points-to relations are introduced by frameworks using particular configuration parameters. The framework semantics are abstracted as mappings from configuration parameter sets to relation types, referred to as *specification* in this paper. Specifications are framework-related, yet application-independent, and can be applied to specific applications to model framework semantics.

---

[1] Techniques described in this paper are applicable to infer other user-defined relations.

Based on the observation that a framework-introduced relation is declared by the *minimal sufficient and necessary set* (MSNS), which is the set of minimum configuration parameters to trigger this relation, we propose a *mutation-based approach* to identifying the MSNS for each relation. Specifically, given an application program construct P with framework-introduced relation R, our approach first identifies the *necessary* condition of relation R by removing configuration parameters from P until R cannot be triggered at runtime. Then, new configuration parameters (mutated from identified necessary conditions) are introduced to further verify that the set of necessary configuration parameters is *sufficient* to trigger relation R.

We develop a prototype tool, AUTOWEB, to demonstrate the effectiveness of our approach. AUTOWEB observes framework-introduced relations during execution, then mutates configuration parameters to identify the MSNS for a relation. We have experimented with AUTOWEB on three popular Java web application frameworks, namely Servlet, Spring, and Apache Struts2. Experimental results demonstrated that AUTOWEB can automatically generate specifications as precise as the state-of-the-art manual approaches. To summarize, this paper makes the following contributions:

- We propose the first automated method to infer web framework semantics, by identifying the MSNS for framework-introduced relations.
- We develop AUTOWEB, utilizing a novel mutation-based approach to deduce the framework specifications automatically.
- We experimented AUTOWEB on three popular Java web frameworks, and experimental results demonstrated that the inferred specifications are comparable with hand-written specifications over precision and soundness.

The rest of the paper is organized as follows. Section 2 motivates our approach with an example, and Sect. 3 describes AUTOWEB in detail. We evaluate the tool AUTOWEB in Sect. 4. Section 5 reviews related work and Sect. 6 concludes this paper.

2   Motivation

In this section, we aim to introduce the dynamic relations facilitated by web frameworks, illustrating their impact on static analysis through an example.

2.1   Motivating Example

Figure 1 gives an example application built on top of the Spring framework. The execution flows for two URLs are illustrated in Fig. 2. The example processes two URLs: /root1/path1 and /root2/path2. URL /root1/path1 is handled by Controller1.handle1 and URL /root2/path2 is firstly processed by doFilterInternal before being handled by Controller2.handle2. In the example shown in Fig. 1, an SQL injection vulnerability exists in line

```
1  @Controller
2  public class Controller1 {
3      @Autowired
4      @Qualifier("service2")
5      ServiceInterface srv;
6
7      @GetMapping("/root1/path1")
8      public String handle1(HttpServletRequest request){
9          ...
10         data = request.getParameter("name");
11         srv.service(data);
12         ...
13     }
14 }
15
16 @Controller
17 @RequestMapping("/root2")
18 public class Controller2 {
19     @RequestMapping("/path2")
20     public String handle2(HttpServletRequest request){
21         ...
22         data = request.getParameter("name");
23         sql = "update users set hit=hit+1 where name='"+data+"'";
24         statement.executeUpdate(sql);
25         ...
26     }
27 }
28
29 public class Filter1 extends OncePerRequestFilter{
30     protected void doFilterInternal(HttpServletRequest request, ...){
31         if(validateSqlCharactor(request)) // Santitizer
32             chain.doFilter(request, response);
33         ...
34     }
35 }
36
37 @Service("service1")
38 public class ServiceImpl1 implements  ServiceInterface{
39     public String service(String name) {
40         // safe SQL operation
41         ...}
42 }
43
44 @Service("service2")
45 public class ServiceImpl2 implements ServiceInterface{
46     public String service(String name) {
47         ...
48         String sql = "select * from users where name='"+name+"'";
49         stmt.executeQuery(sql); // SQL injection
50         ...}
51 }
```

(a) Application code.

```
1  <!--web.xml configuration file-->
2  <web-app>
3      <filter>
4          <filter-name> myFilter </filter-name> <filter-class> Filter1 </filter-class>
5      </filter>
6      <filter-mapping>
7          <filter-name> myFilter </filter-name> <url-pattern> /root2/* </url-pattern>
8      </filter-mapping>
9  </web-app>
```

(b) XML configuration file. "myFilter" aliases the class "Filter1" in Figure 1a.

Fig. 1. Motivating example of a Spring-based application.

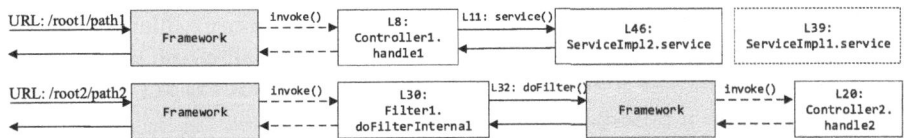

Fig. 2. Execution flow of the example in Fig. 1. Solid lines indicate direct call edges, while dotted lines denote indirect calls. The content above the solid line represents the callsite, with the line number appearing to the left of the colon.

49. This vulnerability occurs because `Controller1.handle1()`(line 8) invokes `ServiceImpl2.service()` (line 46) at line 11, which directly passes input data to a SQL query (line 49). Note that `Controller2.handle2` (line 20) also passes input data to SQL query statements (line 24). However, it is considered safe since the input request is sanitized by `Filter1` at line 31, before being processed by this handler.

The SQL injection in the above example can be detected via a classical taint analysis which computes whether the parameters of SQL queries are input data without being sanitized or not. However, without awareness of framework-introduced relations, traditional static taint analyses often fail to recognize the execution flow as in Fig. 2, resulting in ineffective analysis results.

2.2 Framework-Introduced Relations

Entry Point Relation. Static analysis including taint analysis typically process on a call graph consisting of all reachable methods from *entry points*. In stand-alone Java applications, the `main` method is considered the entry point, whereas in web applications, entry points are often the request-handling methods directly invoked by frameworks. In Fig. 1, `Controller1.handle1()` and `Filter1.doFilterInternal()` are entry points.

Entry point relations are defined by configuring methods and/or their declaring classes with Java annotations or XML configurations. For example, as shown in Fig. 1a, the annotation `@Controller` (line 1) defines the entry class, and the annotation `@GetMapping` (line 7) declares the entry method. Note that the parameter values of annotations also specify corresponding request URLs (lines 7, 17, and 19). Additionally, entry points can be declared via XML configurations as shown in Fig. 1b.

Points-to Relation. Points-to relations denote the set of heap objects referred to by a pointer variable. Frameworks can dynamically introduce points-to relations: frameworks can create objects outside the application code and inject these objects managed by frameworks into particular field references of application code. In our example, an object with type `ServiceImpl2` is managed by the framework and injected into field reference `Controller1.srv` (line 5). Such framework-introduced points-to relations cannot be computed by existing points-to analyses. Consequently, a call graph algorithm based on points-to analysis will miss the call relation from line 11 to `ServiceImpl2.service()` (line 46), resulting in false negatives. On the other hand, a CHA-based call graph construction algorithm will introduce a spurious call relation to `ServiceImpl1.service()` (line 39), resulting in false positives.

Points-to relations are specified by annotating fields and the classes of injected objects. In our example, the field `Controller1.srv` is annotated with `@Autowired` (line 3) and `@Qualifier` (line 4), indicating that the field is injected with framework-managed object `service2`. The `@Service` annotation at line 37 and 44 suggest that object `service1` and `service2` are managed by the framework, with type `ServiceImpl1` and `ServiceImpl2`, respectively. Note that the parameter value of `@Qualifier` matches with that of `@Service` (line 44), indicating that object `service2` is injected into field `Controller1.srv`.

Call Relation. Call relations can be statically computed using techniques such as class hierarchy analysis (CHA) [11] or points-to analysis [18], to establish connections between invocation sites and corresponding callee methods. Nevertheless, these analyses are unable to handle indirect call relations introduced by frameworks. In such cases, applications invoke framework APIs, which, in turn, call back into application methods, making it challenging for static analysis techniques to accurately track and resolve these dynamic interactions.

Indirect call relations can be specified in various ways. In the example shown in Fig. 1a, the method `Controller2.handle2()` at line 20 is indirectly called by `chain.doFilter()` at line 32 because it handles the URL `/root2/path2` which matches with the URL `/root2/*` (line 7) in the XML configuration for filter-mapping (Fig. 1b).

Objective. The framework-introduced relations mentioned above are concrete because they are tied to a particular application. Our objective is to automatically abstract the general framework semantics (e.g., the annotations, `@Controller` and `@GetMapping`, together specify an entry method relation), which referred as *specifications*. Specifications can be leveraged by a static analysis tool to analyze a wide range of applications that are built on similar frameworks with different concrete configuration parameter values. Alternatively, we can manually produce specifications for frameworks, one by one. Nonetheless, such manual work can be error-prone and labor-intensive.

3   Methodology

We present AUTOWEB to automatically infer framework specifications, which describe how relations are introduced by frameworks for given configurations. The key idea of AUTOWEB is to automatically identify the minimal sufficient and necessary set (MSNS) for a relation by mutating configurations. In practice, AUTOWEB takes a set of sample applications as inputs to infer the general framework semantics and the generated specifications can be used when analyzing other web applications.

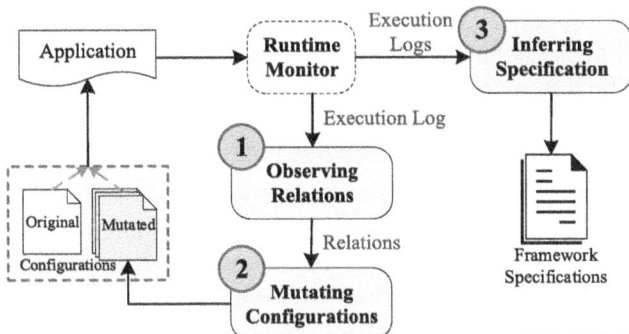

Fig. 3. Overview of AUTOWEB. Black solid lines depict the workflow, while gray dotted lines indicate a singular choice. (Color figure online)

Figure 3 overviews AUTOWEB. The input is a runnable web application, and we partition it into two distinct logical modules: configurations and others in the input application. The steps are shown below:

1. The input application is firstly instrumented and traced to acquire the initial execution log using the original configurations. This step aims to observe the concrete relations including entry points and call/points-to relations.
2. Mutated configurations are generated by removing or adding configuration parameters from the original configurations. The MSNS for each relation can be identified by comparing the dynamic relations of different mutations. The application runs using mutated configurations in this step.
3. Concrete relations and configuration parameters are symbolized to generate specifications that map each relation type to the configuration set.

Next, we will elaborate on each step using the example in Fig. 1.

3.1 Observing Relations

This step aims to collect the concrete framework-introduced relations from the initial execution information. In the Runtime Monitor component, we employed Javaassist [6–8] to instrument the input application, modifying its Java bytecode before class loading. To capture entry point and call relations, we insert logging statements before and after each method call, as well as at the entry and exit of each method. We log field read statements for points-to relations. Additionally, the entry and exit of the request handling methods of web containers (e.g., *Apache Tomcat* [16]) are also logged to track coming requests. These request sequences will later be used to interact with mutated applications.

As web applications handle multiple user requests concurrently, each log statement includes its thread identifier. Relations including entry methods and call/points-to relations can be easily deduced from execution logs within the same thread, as follows:

*Rule*_{entry}: Method m is an entry method if there is a log instance "[mtd] m" immediately following "[Req]".

*Rule*_{call}: Method m is called directly or indirectly at call site c if there is a log instance "[mtd] m" immediately following "[callsite] c".

*Rule*_{ptsto}: Field f references to an object of type t if there is a log instance "[fieldRead] f t".

Listing 1.1 shows the simplified execution logs of our motivating example. The logs are grouped by thread id ㉘ and ㉗, triggered by the coming request "/root1/path1" (line 1–11) and ends with the log instance "/root2/path2" (line 12–21), respectively.

```
 1   ㉘[Req]/root1/path1
 2   ㉘[mtd]Controller1.handle1(...)
 3   ...
 4   ㉘[fieldRead]Controller1.srv:ServiceImpl2
 5   ㉘[callsite]line:11
 6   ㉘[mtd]ServiceImpl2.service(...)
 7   ...
 8   ㉘[mtdEnd]ServiceImpl2.service(...)
 9   ㉘[returnSite]line:11
10   ㉘[mtdEnd]Controller1.handle1(...)
11   ㉘[ReqEnd]/root1/path1
12   ㉗[Req]/root2/path2
13   ㉗[mtd]Filter1.doFilterInternal(...)
14   ...
15   ㉗[callsite]line:32
16   ㉗[mtd]Controller2.handle2(...)
17   ...
18   ㉗[mtdEnd]Controller2.handle2(...)
19   ㉗[returnSite]line:32
20   ㉗[mtdEnd]Filter1.doFilterInternal(...)
21   ㉗[ReqEnd]/root2/path2
```

Listing 1.1. Simplified runtime logs of the example in Fig. 1.

The execution logs precisely capture all triggered relations. To focus solely on the relations introduced by the framework, we exclude the relations that can already be computed by existing static analyses. In Listing 1.1, line 2 and line 13 confirm that `Controller1.handle1` and `Filter1.doFilterInternal` are entry methods (*Rule*_{entry}). Lines 15–16 suggest that `Controller2.handle2` is indirectly invoked at line 32 of application code in Fig. 1a by the invocation `doFilter` (*Rule*_{call}). Line 4 states that field `Controller1.srv` refers to an object of type `ServiceImpl2` (*Rule*_{ptsto}). Column 2 in Table 1 shows the set of concrete relations.

Table 1. Observed relations and corresponding configurations.

Relation Type	Concrete Relation R	Minimal Sufficient and Necessary Set S
Entry point	*entry* : `Controller1.handle1`	$\{$`@Controller`$\prod$`Controller1`, `@GetMapping`$\prod$`handle1`$\}$
	entry : `Filter1.doFilterInternal`	$\{$`@filter`$\prod$`Filter1,Filter1`$\trianglelefteq$ ..., `doFilterInternal`$\trianglelefteq$ ... $\}$
Points-to	`Controller1.srv:ServiceImpl2`	$\{$`@Autowired`$\prod$`srv`, `@Qualifier`$\prod$`srv`, `@Service`$\prod$`ServiceImpl2` $\}$
Call relation	`32:Controller2.handle2` ˙	$\{$`API:doFilter`, C_{32}/m_{32} $\trianglelefteq$ ..., `@filter`$\prod C_{32}$, $S\prod\{$`Controller2,handle2`$\}$ $\}$

3.2   Mutating Configurations

After step 1, we observed the set of dynamically triggered concrete relations. Given a framework-introduced relation R, we try to identify the MSNS S for R, i.e., the minimum set of configuration parameters triggering R. To this end, we mutate configuration parameters based on the following guidelines.

> **Necessity:** Configuration parameter C is a necessary condition of R, i.e., $C \in S$, if the resulting effect of removing C is that R is not triggered, while other relations remain unaffected.
>
> **Sufficiency:** The configuration set S is sufficient to trigger R, if a mutated configuration set S' can trigger a correspondingly mutated relation R'.

Hence, our approach firstly identifies necessary conditions for relation R by removing each configuration parameter one by one until R cannot be triggered. Next, the set of necessary conditions is mutated to further verify whether it is sufficient to trigger a correspondingly mutated relation R'.

In processing relation R, we only need to consider configurations related to R, which is the set of Java annotations or XML attributes attached to related program constructs of R. Hereafter, we use the notation $S\prod p$ to denote configurations S on program construct p. Entry point relation *entry* : $C.m$ involves configurations $S\prod\{C, m\}$, where $C.m$ represents the entry method m inside class C; points-to relation $f : C$ relates to configurations $S\prod\{f, C\}$, where f and C are the field and its type respectively; and Call relation $c : C.m$, denoting that method m inside class C is invoked at callsite c, involves configurations $S\prod\{c, C_c, m_c, C, m\}$, where C_c and m_c refer to the containing class and containing method of callsite c, respectively. The types of configuration parameters include Java annotations and XML configuration files. In addition, we also consider extensions of framework APIs including sub-typing and method overriding as special configurations, denoted by the notation $\trianglelefteq$.

Table 2. Mutation strategy.

original relation	mutated relation
$S\prod\{C, m\} \implies$ *entry* : $C.m$	$S\prod\{C', m'\} \implies$ *entry* : $C'.m'$
$S\prod\{f, C\} \implies f : C$	$S\prod\{f', C'\} \implies f' : C'$
$S\prod\{c, C_c, m_c, C, m\}$ $\implies c : C.m$	$S\prod\{c', C_{c'}, m_{c'}, C', m'\}$ $\implies c' : C'.m'$

Table 2 presents the mutation strategy to verify sufficient conditions. In summary, original program constructs p related to relation R are duplicated and renamed to p'. Configurations S on p are moved to the mutated construct p' instead. Each mutation action generates a *mutated configuration set*, which is used to replace original configurations and then executed to verify whether a correspondingly mutated relation R' on p' can be triggered or not.

Next, we elaborate on how the MSNS for distinct relations are identified for our motivating example. The result for each observed relation is shown in Column 3 of Table 1.

Entry Point Relation. In Fig. 1, `Controller1.handle1` and `Filter1.do-FilterInternal` are entry points. Let us consider the concrete relation *entry* : `Controller1.handle1`. There are two related configuration parameters of this method: `@Controller∏Controller1` and `@GetMapping∏handle1`, and both parameters are necessary since the relation cannot be triggered if either of them is removed. Furthermore, a new relation *entry* : `Controller1'.handle1'` can be triggered by applying our mutation strategy in Table 2. As a result, we have identified the MSNS {`@Controller∏Controller1, @GetMapping∏handle1`} for this concrete relation. The entry method `Filter1.doFilterInternal` involves the XML configuration on class `Filter1` (`filter∏Filter1`) and two additional configurations derived from API extension, sub-typing from `OncePerRequestFilter` (denoted as `Filter1⊴ ...`) and overriding of method `doFilterInternal` (denoted as `doFilterInternal⊴ ...`), which form the MSNS for this relation.

Points-to Relation. We observe a points-to relation `srv:ServiceImpl2` in our motivating example. There are three related configuration parameters, which are `@Autowired∏srv`, `@Qualifier∏srv`, and `@Service∏ServiceImpl2`. Removing either annotation will disable the points-to relation. To mutate the set of configurations, we introduce a new class `ServiceImpl2'` and a new field `srv'`, duplicated from `ServiceImpl` and `srv`, respectively. Next, the set of configurations is removed from the original program constructs and applied to the newly duplicated field and class instead. In another word, the configuration set $S \prod$ {`srv, ServiceImpl2`} is mutated to another set $S \prod$ {`srv', ServiceImpl2'`}. The mutated application will trigger the points-to relation `srv':ServiceImpl2'`. Hence, the three configuration parameters consist of the minimal sufficient and necessary set for the points-to relation `srv:ServiceImpl2`.

Call Relation. Call relation preserves the semantics of indirectly invoking application methods via framework API. That is, application invokes framework APIs, which in turn, call back into application methods. For this, we consider configuration parameters on the callsite (including its containing class and method), as well as configuration parameters on the invoked method (including its containing class). For concrete call relation `32:Controller2.handle2` in our example, the callsite (line 32 in Fig. 1a) is constrained with the following configurations: line 32 invokes API doFilter (`API:doFilter`), m_{32} derived from `doFilterInternal` ($m_{32} \trianglelefteq ...$), C_{32} (class `Filter1`) derived from `OncePerRequestFilter`

$(C_{32} \trianglelefteq \ldots)$, and C_{32} configured as `filter` in XML (`filter` $\prod C_{32}$). These configurations, together with configurations on the invoking method $(\mathcal{S} \prod$ {`Controller2, handle2`}) are the corresponding minimal sufficient and necessary set.

Table 3. Inferred Specifications of the motivating example shown in Fig. 1.

Relation type	Relation R	Specification (content for relation type)
Entry point	$entry : C.m$	{`@Controller`$\prod C$, `@GetMapping`$\prod m$ }
		{`@filter` $\prod C$, $C \trianglelefteq$ `OncePerRequestFilter`, $m \trianglelefteq$ `doFilterInternal` }
Points-to	$f : C$	{`@Autowired`$\prod f$, `@Qualifier`$(S_1) \prod f$, `@Service`$(S_2) \prod C$, $S_1 \sim S_2$}
Call relation	$c : C.m$	{`API:doFilter`, `@filter` $\prod C_c$, $C/m \trianglelefteq \ldots$, `@Controller` $\prod C$, `@RequestMapping` $\prod m$ }

3.3 Inferring Specifications

Until this point, we have obtained a set of relations with their corresponding minimal sufficient and necessary sets. However, the relations and configurations are concrete: they are tied to concrete program constructs and the value (if any) of a configuration parameter is a constant string. In this step, we generalize the relation R and its corresponding configurations to abstract program constructs, according to the following rule.

> **Generalization:** Concrete application program constructs (classes, methods, and fields) are generalized to abstract program constructs. The constant string parameters of a configuration are generalized to a symbolic string with constraints matching the string to the name of a program construct, or to the parameter of another configuration. Symbolic strings with no matching constraints can be discarded.

Table 3 summarizes the specifications inferred from our example in Fig. 1. The specifications look similar to the minimal sufficient and necessary set for a concrete relation, with concrete program constructs and constant configuration parameters symbolized.

As shown in Table 3, we have inferred two rules for entry point relation. The first rule states that the two configurations which are `@Controller`$\prod C$ and `@GetMapping`$\prod m$, collectively declare that $C.m$ is an entry method, regardless of their parameters. The second rule indicates that method $C.m$ is an entry method if $C.m$ extends from `OncePerRequestFilter.doFilterInternal` and C is configured as a filter by `@filter`$\prod C$. Note that in this rule, we only generalize application classes and methods while preserving concrete framework constructs. In the third rule, points-to relation $f : C$ holds under the following conditions: f is annotated with `@Autowired` and `@Qualifier`(S_1), C is annotated with `@Service`(S_2), and the two symbolic string parameter S_1 and S_2 match with each other $(S_1 \sim S_2)$. Here S_1 and S_2 are parameters of configuration `@Qualifier` and `@Service`, respectively. The last rule indicates that if c invokes API `doFilter` in a method derived from `doFilterInternal` $(C/m \trianglelefteq \ldots)$, it may indirectly invoke $C.m$ if $C.m$ is a request handler (`@Controller` $\prod C$, `@RequestMapping` $\prod m$

$\prod m$) and C_c is a filter (`@filter` $\prod C_c$) with parameter matching the name of C ($S \sim C.name$).

Limitations. Points-to/Call relation connects a field/callsite to corresponding class/methods. The connection between distinct program constructs are often indicated by their configuration parameters, where the parameter may match with another parameter or with the name of a program construct. For the points-to relation `srv:ServiceImpl2` in our example, the parameter of configuration `@Qualifer("service2")` on field `srv` matches with the parameter of configuration `@Service("service2")` on class `ServiceImpl2`.

However, it is often challenging to statically determine whether a parameter matches another due to the flexibility provided by frameworks. Parameters can be configured using options such as regular expressions or string manipulation operations. For instance, the call relation `32:Controller2.handle2` happens because the URL processed by the containing class of line 32 (class `MyFilter`) matches with the URL handled by `Controller2.handle2`. However, the URL processed by `MyFilter` is configured as a regular expression `/root/*`. Moreover, we need to join parameters of the two configurations `@RequestMapping(/root2)` and `@RequestMapping(/path2)` together to construct the full URL handled by method `Controller.handle2`. It is rather challenging to recognize the above intricate connection automatically, and our approach will discard the matching constraints between the two URL parameters.

4   Evaluation

4.1   Experimental Setup

Implementation. We implemented a prototype that includes all the components as depicted in Fig. 3, and took the benchmarks as inputs to generate framework specifications. Additionally, we applied the specifications inferred by AUTOWEB to JackEE, a web application analysis engine built on top of Doop [4] for static program analysis.

Platform. The experiments for inferring framework specifications were conducted on an Intel Core(TM) i5-4590 (3.3GHz) laptop equipped with 32 GB of RAM, operating on the Windows 10 Professional version.

Benchmarks. We experimented AUTOWEB on three popular web application frameworks: *Servlet* [13], *Spring*(including *Spring* [25] and *Spring-boot* [24]), and *Apache Struts2* [15]. Our experimental benchmarks comprise two parts.

- A collection of 16 open-source web applications[2], as listed in Table 4, was curated from various open platforms and filtered based on the underlying web frameworks, application categories (such as blogging systems and e-shops), and their respective star ratings. This benchmark encompasses a diverse array of applications, encompassing both popular and lesser-known examples, as well as those with complex and straightforward architectures.

---

[2] https://gist.github.com/menghaining/38286f83c8b674fab771be66d5bf371f.

Table 4. Details of collected open-source benchmarks.

ID	Benchmark	Properties				
		Application Classes	Total Classes	Stars	Forks	Frameworks
1	community	94	19544	2.3k	739	Servlet, Spring
2	halo	425	45359	22.1k	7.5k	Servlet, Spring
3	iCloud	22	9498	183	115	Servlet, Spring, Struts2
4	jpetstore	24	6847	521	745	Servlet, Spring
5	logicaldoc	2013	51494	61	31	Servlet, Spring
6	LMS	33	14329	420	187	Servlet, Spring, Struts2
7	B2CWeb	41	17434	481	343	Servlet, Spring, Struts2
8	newbee-mall	89	13463	9.4k	2.5k	Servlet, Spring
9	NewsSystem	66	20065	19	8	Servlet, Spring, Struts2
10	openkm	2968	88843	527	255	Servlet, Spring
11	RuoYi	290	38320	3k	1k	Servlet, Spring
12	showcase	42	8937	5k	3.8k	Servlet, Spring
13	petclinic	24	29092	395	1.8k	Servlet, Spring
14	WebApp	75	28722	1.3k	610	Servlet, Spring
15	struts-examples	170	13507	405	543	Servlet, Struts2
16	Struts2-Vuln	20	4569	170	38	Servlet, Struts2

The rest of the figures and tables of this paper would use "ID" to represent each benchmark instead of benchmark names.

Benchmarks **struts-examples** and **Struts2-Vuln** are two collections that contain 41 and 16 micro-benchmarks, respectively.

– A collection of 8 web applications from JackEE, which are suggested by experts or top-popularity representatives of major classes of enterprise applications. One is free-binary-only, and the others are open-source.

Our experiments aims to answer the following research questions:

RQ1 Can our approach automatically infer framework-introduced semantics of web applications?
RQ2 How precise are the inferred specifications?
RQ3 How is the quality of generated specifications compared to manually written ones?

4.2   RQ 1: Feasibility

AUTOWEB. AUTOWEB offers an automated and user-friendly solution that minimizes the need for manual setup throughout the workflow. Human involvement is solely required during the initial phase, primarily to interact with the deployed applications, which can be streamlined through using tools like [31]. Then, AUTOWEB automates the following steps leveraging information directly derived from the initial phase. Compared to human learning, which involves static analysis and framework knowledge, the cost is notably lower. Even though resources like StackOverflow [10] and official documentation facilitate rapid initiation, they often lack insights into the correlation between framework usage and static analysis semantics. Establishing these connections requires intricate, labor-intensive manual intervention, making it a challenging and time-consuming endeavor.

The applications in Table 4 are used as the input of AUTOWEB to generate specifications. In Table 5, columns 2–5 detailed the runtime information of AUTOWEB. The size of the execution log produced by each application using the original configurations is outlined in column 2. Column 3 displays the number of mutated configuration sets generated by AUTOWEB. Each mutation concerns one configuration for one relation. Column 4 denotes the success rate of application execution using these mutated configurations. Apart from benchmarks 2 and 5, all other benchmarks achieved success rates exceeding 75% during execution. Even in benchmarks 2 and 5, a significant number of successful runs amounted to 277 and 488, respectively. Column 5 presents the recall $(TP/(TP + FN))$, demonstrating that all benchmarks, except benchmark 9 (to be discussed in 4.3), achieved recall rates surpassing 90%. The extensibility of AUTOWEB lies in two aspects, namely new frameworks and new relation types. New frameworks are already supported by AUTOWEB (discussed shortly), while certain components need to be enhanced to support new relation types.

Table 5. Detailed results of inferred specifications and runtime information.

ID	Instrument	Mutation		Inferring	Configurations			Entry		Inject		Call	
	Log Size(M)	Testcase Number	Trigger Success%	Recall%	All	Reach.	Spec.	FN	FP	FN	FP	FN	FP
1	47.6	167	86.75%	90.00%	19	15	9	1	0	0	0	0	0
2	12.5	398	69.41%	94.44%	80	60	17	0	0	0	0	1	0
3	0.5	242	96.61%	100.00%	6	6	4	0	0	0	0	0	0
4	36.6	104	98.08%	100.00%	6	6	2	0	0	0	0	0	0
5	122	780	62.50%	100.00%	18	9	9	0	0	0	0	0	0
6	0.5	213	84.13%	100.00%	8	7	6	0	0	0	0	0	0
7	0.8	155	82.96%	100.00%	7	4	4	0	0	0	0	0	0
8	172	117	76.92%	100.00%	19	14	8	0	0	0	0	0	0
9	158	93	76.92%	60.00%	9	9	3	0	0	0	0	2	0
10	207	421	98.05%	100.00%	67	30	5	0	0	0	0	0	0
11	2.79	310	75.28%	92.31%	59	27	12	0	0	1	0	0	0
12	1.4	695	78.71%	100.00%	29	28	12	0	0	0	0	0	0
13	1.75	337	75.00%	100.00%	33	31	10	0	0	0	0	0	0
14	1.56	290	80.07%	100.00%	34	19	7	0	0	0	0	0	0
15	11.2	1022	98.33%	100.00%	29	16	8	0	0	0	0	0	0
16	0.3	176	99.43%	100.00%	3	3	2	0	0	0	0	0	0

Columns 2–5 show the details of runtime information. Columns 6–8 show the configuration numbers. Columns 9–14 show the accuracy of inferred specifications.
Reach. denotes reachable configurations at runtime.
Spec. denotes specifications inferred by AUTOWEB (Sect. 3.3)

Specification. The benchmarks encompass a total of 152 configuration parameters: 60 for classes, 39 for methods, and 53 for fields. Among these configuration parameters, 121 are specific to the three frameworks under manual investigation, while the remaining parameters are associated with other frameworks utilized within the applications, such as Stripes [17]. Additional details regarding the count of distinct configuration parameters in each benchmark can be found in columns 6–8 of Table 5.

The specifications derived from the 16 benchmarks encompass 96 entry point types, 9 points-to types, and 17 indirect call types. These specifications involve configuration parameters for 17 classes, 20 methods, 10 fields, as well as 46 additional sub-types within the framework API. To enhance understanding and application of these specifications, we provide an excerpt of inferred entry point relations. Both points-to and call relations exhibit similarities.

Table 6 displays a snip of the results regarding entry point types in the inferred specifications. Each row indicates when a method and its associated class satisfy the specified configuration parameters, the method can be considered as an entry point, and the class becomes the entry class. EP 1, 3, and 6 correspond to annotation configurations, while EP 2, 4, and 5 are associated with XML configurations. In the case of EP6, the method also needs to override the designated method. The results shown are divided by frameworks: EP 1–2, EP 3–4, EP 5, and EP 6 belong to frameworks Spring, Struts2, Servlet, and Strips [17], respectively. It is worth noting that initially, we did not consider Strips but AUTOWEB still inferred the corresponding specification (EP 6), thereby validating our approach's potential for generalization to other novel frameworks.

Table 6. Part of inferred entry-point (EP) specifications.

Line	Class	Method
EP1	@RestController	@PostMapping
EP2	beans->bean[class]	beans->bean[destroy-method]
EP3	-	@Action
EP4	struts->package->action[class]	struts->package->action[method]
EP5	web-app->filter->filter-class	Filter.doFilter(...)
EP6	-	@DefaultHandler

The "-" symbol represents any configuration.

Since specifications are generalized and not specific to any application, they can be utilized by existing static analyzers to analyze any web application built on the frameworks outlined in the specifications. After that, static analyzers can understand framework semantics to facilitate various analyses, such as call graph construction [18], and information analysis [3].

Conclusion. To sum up, our proposed technique is feasible in practice: AUTOWEB can automatically generate precise framework specifications, saving heavy human effort. Moreover, AUTOWEB is readily applicable to other frameworks (e.g., Stripes) and relation types, confirming the generality of our approach.

4.3   RQ 2: Specification Accuracy

The specifications generated by AUTOWEB should exhibit minimal or zero false positives to prevent any adverse impacts for later direct use. To this end, we manually verify the accuracy of all inferred specifications by examining the source code, with a summary of the results provided in columns 9–14 of Table 5. During the verification process, we focus on the following two issues:

- **False Positives.** Are there any incorrectly inferred relations that contradict framework semantics?
- **False Negatives.** Are there any correct relations that were not inferred but were observed during runtime?

False Positives. Table 5 (columns 10, 12, 14) reveals no false positives in any of the benchmarks. This is due to the specifications being inferred from runtime information, which is subsequently verified by the actual execution of the application. As a result, AUTOWEB effectively avoids false positives, ensuring the correctness of the framework semantics introduced in the static analysis.

False Negatives. There are only five false negatives for all benchmarks. These false negatives are hidden behind certain factors during runtime, preventing them from being apparent. One factor is embedding the configuration content directly in the code, rendering the configuration on the code(annotations or XML files) ineffective. For example, in the community project, removing the parameter @RequestMapping does not take effect at runtime because the application has a default response function that produces the same result as the configuration parameter. Another factor is the language feature. For example, in RuoYi, the false negative for the field-inject relation is related to the @Value annotation. This occurs because all fields annotated with this configuration are Java primitive types, which always have default values. Moreover, complex string configurations lead to false negatives. In the NewsSystem, a false negative is caused by intricate string manipulation. Specifically, the configuration <action name= "AdminAction_*" method="1"> utilizes the implicit configuration value "{1}" to represent the method name, relying on the incoming URLs. Another false negative arises from multi-layer references in XML attributes, which are not yet supported.

Note that false negatives in one application may appear as true positives in other applications. For instance, the Entry-point relation @RequestMapping is FN in community appears as true positive in newbee-mall. More input projects would bring more complete specifications, which will be discussed in 4.4. The average false negative rate summarized from all inferred specifications is 8.2%.

Conclusion. Upon analysis, we identified that the false negatives in our inferred specifications were a consequence of configurations outside the scope of our analysis. Importantly, the inferred specifications exhibit no false positives, indicating their direct applicability as framework knowledge for existing static analysis tools.

4.4 RQ 3: Comparison with Existing Work

The state-of-the-art tool, JackEE [1], offers open-source specifications across various web frameworks. JackEE is built on top of Doop [4], which is a collection of various analyses expressed in the form of Datalog [23] rules, and all the framework specifications are written in Datalog rules. To assess the efficacy of the specifications generated by AutoWeb, we transformed them into Datalog rules, replacing all the configuration specifications in JackEE. With JackEE's specifications serving as the baseline, we evaluated the specifications produced by AutoWeb alongside the default specifications from Doop.

Comparison Dimension. Framework knowledge helps static analyses to better understand the application behaviors. To evaluate the quality of specifications, we focus on *reachable application methods* and *call graph edges* (same metrics used in JackEE).

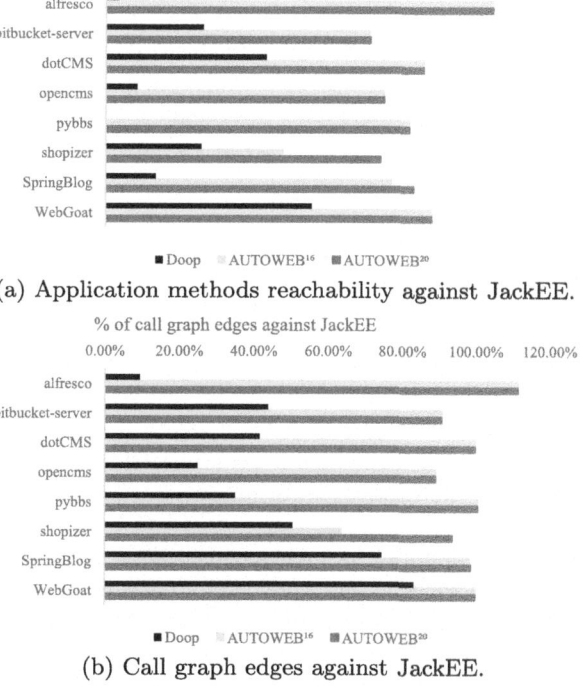

(a) Application methods reachability against JackEE.

(b) Call graph edges against JackEE.

Fig. 4. Reachability for different metrics.

Comparison Method. We adopt JackEE (with its manual specifications) as the baseline for comparison, evaluating the efficacy of JackEE, Doop, and AUTOWEB. We considered the 8 benchmarks from JackEE (mentioned in 4.1) for evaluation. Among these benchmarks, one benchmark entails a prolonged setup duration, two encounter deployment problems, and one is not open-sourced. As a result, we use 4 benchmarks to enrich the specifications generated by AUTOWEB. Hence, we have two sets of specifications: (1) AUTOWEB[16] contains the specifications generated using 16 benchmarks in Table 4. (2) AUTOWEB[20] denotes the enriched specifications using 20 (16 + 4) benchmarks.

Result. Figure 4a and Fig. 4b compare Doop, AUTOWEB[16], and AUTOWEB[20] against JackEE on reachable (application) methods and call graph edges, respectively. When comparing AUTOWEB[16] with JackEE, the average ratio on reachable application methods is 91.85%, while for call graph edges, the average is 94.27%. In some cases, AUTOWEB[16] even outperformed JackEE like `alfresco` and `WebGoat`. When comparing AUTOWEB[20] with JackEE, the ratios range from 83.14% to 121.73% on reachable methods, averaging 96.59%; For call graph edges, the numbers vary from 89.18% to 111.48%, averaging 98.09%. On the other hand, AUTOWEB[20] exhibited improvements over AUTOWEB[16], particularly in benchmarks such as `shopizer` and `SpringBlog`, where previously missing configurations, absent in the initial 16 benchmarks, were later inferred. This underscores that using a more extensive set of inputs can lead to richer specifications, as discussed in Sect. 4.3. Furthermore, our specifications include correct configuration parameters that may have been overlooked manually. For example, we identified annotations like `@PostMapping` and `@ExceptionHandler` as entry-point relations, while JackEE did not recognize them. This highlights the presence of unsystematic and incomplete issues of manually configured specifications.

Conclusion. Concerning the quality of constructed call graphs, the specifications inferred by AUTOWEB are comparable with those manual ones in JackEE. Furthermore, our approach excels at identifying overlooked configuration parameters during manual writing.

5   Related Work

The related work encompasses two main areas: modeling framework behaviors, and automatic summarizing of program semantics.

Modeling Framework Behaviors. As mentioned in Sect. 1, previous works modified each analysis engine with human knowledge on-demand, which suffers from limited reusability. Some studies attempted to develop reusable models for specific frameworks to address this limitation, which still rely on human knowledge. ANTaint [30] needs manually modeled core features of Spring like bean injection and AOP. The Oracle team [12] manually wrote rules to identify entry points only for Java EE Servlet applications. GenCG [19,20], F4F [26,28], and

JackEE [1] all need human effort to obtain knowledge of the framework and static analysis. Static Analysis Refining Language (SARL) [14] can also obtain framework-introduced relations via iterative software analysis. However, the analyzer also needs to point out and add the missing framework knowledge. Unlike AUTOWEB, all these approaches rely on the knowledge of frameworks and static analysis, and require extra manual effort for new frameworks.

Automatic Summarizing Program Semantics. Research on exploiting automatic approaches to summarizing framework library specifications used in static analysis [2,9,22] became more popular. These approaches mined information flow specifications with additional running information over libraries, rather than writing them by hand. However, the purpose of these approaches is to summarize the framework library APIs' semantics, especially for Android, not to deal with complex but frequently used configurations (e.g., XML files). Therefore, these approaches do not apply to web applications that mostly use non-code configurations.

6 Conclusion

Web applications heavily rely on web frameworks, making it imperative to precisely model framework semantics for static analysis. In this paper, we proposed the first automated method to produce specifications that represent general framework semantics. To that end, we identify the minimal necessary and sufficient set for a framework-related relation by mutating configurations. Experimental results on three mainstream Java frameworks demonstrate that our technique is comparable to existing state-of-the-art manual approaches, obtaining a marginal 8.2% false negatives with no false positives.

Acknowledgments. We thank all anonymous reviewers for their valuable feedback which has significantly improved the quality of this manuscript. This work is supported by the National Key R&D Program of China (2022YFB3103900), and the National Natural Science Foundation of China (NSFC) under grant numbers 62132020 and 62202452.

References

1. Antoniadis, A., Filippakis, N., Krishnan, P., Ramesh, R., Allen, N., Smaragdakis, Y.: Static analysis of java enterprise applications: frameworks and caches, the elephants in the room. In: Proceedings of the 41st ACM SIGPLAN Conference on Programming Language Design and Implementation, pp. 794–807 (2020)
2. Arzt, S., Bodden, E.: Stubdroid: Automatic inference of precise data-flow summaries for the android framework. In: 2016 IEEE/ACM 38th International Conference on Software Engineering (ICSE), pp. 725–735. IEEE (2016)
3. Arzt, S., et al.: Flowdroid: precise context, flow, field, object-sensitive and lifecycle-aware taint analysis for android apps. In: Proceedings of the 35th ACM SIGPLAN Conference on Programming Language Design and Implementation, PLDI 2014, pp. 259-269. Association for Computing Machinery, New York, NY, USA (2014). https://doi.org/10.1145/2594291.2594299

4. Bravenboer, M., Smaragdakis, Y.: Strictly declarative specification of sophisticated points-to analyses. In: Proceedings of the 24th ACM SIGPLAN Conference on Object Oriented Programming Systems Languages and Applications, pp. 243–262 (2009)
5. Centonze, P., Naumovich, G., Fink, S.J., Pistoia, M.: Role-based access control consistency validation. In: Proceedings of the 2006 International Symposium on Software Testing and Analysis, pp. 121–132 (2006)
6. Chiba, S.: Javassist, https://www.javassist.org/
7. Chiba, S.: Load-time structural reflection in java. In: Bertino, E. (ed.) ECOOP 2000. LNCS, vol. 1850, pp. 313–336. Springer, Heidelberg (2000). https://doi.org/10.1007/3-540-45102-1_16
8. Chiba, S., Nishizawa, M.: An easy-to-use toolkit for efficient java bytecode translators. In: Pfenning, F., Smaragdakis, Y. (eds.) GPCE 2003. LNCS, vol. 2830, pp. 364–376. Springer, Heidelberg (2003). https://doi.org/10.1007/978-3-540-39815-8_22
9. Clapp, L., Anand, S., Aiken, A.: Modelgen: mining explicit information flow specifications from concrete executions. In: Proceedings of the 2015 International Symposium on Software Testing and Analysis, pp. 129–140 (2015)
10. Community, S.O.: stackoverflow. https://stackoverflow.com/
11. Dean, J., Grove, D., Chambers, C.: Optimization of object-oriented programs using static class hierarchy analysis. In: Tokoro, M., Pareschi, R. (eds.) ECOOP 1995. LNCS, vol. 952, pp. 77–101. Springer, Heidelberg (1995). https://doi.org/10.1007/3-540-49538-X_5
12. Dietrich, J., Gauthier, F., Krishnan, P.: Driver generation for java EE web applications. In: 2018 25th Australasian Software Engineering Conference (ASWEC), pp. 121–125. IEEE (2018)
13. Edition, J.E.: Jakarta EE. https://jakarta.ee/
14. Ferrara, P., Negrini, L.: SARL: OO framework specification for static analysis. In: Software Verification, pp. 3–20. Springer, Cham (2020)
15. Foundation, A.: Apache struts 2. https://struts.apache.org/
16. Foundation, A.S.: Apache tomcat. https://tomcat.apache.org/
17. Framework, S.: Stripes framework. https://stripesframework.atlassian.net/
18. Lhoták, O., Hendren, L.: Scaling java points-to analysis using SPARK. In: Hedin, G. (ed.) CC 2003. LNCS, vol. 2622, pp. 153–169. Springer, Heidelberg (2003). https://doi.org/10.1007/3-540-36579-6_12
19. Luo, L.: A general approach to modeling java framework behaviors. In: Proceedings of the 29th ACM Joint Meeting on European Software Engineering Conference and Symposium on the Foundations of Software Engineering, pp. 1680–1682 (2021)
20. Luo, L.: Improving Real-World Applicability of Static Taint Analysis. Ph.D. thesis, Universität Paderborn, October 2021. https://www.bodden.de/pubs/phdLuo.pdf
21. Martínez, S., Cosentino, V., Cabot, J.: Model-based analysis of java EE web security configurations. In: 2016 IEEE/ACM 8th International Workshop on Modeling in Software Engineering (MiSE), pp. 55–61. IEEE (2016)
22. Nimmer, J.W., Ernst, M.D.: Automatic generation of program specifications. ACM SIGSOFT Softw. Eng. Notes **27**(4), 229–239 (2002)
23. Smaragdakis, Y., Bravenboer, M.: Using Datalog for fast and easy program analysis. In: de Moor, O., Gottlob, G., Furche, T., Sellers, A. (eds.) Datalog 2.0 2010. LNCS, vol. 6702, pp. 245–251. Springer, Heidelberg (2011). https://doi.org/10.1007/978-3-642-24206-9_14
24. Spring: Spring boot. https://spring.io/projects/spring-boot

25. Spring: Spring framework. https://spring.io/projects/spring-framework
26. Sridharan, M., Artzi, S., Pistoia, M., Guarnieri, S., Tripp, O., Berg, R.: F4f: taint analysis of framework-based web applications. In: Proceedings of the 2011 ACM International Conference on Object Oriented Programming Systems Languages and Applications, pp. 1053–1068 (2011)
27. Toman, J., Grossman, D.: Taming the static analysis beast. In: 2nd Summit on Advances in Programming Languages (SNAPL 2017). Schloss Dagstuhl-Leibniz-Zentrum fuer Informatik (2017)
28. Tripp, O., Pistoia, M., Cousot, P., Cousot, R., Guarnieri, S.: ANDROMEDA: accurate and scalable security analysis of web applications. In: Cortellessa, V., Varró, D. (eds.) FASE 2013. LNCS, vol. 7793, pp. 210–225. Springer, Heidelberg (2013). https://doi.org/10.1007/978-3-642-37057-1_15
29. Tripp, O., Pistoia, M., Fink, S.J., Sridharan, M., Weisman, O.: Taj: effective taint analysis of web applications. ACM Sigplan Notices 44(6), 87–97 (2009)
30. Wang, J., Wu, Y., Zhou, G., Yu, Y., Guo, Z., Xiong, Y.: Scaling static taint analysis to industrial SOA applications: a case study at Alibaba. In: Proceedings of the 28th ACM Joint Meeting on European Software Engineering Conference and Symposium on the Foundations of Software Engineering, pp. 1477–1486 (2020)
31. Yu, X., Jin, G.: Dataflow tunneling: mining inter-request data dependencies for request-based applications. In: 2018 IEEE/ACM 40th International Conference on Software Engineering (ICSE), pp. 586–597. IEEE (2018)

SafePtrX: Research on Mitigation of Heap-Based Memory Safety Violations for Intel x86-64

LiLie Chen(✉), JunYu Wu, and Yuan Liu

Jianngnan University, Wuxi, China
6223115010@stu.jiangnan.edu.cn, lyuan1800@jiangnan.edu.cn

Abstract. In contemporary programming languages that lack automatic memory management, such as C/C++, ensuring memory safety remains an unresolved practical challenge. Applications developed in these languages often exhibit various safety vulnerabilities. While numerous solutions have been proposed by both academia and industry, some of which have gained widespread adoption, they commonly present limitations such as the requirement for specialized hardware support, significant runtime or memory overhead, or a limited scope of problem coverage.

In this paper, we present an efficient, software-based memory safety violation mitigation scheme based on intermediate pointers and metadata embedding for the Intel x86-64 platform. The fundamental idea is to insert intermediate pointers to every pointer that points to heap memory and embed tags in the unused bits of the intermediate pointers. By inserting checks on these intermediate pointers, potential memory safety violations can be mitigated. Based on this scheme, we implement SafePtrX, a mitigation solution for heap memory safety with enhanced security properties and improved performance compared to existing methods. We also demonstrate the feasibility of SafePtrX by using publicly disclosed vulnerabilities.

Keywords: memory safety · software system · use-after-free · out-of-bounds

1 Introduction

Nowadays, although the use of automatic memory management languages such as Java and Python for software development has become a popular way, the C/C++ programming language is still frequently used [1], many software systems pursuing high performance, such as browsers [2], databases, operating system kernels, etc., are implemented in C/C++, and memory management needs to be carried out manually when developing with these languages. Developers need to explicitly allocate and free memory (e.g., malloc(), free(), etc.), which may introduce a large number of memory safety violations, and memory safety protection of C/C++ codebases remains a key issue in reality.

In recent years, many efforts have been made in academia and industry to address the issue of memory safety violations. For example, ASLR (Address

© The Author(s), under exclusive license to Springer Nature Switzerland AG 2025
G. Bai et al. (Eds.): ICECCS 2024, LNCS 14784, pp. 390–408, 2025.
https://doi.org/10.1007/978-3-031-66456-4_21

Space Layout Randomization) randomizes the memory layout of an application at runtime, making it difficult for attackers to exploit known memory layout information. Fuzzing [3] is an automated testing technique that involves providing programs with random, invalid, or abnormal input data to check for memory safety issues. AddressSanitizer [4], an open-source tool developed by Google, is used for detecting memory errors and can detect most of them. CETS+Softbound [5,6] is also one of the robust software-based memory safety solutions for the C language. The mentioned efforts either allow attackers aim to code reuse attacks [7], require high-quality test samples, or come with significant runtime overhead.

Generally speaking, memory violations can be divided into two types: temporal violations and spatial violations. Temporal violations typically involve errors in the timing of memory access and usage, such as null pointer dereferences. Spatial violations are always the situations where a program accesses or uses memory beyond the boundaries of allocated space, such as buffer overflow. In the "Top 25 Most Dangerous Software Weaknesses" of Common Weakness Enumeration (CWE) 2023, the Use-After-Free (CWE-416) and Out-of-bounds Write (CWE-787) vulnerabilities are ranked 4th and 1st, respectively [8], as examples of temporal and spatial memory violations. In implemented sanitizers, most of them only provide partial protection against a specific type of error [9]. For instance, LowFat [10] only detects some out-of-bounds errors but cannot detect Use-After-Free errors.

In some hardware architectures, memory tagging mechanisms have been introduced to mitigate the impact of memory violations. For example, ARM's Top Byte Ignore (TBI) utilizes 8 bits from a 64-bit pointer as a tag, and ARM's Memory Tagging Extension (MTE) [13] assigns a 4-bit tag to each memory region. However, Intel x86-64, which is widely used in server and desktop applications, such mechanisms are almost absent, and other memory technologies and software-level protection methods need to be sought.

In view of the above problems and challenges, this paper proposes a memory violation mitigation scheme based on the Intel x86-64 platform, which mainly focuses on heap memory. For each pointer pointing to heap allocation, an intermediate pointer is set, and the original pointer will point to it. The unused bits of the intermediate pointer will be embedded with a tag. When heap memory is freed, the intermediate pointer will be set to null. When dereferencing the original pointer, the tag of the intermediate pointer will be compared with the tag stored in disjoint shadow memory to detect and mitigate the Use-After-Free and heap overflow vulnerabilities.

To demonstrate the feasibility of the proposed solution, we implement the scheme based on LLVM 10 [11] and named it SafePtrX. It consists of an LLVM Pass and a self-implemented memory allocation function library. In our evaluations, we select 4 real-world vulnerabilities involving Use-After-Free and heap overflow to prove the effectiveness of the scheme. In addition, SPEC CPU2006 benchmark suite was used to evaluate the runtime and memory overhead. Experiments have shown that SafePtrX has considerable improvements compared to previous classical works.

The contributions of this paper are as follows:

(1) We propose a Use-After-Free mitigation scheme based on intermediate pointers.
(2) Based on the mitigation scheme above, a memory tagging solution is proposed, which extends the scheme to the mitigation of spatial memory violations, and supplements the detection of Use-After-Free vulnerabilities.
(3) We implement the scheme and demonstrate its effectiveness and performance through experiments, which has shown considerable improvements.

2   Background

In this section, we will review the basic concepts of dangling pointers, buffer overflows, lock-and-key checking mechanisms, and memory tagging. This will provide a better understanding of the content of this paper.

2.1   Dangling Pointers

As programming languages with high performance, C/C++ leaves many memory management to developers. Unlike stack variables, the C standard library provides the malloc() and free() functions for allocating and deallocating heap memory objects, and the timing of memory allocation and deallocation is determined by the programmer. When memory in a program is freed, the pointer still exists. Such a pointer is called a dangling pointer. Dangling pointers are safe if they are not accessed (freed/dereferenced), but developers may inadvertently use dangling pointers, leaving them vulnerable to exploitation by attackers.

An attacker can exploit the Use-After-Free (UAF) vulnerability, and the code in Listing 1.1 illustrates the mechanism of the UAF vulnerability.

Listing 1.1. An example of Use-After-Free error in C source code.

```
int main()
{
int *p = malloc(8);
int *q = p;
*p = 1;
free(p);
int *r = malloc(16);
*q = 10;
......
return 0;
}
```

In the code showed in Listing 1.1, two pointers p and q point to the same memory object. When the memory object is freed through pointer p, pointer q becomes a dangling pointer. An attacker can allocate another object r at the same address as the previously allocated object, and using the freed memory address q can cause undefined behavior. Modifying function pointers in the heap and hijacking control flow is a common attack method [42].

Another vulnerability related to dangling pointers is the Double-Free vulnerability, where memory is freed but the pointer is not set to NULL. Taking the code as an example, the dangling pointer would be p instead of q, and then the same block of memory is freed using the same pointer (free(p)) again. This also result in undefined behavior, as the memory allocator may allocate the same block of memory to another object after the second deallocation.

2.2   Heap Overflow

A heap is a memory region used to store dynamically allocated data, commonly used for dynamic memory allocation and deallocation during program execution. It is a region of memory managed by the operating system or a memory allocator, which allocates and deallocates memory regions as requested by the application.

Heap overflow is a type of out-of-bounds write that occurs due to improper memory operations or miscalculation of data size, where data is written beyond the boundaries of a dynamically allocated memory region on the heap. When data is written to a heap-allocated region exceeding its capacity, the excess data is written into adjacent memory regions on the heap. Attackers can exploit this vulnerability to cause program crashes, data corruption, and even information leakage or arbitrary code execution.

2.3   Lock-and-Key Mechanism and Memory Tagging

The lock-and-key mechanism is a common mechanism used in memory temporal safety checks. It regards the use of memory as authorization. When accessing memory, the visitor must hold the correct key to unlock the corresponding lock, granting access or modification permissions to the corresponding memory region. Oscar [12] summarizes the existing lock-and-key mechanisms into two types: explicit and implicit, with each type further divided into "change the lock" or "revoke the keys".

Memory tagging is widely used in building secure systems as an application of the lock-and-key mechanism. It involves allocating additional tag bits to memory addresses to record relevant information about memory blocks, aiming to prevent certain temporal and spatial security violations. Figure 1 illustrates how memory tagging is used to detect violations. The memory region pointed to by the pointer p is assigned a tag, represented in green color. Different tags, represented in different colors, are placed outside the memory region. When accessing memory, a comparison is made between the tag of p [12] and the tag of the memory region. If the tags do not match (colors are different), it indicates an overflow error. The detection of UAF errors follows a similar process.

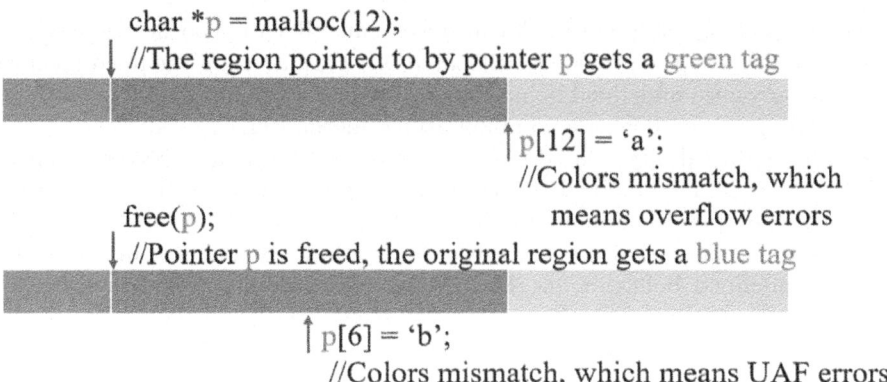

char *p = malloc(12);

↓ //The region pointed to by pointer p gets a green tag

↑ p[12] = 'a';

//Colors mismatch, which

free(p);   means overflow errors

↓ //Pointer p is freed, the original region gets a blue tag

↑ p[6] = 'b';

//Colors mismatch, which means UAF errors

Fig. 1. An example of how memory tagging works, different colors represent different tags. (Color figure online)

3   Design

In this section, the design of SafePtrX in Intel x86-64 will be described in detail, and we will show how it works to mitigate memory violations.

The overall framework of the system is illustrated in Fig. 2. The system takes C/C++ code that may contain memory safety violations as input. Through pointer analysis and transformation, it generates intermediate pointers and modifies the original pointer's references. Instrumentation for tag checking is embedded into the binary file. During runtime, the program is linked to a modified memory allocator. This allocator embeds metadata while allocating memory and sets the intermediate pointers to null when releasing memory. When accessing memory, the stored tag in the intermediate pointer is checked against the shadow memory. If the tags do not match, the pointer is set to null as well. The final output is a protected binary program.

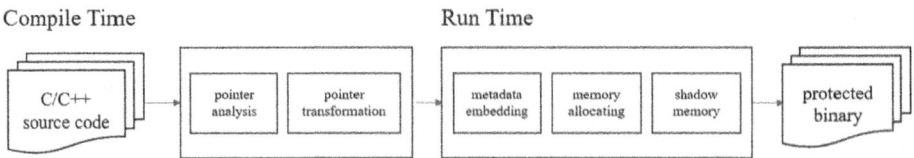

Fig. 2. An overall framework of SafePtrX.

3.1   Intermediate Pointer

This paper mainly focuses on memory safety violations in the heap. Special attention needs to be given to the heap memory region (obtained through functions like malloc()) and the pointers pointing to this memory region. The heap memory region can have multiple pointers pointing to it, and a single pointer can also have multiple aliases. Analyzing each of these pointers individually would incur significant overhead. Hence, we introduce an intermediate pointer between each heap memory region and the pointers pointing to it. Assuming there are three pointers, p, q, and u. p is the pointer that allocates memory, q is an alias for p, and u is another pointer that points to the heap region (derived arithmetically from p). The original pointer relationships in the program are depicted in Fig. 3.

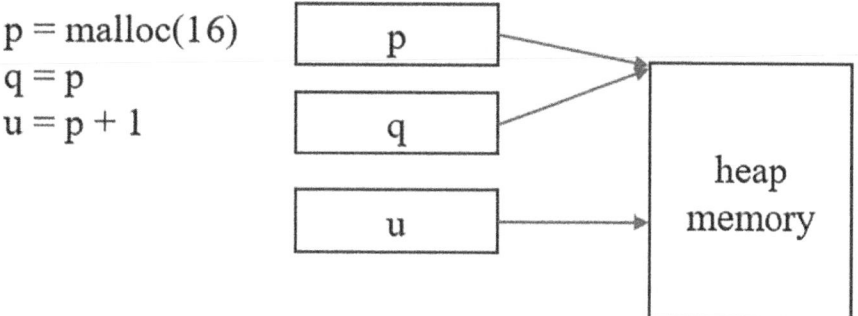

Fig. 3. Pointers in original program.

After pointer transformation, the relationship between pointers and memory regions is depicted in Fig. 4. Pointers that originally pointed to the same address in the heap memory have an intermediate pointer inserted between them. The original pointers are then made to point to the intermediate pointer. Pointers derived arithmetically from the original pointers will now be derived arithmetically from the intermediate pointer in order to access the same memory address.

The intermediate pointer have three main purposes: organizing and managing pointers and their aliases pointing to the heap region, embedding metadata and inserting comparisons, and nullifying the pointer at appropriate times. These will be discussed in detail in the following sections.

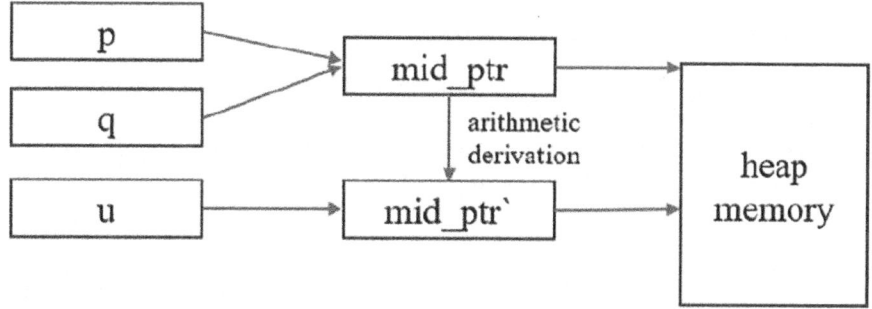

Fig. 4. Pointers after transformation.

3.2   Semantic Changes

Due to the introduction of the intermediate pointer, all first-level pointers in the source code will be transformed into second-level pointers. This will result in changes in the usage of pointers in many situations. Consider the original code in Listing 1.2. In the code, where the pointer p points to the beginning of the 16-byte heap area, and then the pointer q points to the same location, the transformed code is shown in Listing 1.3. The pointer p and q remain unchanged in their declarations. However, when allocating heap space, a two-level pointer is declared and an additional space (equal to 64 in Intel x86-64) is allocated. This space points to the originally allocated heap space. Then, this two-level pointer is pointed to the original heap pointer, pointing to the same memory space. The code will keep unchanged when performing pointer passing.

Listing 1.2. Code in original program.

```
int *p,*q;
p = (int *)malloc(16);
q = p;
```

Listing 1.3. Code in transformed program.

```
int *p,*q;
void **mid_ptr = malloc(sizeof(void *));
*mid_ptr = malloc(16);
p = mid_ptr;
q = p;
```

Similarly, when dereferencing, the pointer must be changed to be consistent with the original program, if there is a dereference "*p" in the original program. After the transformation, the dereferenced statement is converted to "*((int *)*p)".

3.3   Nullification of the Intermediate Pointer

When deallocating, the intermediate pointer will be set to null to prevent memory violations caused by dangling pointers. As shown in Fig. 5, when deallocating heap memory, the program will be transformed into deallocating the memory pointed to by the intermediate pointer and setting the intermediate pointer to null.

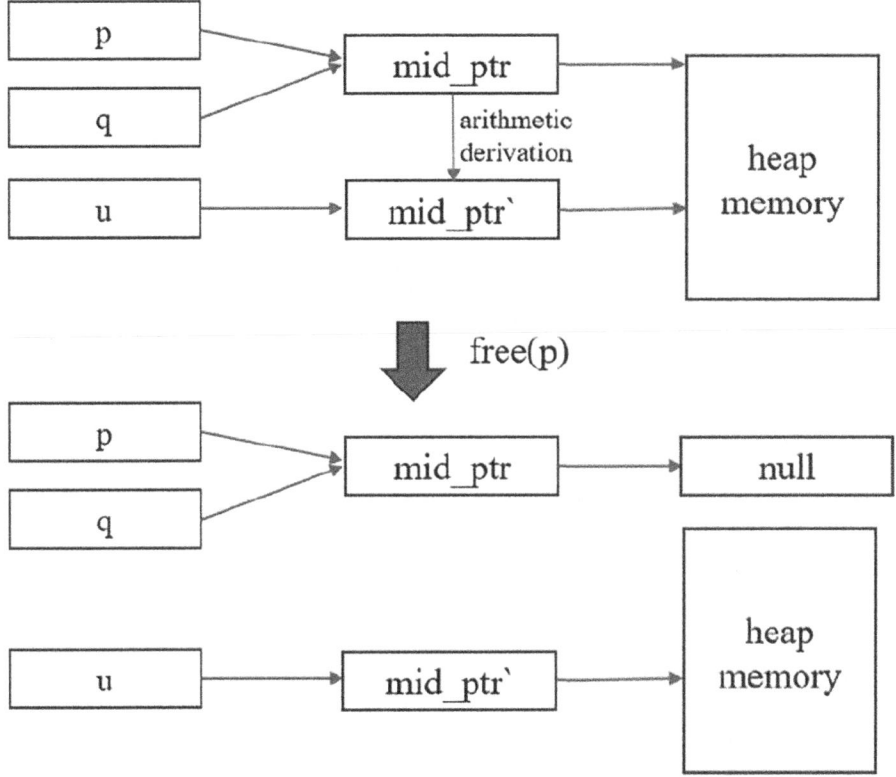

Fig. 5. Nullification of the intermediate pointers.

In the given program, if there is a "free(p);" statement in the original code, it will be transformed into "free((int *)*p);" statement after transformation. Additionally, the statement "(int *)*p = null;" needs to be added to nullify the intermediate pointer. After nullifying the region, accessing the original pointer and its aliases (e.g., pointer q) through dereferencing will access the null region, as they now point to the same intermediate pointer. This approach ensures that when an attacker attempts to dereference a dangling pointer, they will encounter a null pointer directly, making the UAF vulnerability unexploitable. DangDone [15] introduces static analysis to transform only suspicious dangling pointers. In this paper, all pointers related to the heap are transformed because additional safety violations need to be checked. Nullifying the intermediate pointer is not sufficient to prevent UAF errors caused by arithmetic-derived pointers. We address this issue in the out-of-bounds check, as explained in detail in Sect. 3.6.

3.4 Tag Embedding Based on Intermediate Pointer

ARM MTE uses a 4-bit tag to check memory errors in a lock-and-key mechanism. This paper designs a similar mechanism for the Intel x86-64 architecture. In the user space pointers of Intel x86-64, where bits 0–47 are used by the user and bits 48–63 are typically set to 0 [14], known as the canonical address. To make better use of the intermediate pointer, this paper utilizes bits 48–51 as tags, embedding metadata and inserting checks. As shown in Fig. 6, the upper is the canonical address of x86-64, and the lower is a pointer with an embedded tag (assuming it is 0111).

When heap memory is allocated, we introduce a 4-bit randomly generated tag, which will be written to shadow memory that does not intersect with the heap region, and will also be embedded in bits 48–51 of the intermediate pointer. xTag [16] embeds the 4-bit tag into bits 33–36 of the heap pointer, and maps virtual memory using different tags to the same physical page, which requires additional modification of the system function. Our embedding scheme does not involve the paging storage related to the 4-level paging, because in x86-64, only the first 48 bits participate in linear address translation.

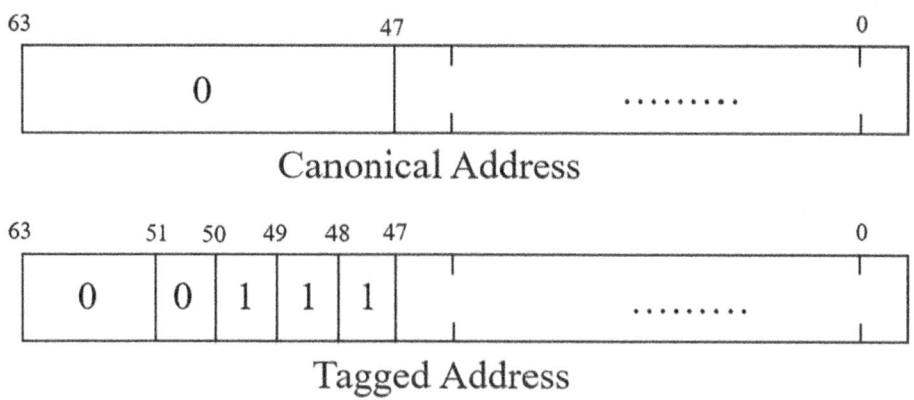

Fig. 6. A canonical address and a tagged address which has a tag of 0111.

3.5 Handling Segmentation Fault

In Intel x86-64, the high 16 bits of user space pointers must be all zeros, otherwise a segmentation fault will occur when accessing the memory. If the high 16 bits of the pointer are masked to zero using the compiler during memory access, the segmentation fault will not occur. Consider the code in Listing 1.4, where mid_ptr is a pointer with a tag of 0111.

Listing 1.4. Mask the high bits of an address to avoid segmentation fault.

```
mid_ptr |= 0x0007000000000000; // tag is 0111
*mid_ptr = ...; // segmentation fault occurs
*(mid_ptr & 0x0000ffffffffffff) =...; // correct, avoids segmentation fault
```

In the given code, the first line embeds the tag into the pointer mid_ptr. Assuming the allocated tag here is 0111, directly dereferencing mid_ptr will result in a segmentation fault. Therefore, when using mid_ptr for memory access, SafePtrX masks the high 16 bits of the pointer to zero to figure prevent segmentation faults. In TailCheck [17], a similar approach is used to store the distance in the high bits of the pointer.

3.6   Out-of-Bounds Checks

Because a 4-bit tag is randomly generated each time the heap memory is allocated, there is a high probability that the tag in the adjacent heap region will be different, and the tag can be used for out-of-bounds checking.

SafePtrX uses shadow memory scheme to store and compare tags. The advantage of this approach is that tag lookup can be done directly through mapping without the need for traversal. In this paper, 16 bytes of heap memory is mapped to 1 byte of shadow memory, as in typical cases heap allocations are larger than 16 bytes, and some high-performance allocators have increased the granularity of heap allocations to this size, such as ptmalloc and mimalloc [43].

When the program accesses a specific address in the heap memory region through the intermediate pointer mid_ptr (note that mid_ptr may be incremented or decremented due to arithmetic derivation, and it may not point to the beginning of the heap), additional checks will be added in Listing 1.5.

Listing 1.5. Additional checks to detect out-of-bounds when accessing memory.

```
tag = (mid_ptr >> 48) & 0x0f;
tag_in_shadow = shadow[mid_ptr/16];
if(tag != tag_in_shadow)
mid_ptr = null;
```

The check first extracts the tag from the intermediate pointer, which can be seen as a shift operation followed by a logical AND operation. The second line of the code retrieves the tag from the shadow memory by using a linear mapping. The next step is to compare the tags, and if they do not match, it is considered a possible error. The pointer at this location will be set to null to prevent exploitation by attackers.

3.7   Optimization

Performance overhead associated with the insertion of the intermediate pointer is small. We mainly focus on optimizations related to out-of-bounds checks. Inserting tag checks every time a pointer is used for memory access would introduce significant overhead. In some cases, it is possible to eliminate unnecessary checks to reduce code instrumentation.

Firstly, the pointer that points to the start of the heap will skip the check because overflow is not possible. Secondly, for the same pointer used for multiple memory accesses, if there are no operations that point to other memory regions and no pointer arithmetic operations, the pointer will only be instrumented with only one check.

4   Implementation

We have implemented the design of SafePtrX on an Intel x86-64 architecture Linux system to protect C/C++ programs running on this system, with both static analysis and dynamic checks. The implementation consists of two parts:

(1) An LLVM compiler framework that utilizes an LLVM Pass [18] to perform intermediate pointer transformation and add nullification operations and additional checks.
(2) A runtime library that replaces the original memory functions related to allocation. This library retains the functionality of the original functions while also handling metadata allocation and management. It works in conjunction with the Pass mentioned in (1).

The pass implemented in (1) traverses the instructions in the Intermediate Representation (IR) of the program. It replaces memory allocation functions with modified functions and records the pointer variables that result from heap memory allocations. It transforms some pointers and analyzes the use points of pointer variables using Use-Def chains and Def-Use chains. It focuses on pointer propagation and arithmetic derivation. Checks will be inserted when suspicious pointer accesses memory and sets the intermediate pointer to null when memory is deallocated. This pass is implemented as an LLVM extension plugin and can be enabled by a flag when compiling C/C++ programs that may have memory safety violations.

The runtime library in (2) refers to the modified version of memory allocation functions. Instead of returning the address of the allocated heap memory, these functions return the address of the intermediate pointer. The generated tag is embedded in the unused bits of the intermediate pointer and mapped to the corresponding range in the disjoint shadow memory. During program execution, the binary file is linked with the runtime library implemented in (2) to work together and provide protection.

5   Evaluation

We evaluate SafePtrX from several aspects: effectiveness, runtime overhead, and memory overhead. Then we discuss potential issues related to SafePtrX. The evaluation was performed on a machine with Intel x86-64 hardware platform and Ubuntu 18.04.6 operating system. The machine has two 4-core CPUs and 8GB of RAM.

5.1 Effectiveness

To evaluate the effectiveness of SafePtrX, we selected several publicly available Proof of Concept (PoC) vulnerabilities from the CVE website [19]. These vulnerabilities included UAF errors and heap overflow errors. The purpose was to assess the detection capabilities of SafePtrX.

CVE-2016-3141 is a UAF vulnerability in the wddx.c file of the WDDX extension, affecting PHP versions before 5.5.33 and 5.6.x before 5.6.19. An attacker can trigger a call to the wddx_deserialize function by providing XML data with a maliciously crafted "var" element. This can lead to Denial of Service (DoS), memory corruption, and program crashes. SafePtrX terminates the execution of the program before accessing the freed memory.

CVE-2017-16943. In versions 4.88 and 4.89 of the Exim SMTP daemon, there is a UAF vulnerability in the receive_msg function of receive.c. This vulnerability allows remote attackers to execute arbitrary code or cause a DoS through a vector involving the BDAT command. We reproduced this vulnerability, and SafePtrX terminated the program's execution before accessing the freed memory, preventing the exploitation of this vulnerability.

CVE-2017-14491. DNSmasq is a tool used for configuring DNS and DHCP in small networks. Prior to version 2.78, there was a heap overflow vulnerability that could lead to remote code execution (RCE) or DoS. We simulated the relevant behavior to trigger the vulnerability, and SafePtrX caused a segmentation fault and stopped the program, thereby preventing the exploitation of the heap overflow.

CVE-2018-18500. This vulnerability exists in the HTML5 parser of Firefox. When parsing HTML5 streams that match custom HTML elements, a UAF vulnerability may occur. This can lead to the release of the stream parser object while it is still in use, resulting in a potentially exploitable crash. SafePtrX successfully terminates the program execution before accessing the freed memory, effectively preventing the exploitation of this vulnerability.

5.2 Runtime Overhead

We measured the runtime overhead of SafePtrX by using the SPEC CPU2006 benchmark. We compared it with two other memory tools, AddressSanitizer (Asan) and CETS+SoftBound, as they, like SafePtrX, are capable of detecting various types of memory errors. To prevent potential termination of the program due to memory safety vulnerabilities during testing, SafePtrX replaces potentially terminating code with no-op instructions (NOP). We compiled the original programs with the default clang -O0 compiler flags as the baseline and enabled Asan with the compilation flag -fsanitize=address. Since CETS+SoftBounds was implemented in LLVM version 2.6, which is different from the version used in this paper, we referred to the data reported in the CETS paper [5]. Due to compilation issues, we only selected a subset of the test programs. The performance overhead of SafePtrX, Asan, and CETS+SoftBounds is shown in Fig. 7, where the numbers on the left represent the percentage of runtime overhead compared to the baseline. Lower values indicate higher performance.

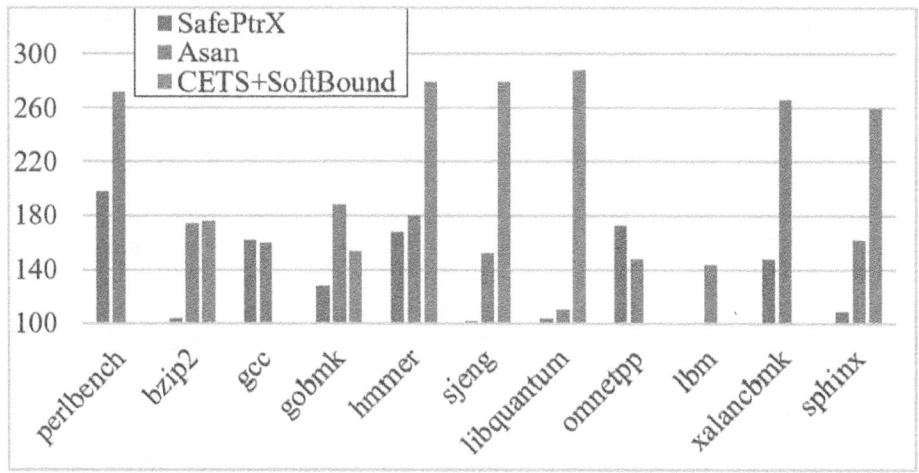

Fig. 7. Runtime overhead of SafePtrX compared with prior work, the vertical axis represents the percentage of runtime overhead.

CETS+SoftBound does not have data for perlbench and omnetpp, so they were not included in the calculation of the geomean overhead. After calculating, SafePtrX has a geomean overhead of 32.07%, which is significantly lower than Asan's 72% and CETS+SoftBound' s 106%. In the majority of programs, SafePtrX has better runtime overhead compared to Asan and CETS+SoftBound, demonstrating its superiority. Among them, perlbench has the highest runtime overhead, reaching 198%. We analyzed perlbench using SafePtrX and found that it has a significant number of heap memory allocation calls (over 15M), which means that more code is instrumented, and more system calls in our allocator, explaining the high overhead.

5.3   Memory Overhead

In theory, the additional memory overhead comes from intermediate pointers and shadow memory. Each heap object contains intermediate pointers of 8 or more bytes, and every 16 bytes of heap space corresponds to 1 byte of shadow memory. We present a script that samples and retrieves the Resident Set Size (RSS) information from /proc/PID to record the physical memory usage of a specific process. Running the SPEC CPU2006 benchmark, the additional memory overhead obtained is shown in Fig. 8.

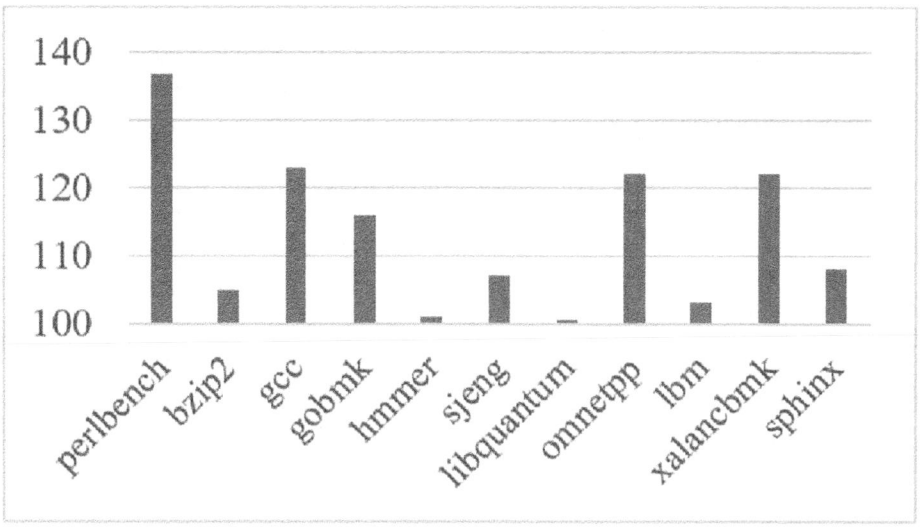

Fig. 8. Memory overhead of SafePtrX, the vertical axis represents the percentage of memory overhead.

After calculation, SafePtrX has a geomean number of 13.7% for memory overhead. Programs with more heap allocations will result in more additional overhead. In typical scenarios, modern complex computer systems have far more memory than the experimental environment (8G). Therefore, the paper suggests that the additional memory overhead can be ignored in high-performance servers. Additionally, embedding metadata in the high 16 bits of intermediate pointers does not affect the allocation of virtual addresses and has no impact on virtual address exhaustion or TLB pressure.

5.4 Discussion

Arithmetic-Derived Pointers. In the mitigation of UAF errors, arithmetic-derived pointers cannot be detected solely by the release of intermediate pointers. This is because the intermediate pointer of an arithmetic-derived pointer is not set to null, as shown in Fig. 5. When using free(p), the intermediate pointer mid_ptr1, which is pointed to by both p and its alias q, is set to null. However, mid_ptr2 still points to the heap memory, allowing an attacker to continue exploiting the UAF vulnerability using the pointer u.

However, this can be supplemented by the use of tags for overflow detection. In the case of mid_ptr2 derived from mid_ptr1 through arithmetic operations, the tag values of these pointers are expected to be the same. When an attacker uses arithmetic-derived pointers to exploit the UAF vulnerability, when the previously freed heap region is reallocated, the tag of the heap region pointed to by the dangling pointer will be randomly generated. It is highly likely to be different from the previous tag. After instrumentation and comparison, this will be recognized as an out-of-bounds error and set to null. In other words, SafeP-trX's out-of-bounds detection can also be applied to detect some UAF errors caused by arithmetic-derived pointers. This is a feature and advantage of the lock-and-key mechanism.

Number of Tag Bits. SafePtrX uses a 4-bit random number as the tag, which means there is a $1/16$ probability of having the same tag in adjacent heap spaces. In practical exploit scenarios, this probability is extremely low. Some hardware architectures, such as ARM MTE and SPARC ADI, also use the same size of tag as SafePtrX.

Stack Safety. SafePtrX does not focus on memory safety violations in the stack because it has a stack frame mechanism, and vulnerabilities on the stack are difficult to exploit [20]. In certain situations, SafePtrX can convert stack memory to heap memory for specially processing at the cost of increased overhead.

6   Related Work

6.1   UAF Detection

Static Analysis. Static analysis is a method of analyzing code without executing the program. Klee [21] uses symbolic execution to detect errors, including UAF vulnerabilities based on event sequences and uses constraint solvers. However, Klee's path coverage is incomplete and can only be applied to small programs. UAFDetector [22] is a binary detection method that establishes an abstract memory model and analyzes pointer aliasing and indirect jumps. However, like KLEE, UAFDetector is only suitable for small programs. SLAyer [23], developed by Microsoft, is an automatic verification tool that detects memory safety issues through static analysis and verification. However, it requires a significant amount of computational resources when dealing with large programs.

Pointer Nullification. Some work prevent UAF exploitation by nullifying dangling pointers explicitly or implicitly. DangSan [24] is a tool inspired by DangNULL [25] and FreeSentry [26]. They perform pointer tracking and record the relationship between pointers and objects using data structures such as red-black trees, logs, and lookup tables. DangSan is more suitable for multi-threaded environments. DangDone [15] and MPChecker [27] use intermediate pointers similar to the ones described in this paper. However, DangDone can only eliminate dangling pointers pointing to the start address of heap objects, and MPChecker has made some supplements for it.

Garbage Collection. Some work optimize garbage collection mechanisms to avoid UAF exploitation. CRCount [28] uses a reference counting mechanism to track the reference count of each memory object but ignores pointers copied in an unsafe type. MineSweeper [29] preserves deallocated allocations in a quarantine until they are no longer pointed to, and then gives permission for reallocation.

MarkUs [30] is implemented based on the Boehm [31] garbage collector and applies garbage collection to all objects marked as reclaimable by calling the standard free() function.

Secure Memory Allocators. This type of work aims to reduce UAF exploitation from the perspective of memory allocation and deallocation. SlimGuard [32] and FreeGuard [33] allow users to specify security levels to provide different levels of entropy. For each memory allocation, SlimGuard randomly selects a memory block from the memory allocation pool to provide stable random entropy. FFmalloc [34] introduces the concept of one-time allocation, where the memory manager always returns an address that will not be reused for each request. Attackers cannot regain access to deallocated objects, thus preventing UAF vulnerabilities from being exploited.

Page-Based Memory Schemes. In recent years, page-based memory schemes have gained popularity in memory safety, not limited to UAF protection, and are often used in combination with the methods mentioned above. Oscar [35] is a protection scheme based on page permissions. It does not require source code modifications and is compatible with standard allocators. It specifies virtual addresses when remapping virtual aliases. xTag [16] embeds metadata in the middle bits of allocated memory pointers and maps them to the same physical page, then checks for a match during dereferencing. DangZero [36] allows the program's allocator to access the page table directly, managing objects and invalidating them.

6.2   Out-of-Bounds Detection

Bounds Checking. Baggy bounds [37] limits the size and alignment of heap allocations, enabling effective boundary checks at runtime. SoftBound [6] maintains a table to store the base and bound addresses of each allocated object. When a pointer in the program accesses memory, SoftBound checks if the pointer is within the boundaries of its allocated object. EffectiveSan [9] transforms C/C++ into a dynamically-typed programming language, binding a dynamic type to each allocated object, which can be retrieved and compared with the declared static type at runtime.

Redzone. This type of work uses additional space added at the boundaries of memory allocations to detect and prevent memory access errors. Purify [38] is a memory and program debugging tool developed by IBM Rational, which was the first tool to utilize redzones. AddressSanitizer [4] and Valgrind [39] detect overflow errors by adding red regions before and after each allocation and checking if these regions have been modified when accessed by the program.

Memory Isolation. It is to isolate different memory regions using specific hardware and operating system features. ASLR randomizes the memory layout of a program to make it unpredictable for attackers to identify critical locations. Data

Execution Protection (DEP) [40] limits the execution of data by marking memory regions as non-executable. PageHeap [41] places allocated objects before protected pages, triggering a hardware page fault in case of out-of-bounds accesses. TailCheck [17] uses pointer tagging and shadow memory access to detect overflows, allowing multiple original objects to share a TailObject, reducing performance and memory overhead.

7   Conclusion and Future Work

In this paper, we present a heap memory violation mitigation solution based on the Intel x86-64 architecture, involving UAF and out-of-bounds violations, called SafePtrX. By introducing intermediate pointers and nullifying them when freeing heap memory, we effectively prevent the UAF exploitation. Additionally, by embedding metadata within the intermediate pointers, we mitigate heap overflow vulnerabilities and further enhance the detection capabilities of UAF. We implement it and demonstrate the effectiveness of our approach in mitigating UAF and heap overflow violations in evaluation. Compared to common methods, our research incurs lower runtime and memory overhead while providing robust security features.

SafePtrX currently focuses on mitigating UAF and heap overflow violations. In the future, it could expand to address a wider range of safety vulnerabilities, such as uninitialized memory usage, to offer more comprehensive security protection. It could also be extended and applied to different operating systems, compilers, and hardware architectures. Additionally, customized optimization strategies can be used for specific application scenarios or workloads. Finally, consideration will be given to integrating SafePtrX with existing security mechanisms like ASLR and DEP to achieve stronger and more comprehensive security protection.

Acknowledgements. This research was funded by the National Natural Science Foundation of China under grant number 61972182.

References

1. TIOBE. 2023. TIOBE Index for November 2023. www.tiobe.com/tiobe-index
2. Mozilla-central:Summary. https://hg.mozilla.org/mozilla-central/
3. American fuzzy lop. https://lcamtuf.coredump.cx/afl/
4. Serebryany, K., Bruening, D., Potapenko, A., Vyukov, D.: Address sanitizer: a fast address sanity checker. In USENIX Annual Technical Conference, pp. 309-318 (2012)
5. Nagarakatte, S., Zhao, J., Martin, M., Zdancewic, S.: CETS: compiler enforced temporal safety for C. In: Proceedings of the 9th International Symposium on Memory Management (ISMM) (2010)
6. Nagarakatte, S., Zhao, J., Martin, M.M.K., Zdancewic, S.: Softbound: highly compatible and complete spatial memory safety for C. In: Proceedings of the 2009 ACM SIGPLAN Conference on Programming Language Design and Implementation (PLDI) (2009)

7. Schuster, F., Tendyck, T., Liebchen, C., Davi, L., Sadeghi, A., Holz, T.: Counterfeit object-oriented programming: on the difficulty of preventing code reuse attacks in C++ applications. In: 2015 IEEE Symposium on Security and Privacy (S&P) (2015)

8. 2023 CWE Top 25 most dangerous software weaknesses. https://cwe.mitre.org/top25/archive/2023/2023_top25_list.html

9. Duck, G.J., Yap, R.H.: EffectiveSan: type and memory error detection using dynamically typed C/C++. In: Proceedings of the 39th ACMSIGPLAN Conference on Programming Language Design and Implementation, PLDI 2018, pp. 181–195. New York, NY, USA, Association for Computing Machinery (2018)

10. Duck, G., Yap, R., Cavallaro, L.: Stack bounds protection with low fat pointers. In: The Internet Society, In Network and Distributed System Security Symposium (2017)

11. The LLVM Compiler Infrastructure. https://llvm.org/

12. Dang, T.H.Y., Maniatis, P., Wagner, D.: Oscar: a practical page-permissions-based scheme for thwarting dangling pointers. In: 26th USENIX Security Symposium (USENIX Security) (2017)

13. ArmDeveloper. https://developer.arm.com/documentation

14. Intel® 64 and IA-32 architectures software developer manuals. https://www.intel.com/content/www/us/en/developer/articles/technical/intel-sdm.html

15. Wang, Y., et al.: DangDone: eliminating dangling pointers via intermediate pointers. In: Proceedings of the Tenth Asia-Pacific Symposium on Internetware, Internetware 2018, pp. 1–10 (2018). https://doi.org/10.1145/3275219.3275231

16. Bernhard, L., Rodler, M., Holz, T., Davit, L.: xTag: mitigating use-after-free vulnerabilities via software-based pointer tagging on intel x86-64. In: 2022 IEEE 7th European Symposium on Security and Privacy (EuroS&P), pp. 502–519 (2022)

17. Gopal, A.U., Soori, R., Ferdman, M., Lee, D.: TAILCHECK: a lightweight heap overflow detection mechanism with page protection and tagged pointers. In: USENIX Symposium on Operating Systems Design and Implementation (2023)

18. Lopes, B.C., Auler, R.: Getting Started with LLVM Core Libraries. Packt Publishing Ltd, Birmingham (2014)

19. CVE® program mission. https://www.cve.org/

20. Lee, B.Y., et al.: Preventing use-after-free with dangling pointers nullification. In: Proceedings of the Network and Distributed System Security Symposium (NDSS) (2015)

21. Cadar, C., Dunbar, D., Engler, D.R.: Klee: unassisted and automatic generation of high-coverage tests for complex systems programs[C]. OSDI 8, 209–224 (2008)

22. Zhu, K., Lu, Y., Huang, H.: Scalable static detection of use-after-free vulnerabilities in binary code. IEEE Access 8, 78713–78725 (2020)

23. Berdine, J., et al.: SLAyer: memory safety for systems-level code. In: International Conference on Computer Aided Verification (2011)

24. Van Der Kouwe, E., Nigade, V., Giuffrida, C.: DangSan: scalable use-after-free detection. In: Proceedings of the European Conference on Computer Systems, pp. 405–419. ACM (2017)

25. Lee, B., Song, C., Jang, Y., et al.: Preventing use-after-free with dangling pointers nullification. In: The Internet Society, NDSS (2015)

26. Younan, Y.: FreeSentry: protecting against use-after-free vulnerabilities due to dangling pointers. In: The Internet Society, NDSS (2015)

27. Qiang, W., Li, W., Jin, H., et al.: Mpchecker: use-after-free vulnerabilities protection based on multi-level pointers. IEEE Access 7, 45961–45977 (2019)

28. Shin, J., Kwon, D., Seo, J., et al.: CRCount: pointer invalidation with reference counting to mitigate use-after-free in legacy C/C++. In: The Internet Society, NDSS (2019)
29. Erdos, M., Ainsworth, S., Jones, T.: MineSweeper: a "clean sweep" for drop-in use-after-free prevention. pp. 212–225 (2022). https://doi.org/10.1145/3503222.3507712
30. Ainsworth, S., Jones, T.M.: MarkUs: drop-in use-after-free prevention for low-level languages. In: SP, 578591 (2020). https://doi.org/10.1109/sp40000.2020.00058
31. A garbage collector for C and C++ https://www.hboehm.info/gc/index.html
32. Liu, B., Olivier, P., Ravindran, B.: SlimGuard: a secure and memory-efficient heap allocator. In: Proceedings of the International Middleware Conference, pp. 1–13. ACM (2019)
33. Silvestro, S., Liu, H., Crosser, C., et al.: Freeguard: a faster secure heap allocator. In: Proceedings of the ACM Conference on Computer and Communications Security (CCS), pp. 2389–2403. ACM (2017)
34. Wickman, B., et al.: Preventing use-after-free attacks with fast forward allocation. In: USENIXSecurity (2021)
35. Dang, T.H., Maniatis, P., Wagner, D.: Oscar: a practical page-permissions-based scheme for thwarting dangling pointers. In: USENIX Security, pp. 815–832 (2017)
36. Gorter, F., et al.: DangZero: efficient use-after-free detection via direct page table access. In: Proceedings of the 2022 ACM SIGSAC Conference on Computer and Communications Security (2022)
37. Akritidis, P., Costa, M., Castro, M., Hand, S.: Baggy bounds checking: an efficient and backwards-compatible defense against out-of-bounds errors. In: USENIX Security Symposium, vol. 10 (2009)
38. Reed, H.: Purify: fast detection of memory leaks and access errors. In: Proceedings of 1992 Winter USENIX Conference, pp. 125–136 (1992)
39. Nethercote, N., Seward, J.: Valgrind: a framework for heavyweight dynamic binary instrumentation. ACM Sigplan Notices **42**(6), 89–100 (2007)
40. Andersen, S., Abella, V.: Data execution prevention. Changes to functionality in microsoft windows xp service pack, vol. 2 (2004)
41. Microsoft Windows. PageHeap. https://learn.microsoft.com/en-us/windows-hardware/drivers/debugger/gflags-and-pageheap (2022). Accessed 21-Nov 2022
42. Xu, W., Li, J., Shu, J., Yang, W., Xie, T., Zhang, Y., Gu, D.; From collision to exploitation: unleashing use-after-free vulnerabilities in linux kernel. In Proceedings of the 22nd ACM SIGSAC Conference on Computer and Communications Security, pp. 414-425, October 2015
43. Mi-malloc Documentation. https://microsoft.github.io/mimalloc/

Towards Efficiently Parallelizing Patch-Space Exploration in Automated Program Repair

Omar I. Al-Bataineh$^{(\boxtimes)}$

Gran Sasso Science Institute, L'Aquila, Italy
omar.bataineh@gssi.it

Abstract. A long-standing open challenge for automated program repair (APR) is search space explosion, which makes the size of the patch space too large to be handled with limited resources of time and memory. This problem is further exacerbated by the accuracy challenges of fault localization techniques, which regularly rank the actual faulty statement low on the list of suspicious statements, resulting in wasted repair effort.

This paper proposes an approach to increase the overall performance of APR for large-scale programs by utilizing parallelism for exploring the patch space, as a promising strategy to parallelize template-based APR and effectively ameliorate the accuracy challenges of fault localization. Our empirical study reveals that a substantial five-fold improvement in performance can be obtained without any degradation in repair quality. We also observe that parallelized APR produces a larger number of plausible patches than sequential APR. These encouraging results show that the proposed approach is both feasible and efficient. Future steps include investigating other promising search space splitting strategies, such as splitting based on the fix templates in template-based APR, and hybrid combinations of such strategies.

1 Introduction

Automated program repair (APR) [1,2] has emerged as an area with continuous research innovations and increasing tool support. However, APR for large-scale programs is a time-consuming activity that requires lots of resources to locate and repair bugs. This paper aims to accelerate the overall performance of APR for large-scale programs by utilizing parallelism in exploring the patch-space.

APR consists of three steps: fault localization (FL for short), patch generation, and patch validation. Despite significant advances [3–11], FL still suffers from accuracy issues [12,13], i.e., the top ranked suspicious statements are often false positives. The poor ranking of faulty statements directly affects APR performance by increasing the number of needlessly examined patch candidates, which increase the overall costs of patch compilation and test execution.

One way to address this issue is to split the patch search space in a way that allows it to be explored more effectively. The challenge is to devise a splitting

© The Author(s), under exclusive license to Springer Nature Switzerland AG 2025
G. Bai et al. (Eds.): ICECCS 2024, LNCS 14784, pp. 409–419, 2025.
https://doi.org/10.1007/978-3-031-66456-4_22

strategy by which the faulty statement corresponding to the target bug can be found as early as possible. The key contribution of this paper is an efficient and scalable splitting strategy for parallel APR that ensures early selection of the faulty statement without degrading repair quality. The algorithm is sound in the sense that when a valid patch can be found by sequential APR, it will also be found by the parallel version.

To demonstrate the feasibility of the proposed search splitting strategies, we consider the state-of-the-art template-based APR tool TBar [14] under two different implementations: the sequential implementation, denoted as $TBar_{seq}$, in which a single repair process is employed to handle the entire patch space, and a parallel implementation, denoted as $TBar_{par}$, in which the patch space is splitted into distinct subspaces that are handled by multiple processes simultaneously. The performance of $TBar_{par}$ has been evaluated and compared to $TBar_{seq}$ on the widely used Defects4J benchmark [15]. Our empirical study shows that $TBar_{par}$ outperforms $TBar_{seq}$ by several orders of magnitude. It also shows the capability of $TBar_{par}$ to produce more plausible repairs than $TBar_{seq}$.

Significance and Impact of Parallel APR: Fixing bugs in large-scale programs can be a non-trivial task that requires significant computational resources. The sequential exploration of the patch space can be a time consuming operation, in particular if the number of suspicious statements to be examined is large and accuracy of the employed FL technique is low. According to many empirical studies [16–18], a valid patch is often only obtained after exploring a large part of the patch space, and thus one needs to allocate a long time budget to be able to find a valid patch. Instead of examining the entire patch space sequentially, this paper proposes an approach to parallelize APR which significantly reduces the repair time of large programs. Most of the currently available APR tools are not constructed in a way that supports parallel exploration of the patch space. However, decomposing the patch space using splitting strategies as discussed in this paper would enable such parallelization and better utilize the multi-core processors that have become the default in today's computers.

2   The Need for Parallel APR

The state-of-the-art template-based APR tool TBar [14] uses 7 h and 19 min[1] to generate a single plausible patch for the bug Lang-10 in Defects4J [15] using its normal sequential repair mode. This time does not include the "debugging time", i.e., the time needed to generate a ranked list of suspicious statements, which is typically obtained by running an FL tool such as GZoltar [19]. The long processing time can be explained by observing that TBar spends a significant time mutating suspicious statements that are not the actual (ground truth) faulty statement responsible for the bug: the FL results for Lang-10 are a ranked list of 287 suspicious statements, in which the ground truth faulty statement is on the 70th position.

---

[1] All results are from an eight-core Macbook Pro M1 with 16 GB memory.

Early selection of the appropriate faulty statement among the suspicious statements is critical for the performance of APR, because if the tool selects a suspicious statement that is not the actual faulty statement and mutates it using the fix templates, a considerable amount of time will be wasted while generating infeasible patches. However, since FL techniques are known to have accuracy issues [12,13], it is desirable to devise an approach that can find the appropriate faulty statement more efficiently, thereby alleviating the impact of FL imperfection. This paper proposes to achieve this goal by means of splitting the patch space and parallelizing the search for the appropriate faulty statement.

There are two advantages of splitting the patch space into smaller disjoint subspaces and process them in parallel. The first is efficiency: as we will see, the speed-up of parallel APR over its sequential counterpart is usually significant. The second is scalability: real-world programs keep growing in size, at a fast rate. As the size of buggy programs increases, generally so does their list of suspicious statements, and thereby the potential of exploiting parallelism in repairs.

3 Patch Space Splitting for Template-Based APR

This section defines a patch space splitting strategy for APR. Let P be a program containing bug b and L be the set of suspicious statements of P computed w.r.t. bug b using some fault localization procedure. Let F be the set of fix templates implemented by a repair system R, and sys be a multi-core system with k processors, where $k \geq 2$. Suppose that S_{seq} is the patch space of P constructed by the sequential mutation of L using the set F. The patch space splitting problem aims at splitting the space S_{seq} into smaller disjoint subspaces $S_1, ..., S_n$ that can be processed simultaneously. Formally, the patch space splitting problem can be described using a split function of the form $split(L, F, sys) = (S_1, ..., S_n)$, where the function $split(L, F, sys,)$ meets the following properties

1. *disjointness property*: $S_i \cap S_j = \emptyset$ for all $i, j = 1...n$,
2. *completeness property*: $S_{seq} \subseteq (S_1 \cup S_2 \cup ... \cup S_n)$.

The above definition gives several ways in which the patch space can be divided into several subspaces. One example is to split suspicious locations L into disjoint subsets and mutate each of them independently. Another way is to split fix templates F into disjoint subsets and apply them in parallel.

The patch space of a buggy program P is typically constructed in an incremental manner starting from the most suspicious statement in the set L. The mutation process continues until either the entire patch space is constructed, a correct patch is found, or the allocated time budget is expired.

Splitting Suspicious Lists: Several splitting strategies can be developed to split the suspicious list L into smaller disjoint subsets. However, it is important to find a splitting strategy S by which the actual faulty statement of the analysed bug can be reached faster. This is crucial in order to ensure that the performance of parallel APR outperforms the corresponding sequential APR. To do so, we select a simple yet effective splitting strategy, as described in formula (1), where n represents the number of involved repair processes, $len(L)$ is a function that returns the length of L, and p_i is a repair process.

$$List(p_i) = \{L_{(i+k \cdot n)}\} \mid k = \{0, 1, 2, ..\} \wedge (i + k \cdot n) \leq len(L) \tag{1}$$

Formula (1) ensures that statements with higher suspiciousness rank will always receive higher priority in the order of mutation. It also ensures the disjointness and completeness properties. To demonstrate the effectiveness of the presented splitting strategy, let us consider the following simple example.

Example 1. Let $L = \{L_1, ..., L_{12}\}$ be a sorted list of suspicious statements for bug b, where $rank(L_i) \geq rank(L_{i+1})$ and $P = \{p_1, p_2, p_3, p_4\}$ is a set of repair processes. Using formula (1) we can split L into four smaller lists which will be assigned to processes p_1, p_2, p_3, and p_4 respectively as follows

$$List(p_1) = \{L_1, L_5, L_9\}, \; List(p_2) = \{L_2, L_6, L_{10}\}$$

$$List(p_3) = \{L_3, L_7, L_{11}\}, \; List(p_4) = \{L_4, L_8, L_{12}\}.$$

For this example, the parallel APR system will be $R_{par} = (p_1 \parallel p_2 \parallel p_3 \parallel p_4)$, where $\bigcup_{i=1...4}(List(p_i)) = L$ and $List(p_i) \cap List(p_j) = \emptyset$ for any two distinct $i, j \in \{1, 2, 3, 4\}$.

Theorem 1. *Soundness of search space splitting strategy. The patch space constructed by a parallel APR algorithm is a superset to the one constructed by a sequential APR algorithm, provided that:*

1. *the same set of fix templates F are used in both the sequential and parallel algorithms, and*
2. *the implemented patch space splitting strategy is complete: each suspicious statement in the set L is guaranteed to occur in at least one subspace.*

Research Questions: To identify the factors that affect the performance of TBar and evaluate the effectiveness of splitting strategies, we consider the following research questions:

RQ1 – How does the outcome of FL analysis (the number of suspicious statements) impact the performance of TBar?

RQ2 – Which factor among the following three factors: the test suite size, FL results, and number of implemented fix templates, has the greatest impact on the performance of TBar?

RQ3 – To what extent does the patch space splitting technique described in Sect. 3 enhance TBar's performance?

4   Evaluation

This section describes the subjects used in the empirical study, as well as the template-based APR tool TBar. Next, we address the three research questions described in the previous section.

Subjects: Our experiments use the Defects4J (v1.2.0) benchmark [15], which is widely used for APR tasks [14,20]. This benchmark consists of 6 open source projects containing 395 real faults. However, for space reasons, we report only the results of one project (the Lang project) and consider the set of bugs that TBar has successfully repaired (see Table 1).

Tool: Among the various available template-based APR tools [14,18,20–28], we select TBar [14], a state-of-the-art repair tool that employs 15 commonly used fix templates and achieves good performance and repair rate on Defects4J.

RQ1: In order to evaluate the impact of FL on the performance of TBar, we conduct experiments under a *perfect fault localization* setup (i.e., the actual ground truth faulty statements are used) and compare them to the repair results with a *normal FL* setup, where the tool GZoltar is invoked to compute a ranked list of suspicious statements for each of the considered bugs.

Tables 1 and 2 summarise our experimental results of running TBar in perfect FL and normal FL modes respectively. As shown in these tables, the number of suspicious statements and the suspiciousness rank of the actual faulty statement have significant impact on the performance of TBar. Our analysis shows that TBar spends a significant amount of time mutating statements incorrectly identified as faulty statements, leading to a patch space with a large number of incorrect patches.

Table 1. Repair times of Lang bugs with Perfect FL

BugID	Repair time (s)
Lang-10	86
Lang-33	45
Lang-39	55
Lang-44	13
Lang-45	52
Lang-51	55
Lang-57	195
Lang-58	96
Lang-59	34
Lang-63	40

Table 2. Sequential and parallel repair times (in seconds) under normal FL setup with complete and reduced test suites

BugID	Fault size	Fault rank	Repair time			
			Sequential		Parallel	
			completed	reduced	completed	reduced
Lang-10	287	70	26340	5400	3240	1320
Lang-33	1061	7	300	73	116	29
Lang-39	75	27	11400	8880	2280	1920
Lang-44	22	7	360	143	138	26
Lang-45	24	16	4080	600	1800	316
Lang-51	19	10	2760	480	960	194
Lang-57	1	1	195	43	195	43 *
Lang-58	36	8	1260	285	557	96
Lang-59	1	1	34	25	34	25 *
Lang-63	1	1	40	29	40	29 *

* equal seq./par. repair times because the fault is ranked @ 1

When analysing the results of FL analysis of bugs Lang-10 and Lang-39, we find that the suspiciousness rank assigned to the actual faulty statements of these bugs is too low. For example, the actual faulty statement of Lang-10 is ranked 70 while the actual faulty statement for Lang-39 is ranked 27.

The low suspiciousness rank assigned to the faulty statements by the FL technique delays the selection of the statement by TBar which leads to a long processing time. It is interesting to note also that the size of the suspicious list does not directly impact the performance of TBar. For example, bug Lang-33 has 1061 candidate suspicious statements but the actual faulty statement is ranked 7th and thus TBar was able to produce a patch for the bug in a short computational time.

> Answer to *RQ1*: The accuracy of FL techniques is critical to the performance of APR. The lower the suspiciousness rank of the faulty statement, the greater the size of patch space and the longer the analysis time.

RQ2: An interesting feature of Defects4J dataset is that it supports a command that can be used to export relevant tests for the target bug. This feature can be used to reduce the size of provided test suite to consider only relevant tests.

In order to evaluate the impact of test suite minimization on the performance of TBar, we conduct first experiments under the normal FL setup with complete test suite (i.e., the complete developer-written test suites). We then compare the results with normal FL setup but with reduced test suites.

The first key observation we made is that TBar is still able to produce the same patches for all considered bugs using reduced test suites but in much shorter computational time (see Table 2). This raises several questions about the optimality of developer-written test suites. Note that when the faulty statement of the bug is ranked too low in the suspicious list, the number of examined patch candidates will increase, which in turn increases the number of patch compilations and test case executions. Thus, by reducing the size of test suites we obtain great reduction on the overall processing time.

> Answer to *RQ2*: The size of test suite has significant impact on the performance of TBar specially when the suspiciousness rank of the faulty statement is too low.

RQ3: In order to evaluate whether or not $TBar_{par}$ is more efficient than $TBar_{seq}$, we conduct experiments by running $TBar_{par}$ and $TBar_{seq}$ under normal FL setup using both complete and reduced test suites. $TBar_{seq}$ is configured as {ochiai, TBar} (the default setup of TBar), where ochiai represents the output ochiai file generated by GZoltar which contains a sorted list of suspicious statements of the target bug. On the other hand, $TBar_{par}$ is configured as {(ochiai1, TBar), (ochiai2, TBar), (ochiai3, TBar), (ochiai4, TBar)}, where (ochiai1 $\cup$ ochiai2 $\cup$ ochiai3 $\cup$ ochiai4) = ochiai and (ochiaiI $\cap$ ochiaiJ = $\emptyset$) for any I, J = 1...4. In our setting, $TBar_{par}$ uses four repair processes to process the patch space.

Several interesting observations can be made based on the results shown in Table 3. First, by using TBar$_{par}$, it is possible to handle large-scale buggy programs and produce patches many orders of magnitude faster than TBar$_{seq}$. Note that by splitting the patch space into smaller disjoint subspaces and involving multiple processes, we reduce the number of patch candidates examined before reaching the faulty statement and producing a repair, thereby we reduce the number of patch compilations and test case executions. Second, TBar$_{par}$ was able to produce more plausible patches than TBar$_{seq}$: TBar$_{par}$ produces two plausible patches for each of Lang-10, Lang-39, and Lang-51 while one patch is produced by TBar$_{seq}$.

Note that once a plausible patch is generated, TBar$_{seq}$ will be terminated, and finally the actual fix templates or the other faulty statements of the target bug have no chance to be selected. The analysis of this RQ led to the following observations. First, we found that some sibling positions to the developer-provided bug positions can be alternative inputs to drive correct fixing. Second, some bugs involve multiple faulty statements and that sequential APR tends to select only one statement and generate a plausible patch (e.g., Lang-10 and Lang-39). Therefore, TBar$_{par}$ increases the chances of trying different fix templates on different faulty statements which increases the chances of producing correct patches. Table 3 summarizes the total repair times of Lang bugs using both sequential and parallel repair modes. TBar$_{seq}$ takes 13 h to produce 10 patches, while TBar$_{par}$ takes 2 h and 32 min to produce 13 patches for the same bugs, where all patches generated by TBar$_{seq}$ were also generated by TBar$_{par}$. It is interesting to note also that combining test suite minimisation techniques with parallel APR based on FL splitting strategy leads to a significant improvement on the repair time: TBar$_{seq}$ takes 13 h to fix the 10 Lang bugs while TBar$_{par}$ with reduced tests takes 1 h and 5 min.

Answer to *RQ3*: TBar$_{par}$ outperforms TBar$_{seq}$ by many orders of magnitudes when handling bugs whose faulty statements are ranked low. TBar$_{par}$ can also increase the chances of producing correct patches.

Table 3. Total repair times of Lang bugs under different repair modes

Test Suite	Sequential repair	Parallel repair
Complete	46826 s ≈ 13h 00	9091 s ≈ 2h 32
Reduced	15958 s ≈ 4h 26	3901 s ≈ 1h 05

5   Related Work

In this section we discuss related work on template-based APR and existing acceleration techniques for APR.

Template-Based APR: Template-based APR are widely used in the APR literature [14,18,20–29], which utilize predefined fix templates to fix specific bugs. However, most of previous work on template-based APR has focused on maximizing the fix-rate by incorporating a large number of useful fix templates. Despite of these great achievements, modern template-based APR tools still suffer from the efficiency issue as demonstrated in [12] and this work. The presented search space splitting strategies would help not only in improving the efficiency of APR by reaching faulty statements earlier, but also increasing their capabilities in producing correct patches.

Accelerating APR: Several techniques have been developed to increase the performance of APR, including regression test selection [30,31], patch filtering [32,33], and on-the-fly patch generation [21] and validation [34,35]. The goal of these techniques is to reduce the patch compilation and test case execution costs, which are the dominant contributors for APR runtime. While these techniques show promise in reducing the repair time of programs, their accuracy and computational overhead are still an issue. The parallel APR approach proposed in this paper has the advantage of accelerating the performance of APR by many orders of magnitude without any degradation in repair quality.

A few attempts have been made to accelerate the performance of APR by utilizing parallelism. For example, Matsumoto et al. aim to accelerate search-based APR with GenProg [36] by building and testing mutated programs in parallel [37]. Yang et al. aim to accelerate semantic-based APR with Nopol [38] by running multiple instances of the so-called suspiciousness-first algorithm (SFA) in parallel [39]. Our work differs from these studies in four key aspects: First, we propose an effective strategy for splitting the patch search space that guarantees both disjointness and completeness of the constructed subspaces. In fact, the algorithm described by Yang et al. does not guarantee the absence of search overlap among parallel processes. Second, the performance and efficiency of our algorithm greatly outperforms the earlier results [37,39], achieving a five-fold speedup in repair time. Third, our approach parallelizes all aspects of the APR process including patch generation, patch compilation, and patch validation. Fourth, we target a different class of APR (template-based APR), and to our knowledge, this is the first work that studies effective patch space splitting and develops a parallel repair algorithm for template-based APR.

6   Concluding Remarks

The patch explosion problem has been, and is likely to remain one of the main challenges faced by APR. One possible technique to combat the patch explosion problem is to apply patch splitting strategies by which the patch space can be divided into smaller disjoint spaces that can be processed simultaneously. To develop an effective patch splitting strategy, we first conducted an empirical study to determine the factors that have the greatest impact on the performance of TBar. The analysis showed that the size of the suspicious list, the rank of faulty statements, and the size of test suite, have great impact on the performance of APR. Based on this observation, we developed an efficient parallel APR algorithm by which the faulty statements responsible for the bug being repaired can be reached much faster. The algorithm is sound in the sense that when a valid patch is found by the sequential APR it can be also found by the parallel APR. Our results show that parallel APR significantly outperforms sequential APR (on average five-fold speedup in repair time), suggesting that parallel APR algorithms may be a desirable way to fix bugs in large-scale programs.

Future Plan: Aside from broadening our empirical study to confirm our initial findings on a wider range of bugs, there is a need to further investigate the relation between the size of test suites, the accuracy of fault localization, and their impact on APR performance. This is not just because the knowledge and insights are interesting in their own right, but because they may be leveraged to further improve the performance of APR. With a deeper understanding of these relations, we can advance the study of parallel APR by integrating effective search space splitting strategies with test case minimisation techniques.

One particular direction of interest is *hybrid parallel APR*. We aim to investigate the effectiveness of hybrid splitting strategy for template-based APR: instead of splitting the patch space using only the suspicious lists, one may also split the 15 fix templates into smaller disjoint sets and apply them on parallel on the resultant smaller suspicious lists. With the hybrid patch splitting strategy, the chance of selecting the correct fix template earlier would increase and thus the repair rate of APR may improve, in particular when dealing with bugs whose faulty statements are ranked too low.

Acknowledgements. The author would like to thank Leon Moonen and anonymous reviewers for comments on an earlier version of the paper.

References

1. Monperrus, M.: Automatic software repair: a bibliography. ACM Comput. Surv. **51**(1), 1–24 (2018)
2. Goues, C.L., Pradel, M., Roychoudhury, A.: Automated program repair. Commun. ACM **62**(12), 56–65 (2019)
3. Abreu, R., Zoeteweij, P., van Gemund, A.J.: On the accuracy of spectrumbased fault localization. In: Testing: Academic & Industrial Conference Practice & Research Techniques - MUTATION, pp. 89–98 (2007)
4. Zhang, X., Gupta, N., Gupta, R.: Locating faulty code by multiple points slicing. Softw. Pract. Exp. **37**(9), 935–961 (2007)
5. Zhang, X., He, H., Gupta, N., Gupta, R.: Experimental evaluation of using dynamic slices for fault location. In: Proceedings of the Sixth International Symposium on Automated Analysis-Driven Debugging, AADEBUG, pp. 33–42 (2005)
6. Zhang, L., Kim, M., Khurshid, S.: Localizing failure-inducing program edits based on spectrum information. In: IEEE International Conference on Software Maintenance, pp. 23–32 (2011)

7. Xuan, J., Monperrus, M.: Learning to combine multiple ranking metrics for fault localization. In: IEEE International Conference on Software Maintenance & Evolution, pp. 191–200 (2014)
8. Li, X., d'Amorim, M., Orso, A.: Iterative user-driven fault localization. In: Bloem, R., Arbel, E. (eds.) HVC 2016. LNCS, vol. 10028, pp. 82–98. Springer, Cham (2016). https://doi.org/10.1007/978-3-319-49052-6_6
9. Pearson, S., et al.: Evaluating and improving fault localization. In: IEEE/ACM 39th International Conference on Software Engineering, pp. 609–620 (2017)
10. Lou, Y., et al.: Can automated program repair refine fault localization? a unified debugging approach. In: International Symposium on Software Testing and Analysis ISSTA, pp. 75–87 (2020)
11. Li, Y., Wang, S., Nguyen, T.N.: Fault localization with code coverage representation learning. In: 43rd IEEE/ACM International Conference on Software Engineering, ICSE, pp. 661–673 (2021)
12. Liu, K., et al.: On the efficiency of test suite based program repair a systematic assessment of 16 automated repair systems for java programs. In: IEEE/ACM 42nd International Conference on Software Engineering, pp. 615–627 (2020)
13. Liu, K., Koyuncu, A., Bissyandé, T.F., Kim, D., Klein, J., Le Traon, Y.: You Cannot fix what you cannot find! an investigation of fault localization bias in benchmarking automated program repair systems. In: Conference, I.E.E.E. (ed.) Software, pp. 102–113. Validation & Verif, Testing (2019)
14. Liu, K., Koyuncu, A., Kim, D., Bissyandé, T.F.: TBar: revisiting templatebased automated program repair. In: ACM SIGSOFT International Symposium on Software Testing and Analysis, ISSTA, pp. 31–42 (2019)
15. Just, R., Jalali, D., Ernst, M.D.: Defects4J: a database of existing faults to enable controlled testing studies for Java programs. In: International Symposium on Software Testing& Analysis, ISSTA, pp. 437–440 (2014)
16. Weimer, W., Fry, Z.P., Forrest, S.: Leveraging program equivalence for adaptive program repair: models and first results." In: IEEE/ACM International Conference on Automated Software Engineering, pp. 356–366 (2013)
17. Goues, C.L., Forrest, S., Weimer, W.: Current challenges in automatic software repair. Softw. Qual. 62(12), 421–443 (2013)
18. Le, X.B.D., Lo, D., Le Goues, C.: History driven program repair. In: IEEE 23rd International Conference on Software Analysis, Evolution, & Reengineering, pp. 213–224 (2016)
19. Campos, J., Riboira, A., Perez, A., Abreu, R.: GZoltar: an eclipse plug-in for testing and debugging. In: IEEE/ACM International Conference on Automated Software Engineering, pp. 378–381 (2012)
20. Jiang, J., Xiong, Y., Zhang, H., Gao, Q., Chen, X.: Shaping program repair space with existing patches and similar code. In: International Symposium on Software Testing and Analysis, ISSTA, pp. 298–309 (2018)
21. Hua, J., Zhang, M., Wang, K., Khurshid, S.: Towards practical program repair with on-demand candidate generation. In: IEEE/ACM 40th International Conference on Software Engineering, pp. 12–23 (2018)
22. Wen, M., Chen, J., Wu, R., Hao, D., Cheung, S.C.: Context-aware patch generation for better automated program repair. In: IEEE/ACM 40th International Conference on Software Engineering, pp. 1–11 (2018)
23. Xin, Q., Reiss, S.P.: Leveraging syntax-related code for automated program repair. In: IEEE/ACM International Conference on Software Engineering, pp. 660–670 (2017)

24. Koyuncu, A., et al.: FixMiner: mining relevant fix patterns for automated program repair. Empirical Softw. Eng. **25**(3), 1980–2024 (2020). https://doi.org/10.1007/s10664-019-09780-z

25. Liu, K., Koyuncu, A., Kim, D., Bissyand'e, T.F.: AVATAR: fixing semantic bugs with fix patterns of static analysis violations. In: IEEE 26th International Conference on Software Analysis, Evolution & Reengineering, pp. 1–12 (2019)

26. Liu, X., Zhong, H.: Mining stackoverflow for program repair. In: IEEE 25th International Conference on Software Analysis, Evolution & Reengineering, pp. 118–129 (2018)

27. Long, F., Amidon, P., Rinard, M.C.: Automatic inference of code transforms for patch generation. In: Joint Meeting on Foundations of Software Engineering, ESEC/FSE, pp. 727–739 (2017)

28. Saha, R.K., Lyu, Y., Yoshida, H., Prasad, M.R.: Elixir: effective objectoriented program repair. In: IEEE/ACM International Conference on Automated Software Engineering, pp. 648–659 (2017)

29. Durieux, T., Cornu, B., Seinturier, L., Monperrus, M.: Dynamic patch generation for null pointer exceptions using metaprogramming. In: Conference, I. (ed.) Software, pp. 349–358. Evolution & Reengineering, Analysis (2017)

30. Yoo, S., Harman, M.: Regression testing minimization, selection and prioritization: a survey. In: Softw. Test. Verification Reliab. **22**(2), 67-120 (2012)

31. Mehne, B., Yoshida, H., Prasad, M.R., Sen, K., Gopinath, D., Khurshid, S.: Accelerating search-based program repair. In: IEEE 11th International Conference on Software Testing, Verification and Validation, pp. 227–238 (2018)

32. Liang, J., et al.: Interactive patch filtering as debugging aid. In: IEEE International Conference on Software Maintenance & Evolution, pp. 239–250 (2021)

33. Yang, C.: Accelerating redundancy-based program repair via code representation learning and adaptive patch filtering. In: Joint Meeting Foundations of Software Engineering, ESEC/FSE, pp. 1672–1674 (2021)

34. Chen, L., Ouyang, Y., Zhang, L.: Fast and precise on-the-fly patch validation for all. In: IEEE/ACM 43rd International Conference on Software Engineering, pp. 1123–1134 (2021)

35. Benton, S., Xie, Y., Lu, L., Zhang, M., Li, X., Zhang, L.: Towards boosting patch execution on-the-fly. In: International Conference on Software Engineering, pp. 2165–2176 (2022)

36. Le Goues, C., Nguyen, T., Forrest, S., Weimer, W.: GenProg: a generic method for automatic software repair. IEEE Trans. Softw. Eng. **38**(1), 54–72 (2011)

37. Matsumoto, J., Higo, Y., Matsuo, H., Arima, R., Matsumoto, S., Kusumoto, S.: GenProg meets cluster computing. In: International Workshop on Empirical Software Engineering in Practice, pp. 37–375 (2019)

38. Xuan, J., et al.: Nopol: automatic repair of conditional statement bugs in java programs. IEEE Trans. Softw. Eng. **43**(1), 34–55 (2016)

39. Yang, D., Qi, Y., Mao, X., Lei, Y.: Evaluating the usage of fault localization in automated program repair: an empirical study. Front. Comput. Sci. **15**(1), 1–15 (2020). https://doi.org/10.1007/s11704-020-9263-1

Correction to: Validation of railML Using ProB

Jan Gruteser⑩ and Michael Leuschel⑩

Correction to:
Chapter 13 in: G. Bai et al. (Eds.):
Engineering of Complex Computer Systems, **LNCS 14784,**
https://doi.org/10.1007/978-3-031-66456-4_13

In an older version of this paper, the casing "railML" was incorrect. This has been corrected.

The updated version of this chapter can be found at
https://doi.org/10.1007/978-3-031-66456-4_13

Author Index

G. Bai et al. (Eds.): ICECCS 2024, LNCS 14784, pp. 421–422, 2025.
https://doi.org/10.1007/978-3-031-66456-4

The manufacturer's authorised representative in the EU is Springer
Nature Customer Service Centre GmbH, Europaplatz 3, 69115 Heidelberg,
Germany. If you have any concerns regarding our products, please
contact ProductSafety@springernature.com

Printed and bound by CPI Group (UK) Ltd, Croydon, CR0 4YY
06/05/2026
02104370-0001